T0393488

Reviews of Environmental Contamination and Toxicology

VOLUME 257

More information about this series at https://link.springer.com/bookseries/398

Reviews of Environmental Contamination and Toxicology Volume 257

Editor
Pim de Voogt

Volume 257

ISSN 0179-5953 ISSN 2197-6554 (electronic)
Reviews of Environmental Contamination and Toxicology
ISBN 978-3-030-88216-7 ISBN 978-3-030-88217-4 (eBook)
https://doi.org/10.1007/978-3-030-88217-4

This Springer imprint is published by the registered company Springer Nature Switzerland AG
The registered company address is: Gewerbestrasse 11, 6330 Cham, Switzerland

Foreword

International concern in scientific, industrial, and governmental communities over traces of xenobiotics in foods and in both abiotic and biotic environments has justified the present triumvirate of specialized publications in this field: comprehensive reviews, rapidly published research papers and progress reports, and archival documentations These three international publications are integrated and scheduled to provide the coherency essential for nonduplicative and current progress in a field as dynamic and complex as environmental contamination and toxicology. This series is reserved exclusively for the diversified literature on "toxic" chemicals in our food, our feeds, our homes, recreational and working surroundings, our domestic animals, our wildlife, and ourselves. Tremendous efforts worldwide have been mobilized to evaluate the nature, presence, magnitude, fate, and toxicology of the chemicals loosed upon the Earth. Among the sequelae of this broad new emphasis is an undeniable need for an articulated set of authoritative publications, where one can find the latest important world literature produced by these emerging areas of science together with documentation of pertinent ancillary legislation.

Research directors and legislative or administrative advisers do not have the time to scan the escalating number of technical publications that may contain articles important to current responsibility. Rather, these individuals need the background provided by detailed reviews and the assurance that the latest information is made available to them, all with minimal literature searching. Similarly, the scientist assigned or attracted to a new problem is required to glean all literature pertinent to the task, to publish new developments or important new experimental details quickly, to inform others of findings that might alter their own efforts, and eventually to publish all his/her supporting data and conclusions for archival purposes.

In the fields of environmental contamination and toxicology, the sum of these concerns and responsibilities is decisively addressed by the uniform, encompassing, and timely publication format of the Springer triumvirate:

Reviews of Environmental Contamination and Toxicology [Vol. 1 through 97 (1962–1986) as Residue Reviews] for detailed review articles concerned with any aspects of chemical contaminants, including pesticides, in the total environment with toxicological considerations and consequences.

Bulletin of Environmental Contamination and Toxicology (Vol. 1 in 1966) for rapid publication of short reports of significant advances and discoveries in the fields of air, soil, water, and food contamination and pollution as well as methodology and other disciplines concerned with the introduction, presence, and effects of toxicants in the total environment.

Archives of Environmental Contamination and Toxicology (Vol. 1 in 1973) for important complete articles emphasizing and describing original experimental or theoretical research work pertaining to the scientific aspects of chemical contaminants in the environment.

The individual editors of these three publications comprise the joint Coordinating Board of Editors with referral within the board of manuscripts submitted to one publication but deemed by major emphasis or length more suitable for one of the others.

Coordinating Board of Editors

Preface

The role of *Reviews* is to publish detailed scientific review articles on all aspects of environmental contamination and associated (eco)toxicological consequences. Such articles facilitate the often complex task of accessing and interpreting cogent scientific data within the confines of one or more closely related research fields.

In the 50+ years since *Reviews of Environmental Contamination and Toxicology* (formerly *Residue Reviews)* was first published, the number, scope, and complexity of environmental pollution incidents have grown unabated. During this entire period, the emphasis has been on publishing articles that address the presence and toxicity of environmental contaminants. New research is published each year on a myriad of environmental pollution issues facing people worldwide. This fact, and the routine discovery and reporting of emerging contaminants and new environmental contamination cases, creates an increasingly important function for *Reviews.* The staggering volume of scientific literature demands remedy by which data can be synthesized and made available to readers in an abridged form. *Reviews* addresses this need and provides detailed reviews worldwide to key scientists and science or policy administrators, whether employed by government, universities, nongovernmental organizations, or the private sector.

There is a panoply of environmental issues and concerns on which many scientists have focused their research in past years. The scope of this list is quite broad, encompassing environmental events globally that affect marine and terrestrial ecosystems; biotic and abiotic environments; impacts on plants, humans, and wildlife; and pollutants, both chemical and radioactive; as well as the ravages of environmental disease in virtually all environmental media (soil, water, air). New or enhanced safety and environmental concerns have emerged in the last decade to be added to incidents covered by the media, studied by scientists, and addressed by governmental and private institutions. Among these are events so striking that they are creating a paradigm shift. Two in particular are at the center of ever increasing media as well as scientific attention: bioterrorism and global warming. Unfortunately, these very worrisome issues are now superimposed on the already extensive list of ongoing environmental challenges.

The ultimate role of publishing scientific environmental research is to enhance understanding of the environment in ways that allow the public to be better informed or, in other words, to enable the public to have access to sufficient information. Because the public gets most of its information on science and technology from internet, TV news, and reports, the role for scientists as interpreters and brokers of scientific information to the public will grow rather than diminish. Environmentalism is an important global political force, resulting in the emergence of multinational consortia to control pollution and the evolution of the environmental ethic. Will the new politics of the twenty-first century involve a consortium of technologists and environmentalists, or a progressive confrontation? These matters are of genuine concern to governmental agencies and legislative bodies around the world.

For those who make the decisions about how our planet is managed, there is an ongoing need for continual surveillance and intelligent controls to avoid endangering the environment, public health, and wildlife. Ensuring safety-in-use of the many chemicals involved in our highly industrialized culture is a dynamic challenge, because the old, established materials are continually being displaced by newly developed molecules more acceptable to federal and state regulatory agencies, public health officials, and environmentalists. New legislation that will deal in an appropriate manner with this challenge is currently in the making or has been implemented recently, such as the REACH legislation in Europe. These regulations demand scientifically sound and documented dossiers on new chemicals.

Reviews publishes synoptic articles designed to treat the presence, fate, and, if possible, the safety of xenobiotics in any segment of the environment. These reviews can be either general or specific, but properly lie in the domains of analytical chemistry and its methodology, biochemistry, human and animal medicine, legislation, pharmacology, physiology, (eco)toxicology, and regulation. Certain affairs in food technology concerned specifically with pesticide and other food-additive problems may also be appropriate.

Because manuscripts are published in the order in which they are received in final form, it may seem that some important aspects have been neglected at times. However, these apparent omissions are recognized, and pertinent manuscripts are likely in preparation or planned. The field is so very large and the interests in it are so varied that the editor and the editorial board earnestly solicit authors and suggestions of underrepresented topics to make this international book series yet more useful and worthwhile.

Justification for the preparation of any review for this book series is that it deals with some aspect of the many real problems arising from the presence of anthropogenic chemicals in our surroundings. Thus, manuscripts may encompass case studies from any country. Additionally, chemical contamination in any manner of air, water, soil, or plant or animal life is within these objectives and their scope.

Manuscripts are often contributed by invitation. However, nominations for new topics or topics in areas that are rapidly advancing are welcome. Preliminary communication with the Editor-in-Chief is recommended before volunteered review manuscripts are submitted. *Reviews* is registered in WebofScience™.

Inclusion in the Science Citation Index serves to encourage scientists in academia to contribute to the series. The impact factor in recent years has increased from 2.5 in 2009 to 7.0 in 2017. The Editor-in-Chief and the Editorial Board strive for a further increase of the journal impact factor by actively inviting authors to submit manuscripts.

Amsterdam, The Netherlands Pim de Voogt
February 2020

Contents

Trends and Sources of Heavy Metal Pollution in Global River and Lake Sediments from 1970 to 2018 1
Yandong Niu, Falin Chen, Youzhi Li, and Bo Ren

A Review on Prediction Models for Pesticide Use, Transmission, and Its Impacts 37
Edwin Prem Kumar Gilbert and Lydia Edwin

Effects of Dissolved Organic Matter on the Bioavailability of Heavy Metals During Microbial Dissimilatory Iron Reduction: A Review 69
Yuanhang Li and Xiaofeng Gong

The Toxic Effect of Silver Nanoparticles on Nerve Cells: A Systematic Review and Meta-Analysis 93
Atousa Janzadeh, Michael R. Hamblin, Narges Janzadeh, Hossein Arzani, MahsaTashakori-Miyanroudi, Mahmoud Yousefifard, and Fatemeh Ramezani

A Systematic Review on Occurrence and Ecotoxicity of Organic UV Filters in Aquatic Organisms 121
Ved Prakash and Sadasivam Anbumani

Micro and Nano-Plastics in the Environment: Research Priorities for the Near Future 163
Marco Vighi, Javier Bayo, Francisca Fernández-Piñas, Jesús Gago, May Gómez, Javier Hernández-Borges, Alicia Herrera, Junkal Landaburu, Soledad Muniategui-Lorenzo, Antonio-Román Muñoz, Andreu Rico, Cristina Romera-Castillo, Lucía Viñas, and Roberto Rosal

List of Contributors

Sadasivam Anbumani Ecotoxicology Laboratory, Regulatory Toxicology Group, CSIR-Indian Institute of Toxicology Research, Lucknow, India

Academy of Scientific and Innovative Research (AcSIR), Ghaziabad, India

Hossein Arzani Department of Medical Physics and Biomedical Engineering, Shahid Beheshti University of Medical Sciences, Tehran, Iran

Javier Bayo Department of Chemical and Environmental Engineering, Technical University of Cartagena, Cartagena, Spain

Falin Chen College of Resources and Environment, Hunan Agricultural University, Changsha, China

Lydia Edwin Department of Mechatronics Engineering, Sri Krishna College of Engineering and Technology, Coimbatore, Tamil Nadu, India

Francisca Fernández-Piñas Departamento de Biología, Facultad de Ciencias, Universidad Autónoma de Madrid, Madrid, Spain

Jesús Gago Instituto Español de Oceanografía (IEO), Vigo, Spain

Edwin Prem Kumar Gilbert Department of Information Technology, Sri Krishna College of Engineering and Technology, Coimbatore, Tamil Nadu, India

May Gómez EOMAR: Marine Ecophysiology Group, IU-ECOAQUA, Universidad de Las Palmas de Gran Canaria, Las Palmas de Gran Canaria, Spain

Xiaofeng Gong School of Resources, Environmental and Chemical Engineering, Nanchang University, Nanchang, China

Key Laboratory of Poyang Lake Environment and Resource Utilization, Ministry of Education, Nanchang University, Nanchang, China

Michael R. Hamblin Wellman Center for Photomedicine, Massachusetts General Hospital, Harvard Medical School, Boston, MA, USA

Laser Research Centre, Faculty of Health Science, University of Johannesburg, Doornfontein, South Africa

Javier Hernández-Borges Departamento de Química, Unidad Departamental de Química Analítica, Facultad de Ciencias, Universidad de La Laguna, San Cristóbal de La Laguna, Spain

Instituto Universitario de Enfermedades Tropicales y Salud Pública de Canarias, Universidad de La Laguna, San Cristóbal de La Laguna, Spain

Alicia Herrera EOMAR: Marine Ecophysiology Group, IU-ECOAQUA, Universidad de Las Palmas de Gran Canaria, Las Palmas de Gran Canaria, Spain

Atousa Janzadeh Radiation Biology Research Center, Iran University of Medical Sciences, Tehran, Iran

Narges Janzadeh Occupational Medicine Research Center (OMRC), Iran University of Medical Sciences (IUMS), Tehran, Iran

Junkal Landaburu IMDEA-Water Institute, Madrid, Spain

Youzhi Li College of Resources and Environment, Hunan Agricultural University, Changsha, China

Yuanhang Li School of Resources, Environmental and Chemical Engineering, Nanchang University, Nanchang, China

Key Laboratory of Poyang Lake Environment and Resource Utilization, Ministry of Education, Nanchang University, Nanchang, China

Soledad Muniategui-Lorenzo Grupo Química Analítica Aplicada, Instituto Universitario de Medio Ambiente (IUMA), Centro de Investigaciones Científicas Avanzadas (CICA), Facultade de Ciencias, Universidade da Coruña, A Coruña, Spain

Antonio-Román Muñoz Departamento de Biología Animal, Facultad de Ciencias, Universidad de Málaga, Málaga, Spain

Yandong Niu College of Resources and Environment, Hunan Agricultural University, Changsha, China

Hunan Academy of Forestry, Changsha, China

Ved Prakash Ecotoxicology Laboratory, Regulatory Toxicology Group, CSIR-Indian Institute of Toxicology Research, Lucknow, India

Academy of Scientific and Innovative Research (AcSIR), Ghaziabad, India

Fatemeh Ramezani Physiology Research Center, Iran University of Medical Sciences, Tehran, Iran

Bo Ren College of Resources and Environment, Hunan Agricultural University, Changsha, China

Andreu Rico IMDEA-Water Institute, Madrid, Spain

Cavanilles Institute of Biodiversity and Evolutionary Biology, University of Valencia, Valencia, Spain

Cristina Romera-Castillo Department of Marine Biology and Oceanography, Institut de Ciències del Mar-CSIC, Barcelona, Spain

Roberto Rosal Department of Chemical Engineering, University of Alcalá, Madrid, Spain

Mahsa Tashakori-Miyanroudi Department of Medical Physiology, Faculty of Medicine, Iran University of Medical Sciences, Tehran, Iran

Marco Vighi IMDEA-Water Institute, Madrid, Spain

Lucía Viñas Instituto Español de Oceanografía (IEO), Vigo, Spain

Mahmoud Yousefifard Physiology Research Center, Iran University of Medical Sciences, Tehran, Iran

Trends and Sources of Heavy Metal Pollution in Global River and Lake Sediments from 1970 to 2018

Yandong Niu, Falin Chen, Youzhi Li, and Bo Ren

Contents

1 Introduction 2
2 Materials and Methods 4
  2.1 Data Collection 4
  2.2 Trend Assessment 4
  2.3 Source Apportionment 11
3 Results 12
  3.1 Trends of Heavy Metal Pollution in River Sediments 12
  3.2 Trends of Heavy Metal Pollution in Lake Sediments 14
  3.3 Sources of Sediment Heavy Metal Pollution Over Time 14
  3.4 Sources of Sediment Heavy Metal Pollution by Continent 17
4 Discussion 17
  4.1 Trends in Heavy Metal Pollution of River and Lake Sediments 17
  4.2 Heavy Metal Concentrations in Sediments by Continents 17
  4.3 Source of Heavy Metal Pollution in River and Lake Sediments 19
  4.4 Control Measures for Heavy Metal Pollution in Sediments 19
5 Conclusions 22
References 22

Abstract Heavy metal pollution is a global problem although its sources and trends differ by region and time. To data, no published research has reported heavy metal pollution in global rivers and lakes. This study reviewed past sampling data across six continents from 1970 to 2018 and analyzed the trends and sources of 10 heavy metal species in sediments from 289 rivers and 133 lakes. Collectively, river sediments showed increasing trends in Cd, Cr, Ni, Mn, and Co and decreasing trends in Hg, indicating that rivers acted as a sink for the former and a source for

Y. Niu
College of Resources and Environment, Hunan Agricultural University, Changsha, China

Hunan Academy of Forestry, Changsha, China

F. Chen · Y. Li (✉) · B. Ren
College of Resources and Environment, Hunan Agricultural University, Changsha, China
e-mail: liyouzhi2004@163.com

P. de Voogt (ed.), *Reviews of Environmental Contamination and Toxicology Volume 257*,
Reviews of Environmental Contamination and Toxicology 257,
https://doi.org/10.1007/398_2020_59

the latter. Lake sediments showed increasing trends in Pb, Hg, Cr, and Mn, and decreasing trends in Cd, Zn, and As, indicating that lakes acted as a sink for the former and a source for the latter. Due to difference in natural backgrounds and development stage in continents, mean metal concentrations were generally higher in Europe and North America than in Africa, Asia, and South America. Principal component analysis showed that main metal source was mining and manufacturing from the 1970s to 1990s and domestic waste discharge from the 2000s to 2010s. Metal sources in sediments differed greatly by continent, with rock weathering dominant in Africa, mining and manufacturing dominant in North America, and domestic waste discharge dominant in Asia and Europe. Global trends in sediment metal loads and pollution-control measures suggest that the implementation of rigorous standards on metal emissions, limitations on metal concentrations in manufactured products, and the pretreatment of metal-contaminated waste have been effective at controlling heavy metal pollution in rivers and lakes. Thus, these efforts should be extended globally.

Keywords Cd, Pb, Hg, Cr, Zn, Cu, Ni, Mn, As, Co · Environmental regulation · Fertilizer and pesticide use · Global pollution · Heavy metal pollution · Lakes · Mining and manufacturing · Pollution source · Pollution trend · Pollution-control measures · Rivers · Rock weathering · Sediment · Waste discharge

Abbreviations

CEC	Council of the European Communities
EPA	Environmental Protection Agency
EU	European Union
M-K	Mann–Kendall
MLR	Multiple linear regression
PCA	Principal component analysis
US	United States
USA	United States of America

1 Introduction

Heavy metal pollution is a global environmental problem caused by increasing mining and refining, the manufacture of metal-contaminated products, the usage of fertilizers and pesticides, and the discharge of domestic waste (Facchinelli et al. 2001; Muhammad et al. 2011; Hu et al. 2015; Huang et al. 2015; Ren et al. 2015). These processes release major heavy metals, such as Cd, Pb, Hg, and Cr, into the atmosphere, water, and soil. In the last half of the twentieth century, the total worldwide release of heavy metals to atmosphere, water, and soil reached 22,000 t for Cd, 939,000 t for Cu, 783,000 t for Pb, and 1,350,000 t for Zn (Singh et al. 2003).

These substances are mainly concentrated in rivers and lakes, ultimately enriching sediments that act as a long-term sinks for heavy metals (Von Gunten et al. 1997; Audry et al. 2004; Peng et al. 2009; Varol 2011; Hu et al. 2015; Huang et al. 2015). For example, in 1974, the concentrations of Cd and Pb in sediments of the Coeur d'Alene River, Idaho, United States of America (USA), reached 17.6 mg kg^{-1} and 2,580.2 mg kg^{-1}, respectively (Reece et al. 1978). In 2008, the concentration of Hg in sediments of the Kızılırmak River, Turkey, was as high as 9.1 mg kg^{-1} (Akbulut and Akbulut 2010). The heavy metals in sediments would be released to the water body and be accumulated in aquatic organisms, such as fish and shrimp (Yu et al. 2012; Liu et al. 2015a, b). A study on India's Jamshedpur Urban Agglomeration showed that heavy metal concentrations in sediments reached 8.1 mg kg^{-1} for Cd (background value 0.3 kg^{-1}) and 135.9 mg kg^{-1} for Pb (background value 20.0 kg^{-1}) and metal concentrations in the fish were as high as 0.8 mg kg^{-1} for Cd and 10.2 mg kg^{-1} for Pb (Kumari et al. 2018). Therefore, the heavy metals in aquatic sediments pose a hazard to human health through food chain and require urgent research attention (Williams et al. 1978; Zingde et al. 1988; Taher and Soliman 1999; Audry et al. 2004).

Many processes contribute to heavy metal pollution in sediments; among them, rock weathering is a natural source, while mining and manufacturing activities, fertilizer and pesticide use, and domestic waste discharge are anthropogenic sources (Mortvedt 1996; Wei and Yang 2010; Muhammad et al. 2011). Global actions have been taken to control the rising trend of heavy metal pollution. For example, since the 1970s, the Congress of the US has mandated that the federal Environmental Protection Agency (EPA) regulates the manufacture, processing, commercial use, labeling, and disposal of harmful substances (Babich and Stotzky 1985). During the 1980s, the EPA's attention was directed towards the regulation of maximum metal concentrations permitted in fertilizers and maximum metal loading in agricultural lands (Mortvedt 1996). In the 1990s, the European Community made the collection and treatment of municipal wastewater compulsory and final disposal to surface water was forbidden (CEC 1991). In the 2000s, the Chinese government prohibited leaded gasoline nationwide and strengthened local emission standards for coal combustion (Duan and Tan 2013). These policies and measures have produced meaningful effects on the source control of heavy metal pollution (Mortvedt 1996; Kelessidis and Stasinakis 2012; Duan and Tan 2013).

However, although the external input of heavy metals is decreasing, sediments can release stored metals into overlying water bodies and thus remain a pollution source (Zoumis et al. 2001; Peng et al. 2009). Whether sediments are acting as a sink or source of heavy metals can influence pollution in waterways and reflect the effects of pollution-control measures (Förstner 1976; Neumann et al. 1998). To date, no published research has reported heavy metal pollution in global rivers and lakes. Therefore, this study investigated the trends and sources of heavy metal pollution in sediments associated with global rivers and lakes from 1970 to 2018, in order to assess the effects of control measures on heavy metal pollution and to propose effective remediation strategies for metal-polluted rivers and lakes.

2 Materials and Methods

2.1 Data Collection

Concentrations of 10 heavy metal species (Cd, Pb, Hg, Cr, Zn, Cu, Ni, Mn, As, and Co) in the surface sediments (0–30 cm depth) of rivers and lakes around the world were collected from published papers; this search was conducted using Google Scholar and Web of Science. Each sample was assigned a specific year by the reported sampling date as follows: for a single sampling year, that year was used; for a sampling range of 1–2 years, the first year was used; and for a sampling range of more than 2 years, the middle year was used. When the sampling date was not provided, the year prior to publication was used. The samples reviewed were collected from a total of 289 rivers and 133 lakes in Africa, Asia, Europe, North America, Oceania, and South America and selected from pristine areas and polluted areas from 1970 to 2018 (Tables 1, 2, 3, 4, and 5).

2.2 Trend Assessment

The Mann–Kendall (M–K) test (Mann 1945; Kendall 1975) has been extensively used to detect change trends in heavy metal pollution over time (Gao et al. 2016; Sharley et al. 2016). In this test, the null (H_0) and alternative hypotheses (H_1) denote the nonexistence and existence of a trend in the time series of the observational data, respectively. The equations for calculating the M–K test statistic S and the standardized test statistic Z_{M-K} are as follows (Kisi and Ay 2014):

$$sgn\ (x_j - x_i) = \begin{cases} +1; & if\ \ x_j > x_i \\ 0; & if\ \ x_j = x_i \\ -1; & if\ \ x_j < x_i \end{cases} \tag{1}$$

$$S = \sum_{i=1}^{n-1} \sum_{j=i+1}^{n} sgn\ (x_j - x_i) \tag{2}$$

$$Var\ (S) = \frac{1}{18}\left[n(n-1)(2n+5) - \sum_{p=1}^{q} t_p(t_p - 1)(2t_p + 5)\right] \tag{3}$$

Table 1 Numbers and regional distributions of rivers and lakes reviewed in this study

	Rivers		Lakes	
Continent	No. of rivers	Typical river examples	No. of lakes	Typical lake examples
Africa	18[a]	Awach River, Congo River, Luangwa River, Nile River, Niger River, Nyando River, Nzoia River, Zambezi River	17[b]	Lake Edku, Jebba Lake, Kainji Lake, Lake Kariba, Lake Manzala, Lake Nakuru, Qarun Lake, Lake Tanganyika, Lake Victoria
Asia	110[c]	Brahmaputra River, Ganges River, Cauvery River, Mekong River, Narmada River, Ob River, Pearl River, Tigris River, Yangtze River, Yellow River,	80[d]	Atatürk Dam Lake, Beyşehir Lake, Chaohu Lake, Dongting Lake, Habbaniyah Lake, Kovada Lake, Poyang Lake, Lake Taihu, Uluabat Lake, Veeranam Lake
Europe	24[e]	Danube River, Elbe River, Guadiamar River, Lanh River, Po River, Rhine River, Scheldt River, Tees River, Tinto River, Tisza River	11[f]	Almadén Lake, Lake Constance, Hjalmaren Lake, Siilinjärvi Lake, Velenjsko Jezero Lake, Venice Lagoon, Zurich Lake
North America	125[g]	Grand River, Maumee River, Mississippi River, Portage River, Raisin River, Saginaw River, South Platte River, Saint Clair River, Tippecanoe River, Trent River	24[h]	Lake Erie, Fontana Lake, Lake Michigan, Lake Monona, Lake Ontario, Lake Palestine, Lake Washington, Yojoa Lake
South America	10[i]	Kaw River, Guandu River, Salado River, Sinnamary River, Suquía River	–	–
Oceania	2[j]	Brisbane River, Yarra River	1[k]	Lake Illawarra
Total	289		133	

[a]Ndiokwere (1984), Nriagu (1986), Zayed et al. (1994), Dupré et al. (1996), Elghobashy et al. (2001), Milenkovic et al. (2005), Rifaat (2005), Lalah et al. (2008), Lasheen and Ammar (2009), Ikenaka et al. (2010), Osman and Kloas (2010), Sekabira et al. (2010), Phillips et al. (2015), Addo-Bediako et al. (2018) and Duncan et al. (2018)
[b]Mothersill (1976), Onyari and Wandiga (1989), Abdel-Moati and El-Sammak (1997), Taher and Soliman (1999), Elghobashy et al. (2001), Rashed (2001), Chale (2002), Kishe and Machiwa (2003), Mavura and Wangila (2003), Oyewale and Musa (2006), Saeed and Shaker (2008), Nakayama et al. (2010), El-Sayed et al. (2015), Maanan et al. (2015) and El-Amier et al. (2017)
[c]Subramanian et al. (1985, 1987), Ajmal et al. (1988), Biksham and Subramanian (1988), Zingde et al. (1988), Ramesh et al. (1989), Shen et al. (1989), Sabri et al. (1993), Prudente et al. (1994), Chen and Wu (1995), Bradley and Woods (1997), Singh et al. (1997), Datta and Subramanian (1998), Ranjbar (1998), Zhang (1999), Dauvalter and Rognerud (2001), Sin et al. (2001), Zhang et al. (2001), Akcay et al. (2003), Che et al. (2003), Liu et al. (2003), Dalai et al. (2004), Feng et al. (2004), Gorenc et al. (2004), Singh et al. (2005), Demirak et al. (2006), Prasad et al. (2006), Dundar and Altundag (2007), Ip et al. (2007), Li et al. (2007), Karadede-Akin and Ünlü (2007), Chaparro et al. (2008), Karbassi et al. (2008), Ruilian et al. (2008), Wang et al. (2008), Gupta et al. (2009), Liu et al.

(continued)

(2009a, b), Niu et al. (2009), Yang et al. (2009), Zhang et al. (2009), Zorer et al. (2009), Akbulut and Akbulut (2010), Kim et al. (2010), Mendil et al. (2010), Mohiuddin et al. (2010), Salati and Moore (2010), Bai et al. (2011b), Chaparro et al. (2011), Cui et al. (2011), Saha and Hossain (2011), Suresh et al. (2011), Varol (2011), Bai et al. (2012), Raju et al. (2012), Yang et al. (2012), Shafie et al. (2013), Fu et al. (2014), Rahman et al. (2014), Yuan et al. (2014), Cheng et al. (2015b), Dhanakumar et al. (2015), Islam et al. (2015a, b), Li et al. (2015), Paramasivam et al. (2015), Zhang et al. (2015), Ali et al. (2016), Islam et al. (2016), Liang et al. (2016), Ma et al. (2016), Nguyen et al. (2016), Yan et al. (2016), Zhang et al. (2016), Zhuang et al. (2016), Ke et al. (2017), Li et al. (2017), Malvandi (2017), Pandey and Singh (2017), Wang et al. (2017), Wong et al. (2017), Xu et al. (2017), Zhang et al. (2017a, b, c), Zhao et al. (2017), Chen et al. (2018), Islam et al. (2018), Patel et al. (2018), Xia et al. (2018), Yan et al. (2018), Zhang et al. (2018), Kang et al. (2019), Mariyanto et al. (2019) and Siddiqui and Pandey (2019)
[d]Chen et al. (1989), Saeki et al. (1993), Karadede and Ünlü (2000), Wenchuan et al. (2001), Al-Saadi et al. (2002), Altındağ and Yiğit (2005), Barlas et al. (2005), Qian et al. (2005), Tekin-Zan and Kir (2005), Yang et al. (2007), Tekin-Zan (2008), Huang et al. (2009), Pathiratne et al. (2009), Peng et al. (2009), Yang et al. (2009), Zheng et al. (2010), Bai et al. (2011a), Bing et al. (2011), Yuan et al. (2011), Zan et al. (2011), Selvam et al. (2012), Suresh et al. (2012), Tao et al. (2012), Huang et al. (2013), Li et al. (2013), Rahman et al. (2014), Yuan et al. (2014), Cheng et al. (2015a), Guo et al. (2015), Hu et al. (2015), Liang et al. (2015), Liu et al. (2015a, b), Swarnalatha et al. (2015), Wang et al. (2015), Zhang et al. (2015), Mamat et al. (2016), Wen et al. (2016), Zhang et al. (2017c), Hu et al. (2018), Jiang et al. (2018), Ramachandra et al. (2018), Xiong et al. (2018) and Bing et al. (2019)
[e]Förstner and Müller (1973), Müller and Förstner (1975), Yim (1976), Bryan and Hummerstone (1977), Guerzoni et al. (1984), Reboredo and Ribeiro (1984), Hudson-Edwards et al. (1997), Ranjbar (1998), Morillo et al. (2002a, b), Bermejo et al. (2003), Galán et al. (2003), Lacal et al. (2003), Vandecasteele et al. (2003), Viganò et al. (2003), Woitke et al. (2003), Audry et al. (2004), Martin (2004), Farkas et al. (2007), Sakan et al. (2009), Beltrán et al. (2010), Bonanno and Giudice (2010), Mendil et al. (2010), Botsou et al. (2011) and Rahman et al. (2014)
[f]Tenhola (1993), Martin et al. (1994), Birch et al. (1996), Müller et al. (1977), Von Gunten et al. (1997), Ranjbar (1998), Müller et al. (2000), Mazej et al. (2010), Hahladakis et al. (2013), Begy et al. (2016) and García-Ordiales et al. (2016)
[g]Oliver (1973), Walker and Colwell (1974), Fitchko and Hutchinson (1975), Grieve and Fletcher (1976), Williams et al. (1978), Wilber and Hunter (1979), Adams et al. (1980), Simpson et al. (1983), White and Tittlebaum (1985), Davies et al. (1991), Murray (1996), Trocine and Trefry (1996), Heiny and Tate (1997), Grabowski et al. (2001), Olivares-Rieumont et al. (2005) and Marrugo-Negrete et al. (2017)
[h]Förstner (1976), Wentsel et al. (1977a, b, c), Jackson (1979), Nriagu et al. (1979), Adams et al. (1980), Glooschenko et al. (1981), Glooschenko (1982), Taylor and Crowder (1983), Abernathy et al. (1984), Tessier et al. (1984), Yousef et al. (1984), White and Tittlebaum (1985), De Vevey et al. (1993), Rowan and Kalff (1993), Ranjbar (1998), Aucoin et al. (1999), An and Kampbell (2003), Arnason and Fletcher (2003), Gewurtz et al. (2007) and Opfer et al. (2011)
[i]Malm et al. (1989), Marchand et al. (2006), Gagneten and Paggi (2009), Harguinteguy et al. (2014), de Paula Filho et al. (2015) and Pesantes et al. (2019)
[j]Duodu et al. (2016) and Lintern et al. (2016)
[k]Chenhall et al. (1992)

Table 2 Metal concentrations (mean and standard deviation [SD]; the first quartile [FQ], median, and third quartile [TQ]; g kg^{-1} for Fe and Al, mg kg^{-1} for other metals), pristine sample number (PRSN), polluted sample number (POSN), and total selected sample number (TSN) for global river sediments and Mann–Kendall (M–K) test results from the 1970s to 2010s

	1970s				1980s				1990s				2000s				2010s				1970–2018							
Metals	Mean ± SD	PRSN	POSN	TSN	Mean ± SD	PRSN	POSN	TSN	Mean ± SD	PRSN	POSN	TSN	Mean ± SD	PRSN	POSN	TSN	Mean ± SD	PRSN	POSN	TSN	M-K test	Mean ± SD	FQ	Median	TQ	PRSN	POSN	TSN
Cd	3.44 ± 6.34	2	127	127	4.00 ± 6.51	0	8	9	1.72 ± 1.86	5	16	20	3.08 ± 7.76	15	59	61	1.47 ± 3.10	15	50	60	0.24	2.83 ± 5.97	0.38	1.20	2.17	37	260	277
Pb	71.62 ± 233.35	9	128	135	43.06 ± 78.00	12	16	24	121.24 ± 390.42	9	26	30	135.45 ± 385.06	20	78	80	72.01 ± 280.13	16	56	70	0.00	89.08 ± 294.39	14.00	26.14	55.59	66	304	339
Hg	0.55 ± 1.51	1	124	125	0.41 ± 0.31	0	2	2	0.10 ± 0.06	1	2	3	0.65 ± 1.96	6	20	21	1.01 ± 0.97	3	7	9	−0.73	0.58 ± 1.52	0.05	0.15	0.44	11	155	160
Cr	56.64 ± 147.86	6	120	125	65.77 ± 34.13	16	11	23	107.67 ± 100.30	6	16	19	73.96 ± 82.86	14	59	62	77.89 ± 37.96	16	48	63	1.22	68.97 ± 109.19	13.45	46.48	84.88	58	254	292
Zn	173.69 ± 332.99	9	133	140	89.27 ± 121.48	20	17	33	272.25 ± 305.47	8	22	26	277.77 ± 452.52	20	78	81	186.83 ± 382.22	16	54	68	0.00	199.81 ± 362.28	43.25	84.00	177.36	73	304	348
Cu	46.44 ± 110.94	10	128	136	40.52 ± 44.82	20	16	31	94.42 ± 181.11	9	25	29	296.76 ± 1155.67	20	78	80	153.51 ± 588.14	15	55	69	0.00	129.47 ± 626.74	13.32	28.10	66.16	74	302	345
Ni	26.88 ± 34.36	9	123	131	33.90 ± 24.90	21	15	32	45.67 ± 67.15	6	18	20	62.61 ± 131.03	10	53	55	47.03 ± 103.20	15	38	50	1.71	39.31 ± 78.46	13.33	23.23	38.50	61	247	288
Mn	297.38 ± 532.78	8	119	127	590.64 ± 347.39	20	11	26	914.36 ± 1018.88	6	16	19	901.71 ± 1139.13	9	33	35	637.61 ± 420.83	7	13	21	0.73	506.34 ± 726.63	150.71	270.00	671.75	50	192	228
As	102.41 ± 272.20	6	2	8	7.35 ± 9.40	0	2	2	43.58 ± 70.97	1	2	4	82.72 ± 239.53	8	24	25	64.73 ± 172.24	10	29	37	0.00	71.90 ± 199.75	5.57	11.00	21.95	25	59	76
Co	15.70 ± 18.28	7	122	128	33.41 ± 15.66	13	8	16	17.97 ± 11.12	2	8	9	21.57 ± 50.58	3	28	29	12.87 ± 7.18	7	7	13	0.24	17.94 ± 25.34	7.72	11.73	18.00	32	173	195

Table 3 Metal concentrations (mean and standard deviation [SD]; g kg^{-1} for Fe and Al, mg kg^{-1} for other metals), pristine sample number (PRSN), polluted sample number (POSN), and total selected sample number (TSN) for the river sediments of five selected continents from 1970 to 2018

	Africa				Asia				Europe				North America				South America			
Metals	Mean ± SD	PRSN	POSN	TSN	Mean ± SD	PRSN	POSN	TSN	Mean ± SD	PRSN	POSN	TSN	Mean ± SD	PRSN	POSN	TSN	Mean ± SD	PRSN	POSN	TSN
Cd	1.63 ± 3.21	5	16	20	1.87 ± 4.76	30	85	96	7.40 ± 10.58	2	33	33	2.56 ± 5.01	7	119	122	2.27 ± 3.94	0	5	5
Pb	46.01 ± 70.38	6	18	23	39.81 ± 45.89	57	121	146	456.63 ± 706.68	3	35	35	57.09 ± 233.77	12	120	123	19.93 ± 7.10	1	10	11
Hg	0.13 ± 0.21	1	7	8	0.83 ± 2.29	8	13	15	2.83 ± 3.43	2	16	16	0.24 ± 0.35	8	114	116	0.66 ± 0.71	0	4	4
Cr	36.85 ± 51.33	6	14	20	82.43 ± 54.80	49	98	125	160.37 ± 137.72	3	26	26	39.46 ± 139.47	4	112	115	67.01 ± 35.72	1	4	5
Zn	120.80 ± 98.26	6	19	24	134.63 ± 172.46	65	116	150	822.02 ± 696.08	3	36	36	121.49 ± 258.51	10	124	127	120.94 ± 51.49	1	9	10
Cu	254.78 ± 1005.50	5	18	22	156.53 ± 832.29	66	120	155	268.24 ± 356.40	3	35	35	27.84 ± 37.30	13	120	122	202.39 ± 274.10	1	9	10
Ni	66.66 ± 164.21	4	16	20	46.94 ± 84.50	54	82	116	86.58 ± 103.64	3	25	25	17.44 ± 19.47	10	115	117	35.60 ± 21.39	0	9	9
Mn	1576.00 ± 1814.76	5	10	15	650.23 ± 487.66	44	47	76	888.65 ± 599.16	1	13	13	232.31 ± 436.01	11	115	117	535.79 ± 155.30	1	5	6
As	3.42 ± 2.78	1	7	8	13.61 ± 11.76	24	39	53	340.60 ± 420.69	0	8	8	5.68 ± 0.95	2	0	2	505.06 ± 274.02	0	4	4
Co	25.86 ± 79.19	0	12	12	27.72 ± 24.39	31	32	53	34.19 ± 18.28	1	13	13	10.85 ± 6.97	5	112	113	9.64 ± 8.04	0	3	3

Table 4 Metal concentrations (mean and standard deviation [SD]; the first quartile [FQ], median, and third quartile [TQ]; g kg^{-1} for Fe and Al, mg kg^{-1} for other metals), pristine sample number (PRSN), polluted sample number (POSN), and total selected sample number (TSN) for global lake sediments and Mann–Kendall (M–K) test results from the 1970s to 2010s

	1970s				1980s				1990s				2000s				2010s				1970–2018							
Metals	Mean ± SD	PRSN	POSN	TSN	Mean ± SD	PRSN	POSN	TSN	Mean ± SD	PRSN	POSN	TSN	Mean ± SD	PRSN	POSN	TSN	Mean ± SD	PRSN	POSN	TSN	M-K test	Mean ± SD	FQ	Median	TQ	PRSN	POSN	TSN
Cd	122.82 ± 212.04	1	15	15	2.24 ± 3.49	2	8	9	8.74 ± 11.39	5	13	16	2.87 ± 11.56	14	43	55	1.20 ± 2.16	33	34	59	−0.73	14.49 ± 73.86	0.23	0.61	2.09	52	113	154
Pb	42.66 ± 43.19	1	11	11	167.33 ± 338.87	3	12	14	253.63 ± 841.85	6	14	18	36.20 ± 25.25	14	47	59	41.81 ± 22.43	34	37	62	1.22	73.81 ± 298.15	20.05	35.70	56.23	57	121	164
Hg	0.56 ± 0.78	1	11	11	–	–	–	–	4.63 ± 2.22	0	2	2	0.09 ± 0.07	12	21	33	1.20 ± 3.51	8	9	11	0.34	0.56 ± 1.79	0.04	0.10	0.19	21	43	57
Cr	353.00 ± 536.91	1	11	11	33.58 ± 11.07	2	4	6	53.61 ± 70.66	2	5	7	69.11 ± 37.50	13	35	47	83.64 ± 34.18	33	36	61	0.24	97.04 ± 171.52	42.19	74.98	91.80	51	91	132
Zn	1649.99 ± 3124.31	1	19	19	253.32 ± 286.45	3	13	15	109.43 ± 80.63	8	13	19	168.92 ± 480.61	14	45	58	131.06 ± 87.78	29	32	57	−1.22	324.38 ± 1169.58	61.14	109.16	166.65	52	122	168
Cu	36.08 ± 29.94	1	15	15	90.56 ± 100.07	3	12	14	62.05 ± 124.62	9	14	21	40.62 ± 44.49	14	46	59	52.28 ± 62.31	33	36	61	0.00	51.16 ± 70.31	20.18	35.82	50.39	56	123	170
Ni	38.07 ± 30.34	1	10	10	93.98 ± 171.47	2	5	7	50.91 ± 39.33	3	6	8	36.74 ± 20.99	13	25	37	43.54 ± 18.53	25	29	52	0.00	44.47 ± 46.95	26.40	39.84	49.27	44	75	114
Mn	294.94 ± 282.5	0	4	4	397.43 ± 300.13	2	2	4	578.40 ± 404.5	4	10	12	638.46 ± 406.14	1	15	16	1004.02 ± 530.49	21	19	40	2.20	790.61 ± 517.59	459.52	740.00	1017.50	28	50	76
As	5.90 ± 5.36	0	4	4	5.62	1	0	1	4.73	1	0	1	15.71 ± 18.60	14	24	36	27.73 ± 46.99	9	14	16	−0.73	17.98 ± 29.00	5.83	11.45	17.86	25	42	58
Co	20.65 ± 17.84	1	5	5	12.10	1	0	1	18.63 ± 5.30	2	1	3	29.48 ± 27.37	1	8	8	17.59 ± 7.65	22	17	39	0.00	19.52 ± 13.35	11.81	18.59	22.05	27	31	56

Table 5 Metal concentrations (mean and standard deviation [SD]; g kg^{-1} for Fe and Al, mg kg^{-1} for other metals), pristine sample number (PRSN), polluted sample number (POSN), and total selected sample number (TSN) for the lake sediments of four selected continents from 1970 to 2018

	Africa				Asia				Europe				North America			
Metals	Mean ± SD	PRSN	POSN	TSN	Mean ± SD	PRSN	POSN	TSN	Mean ± SD	PRSN	POSN	TSN	Mean ± SD	PRSN	POSN	TSN
Cd	7.32 ± 18.43	3	21	22	1.31 ± 2.51	46	64	99	5.68 ± 14.03	2	6	7	76.01 ± 172.25	4	22	25
Pb	37.15 ± 41.57	4	24	25	39.76 ± 17.89	46	66	101	71.23 ± 127.86	5	8	12	249.58 ± 774.78	3	23	25
Hg	0.87 ± 1.46	0	4	4	0.10 ± 0.08	20	24	37	3.09 ± 5.80	1	4	4	1.02 ± 1.79	0	12	12
Cr	50.23 ± 32.54	2	12	13	81.48 ± 35.07	44	59	94	70.28 ± 61.31	4	5	8	244.63 ± 468.29	1	15	16
Zn	128.37 ± 141.03	4	26	27	157.26 ± 374.44	41	60	95	117.05 ± 66.34	5	7	11	1012.61 ± 2427.30	5	30	34
Cu	64.97 ± 119.93	4	26	27	46.25 ± 49.99	45	66	101	44.06 ± 52.46	5	7	11	58.20 ± 77.53	6	25	30
Ni	45.33 ± 30.26	2	10	10	42.28 ± 20.99	37	45	79	43.33 ± 29.16	4	5	8	54.65 ± 117.34	1	16	17
Mn	531.09 ± 392.46	2	13	13	924.54 ± 532.00	22	30	50	744.67 ± 317.29	3	3	6	393.46 ± 374.43	1	5	6
As	11.46 ± 14.26	1	5	5	21.67 ± 33.50	22	28	41	9.03 ± 5.99	2	3	5	7.46 ± 4.63	0	7	7
Co	39.79 ± 27.25	2	5	6	17.60 ± 7.38	21	19	39	20.34 ± 12.77	3	4	6	9.15 ± 3.46	1	4	5

$$Z_{M-K} = \begin{cases} \dfrac{S-1}{\sqrt{Var(S)}}; & if \ S > 0 \\ 0; & if \ S = 0 \\ \dfrac{S+1}{\sqrt{Var(S)}}; & if \ S < 0 \end{cases} \tag{4}$$

where x_i and x_j are the sequential data values of time series at times i and j, respectively; n is the length of the time series; q is the number of tied groups (data with the same value); and t_p is the number of data values in the pth group. Positive values of Z_{M-K} indicate increasing trends, while negative values of Z_{M-K} indicate decreasing trends in the time series. Then, the Z_{M-K} value is compared with the standard normal distribution table with two-tailed confidence levels ($\alpha = 10\%$, 5%, and 1%). When $|Z_{M-K}| > Z_{1-\alpha/2}$, the null hypothesis ($H_0$) is rejected and a significant trend exists in the time series. Otherwise, the H0 hypothesis is accepted and the trend is not statically significant in the time series. $Z_{1-\alpha/2}$ is the critical value for Z_{M-K} and equals 1.96 for the 5% significant level.

As the volume of collected data (number of rivers or lakes) changed by year, the data were classified into five decadal groups (1970–1979, 1980–1989, 1990–1999, 2000–2009, and 2010–2018) to better explore the trends in sediment-borne heavy metal pollution. Average metal concentrations in each decadal group were determined as the mean of all collected data in the decadal group. Average metal concentrations for each continent were determined as the mean of all collected data for that continent. As the data from Oceania contained only two rivers (Brisbane River, Yarra River) and one lake (Lake Illawarra), these were excluded, and only data from Africa, Asia, Europe, North America, and South America were selected to compare mean metal concentrations.

2.3 Source Apportionment

Principal component analysis (PCA) followed by multiple linear regression (MLR) is a useful method for source apportionment (Yang et al. 2017; Ashayeri et al. 2018; Li et al. 2020). In this study, PCA-MLR was used to determine the contribution percentages of estimated metal sources to sediment pollution using the methods described by Larsen and Baker (2003). First, PCA was used to represent the total variability of the original metal data in a minimum number of factors. Each factor is orthogonal to all others, which results in the smallest possible covariance. The first factor represents the weighted (factor loadings) linear combination of the original variables that account for the greatest variability. Each subsequent factor accounts for less variability than the previous one. In this study, factors with an eigenvalue of greater than one were extracted. By critically evaluating the factor loadings, the metal source responsible for each factor could be identified (Larsen and Baker 2003).

Table 6 The potential sources associated with heavy metals

Metals	Potential sources
Cd	Mining, manufacturing wastewater, pesticide use, fertilizer use, battery use, coal combustion, automobile exhaust discharge, colorants use
Pb	Mining, metal smelting, industrial waste gas, pesticide use, fertilizer use, battery use, coal combustion, automobile exhaust discharge
Hg	Industrial waste water, fertilizer use, coal combustion
Cr	Mining, electroplating industry
Zn	Mining, rock weathering, pesticide use
Cu	Rock weathering, pesticide use,
Ni	Mining, rock weathering, electroplating industry
Mn	Rock weathering
As	Rock weathering, coal combustion
Co	Rock weathering

For example, it is assumed that Cd, Cr, and Zn had high loadings in a principal component. According to the previous studies, the Cd mainly originates from mining and manufacturing and domestic waste discharge, Cr from mining and manufacturing, and Zn from mining and manufacturing and rock weathering (Pekey et al. 2004; Bing et al. 2011; Hu et al. 2014; Ma et al. 2016). Therefore, the principal component may be mining and manufacturing. Next, MLR was carried out using the standardized PCA scores and the standardized normal deviation of total metal concentrations as the independent and dependent variables, respectively. Regression coefficients were then applied to estimate the contribution percentage from various metal sources.

As fewer data were available for Hg, As, and Co compared with the other metals (Tables 2 and 4), these three metals were subsequently removed from the database to increase the accuracy of the PCA-MLR results. In order to compare changes in pollution source over time, potential sources in combined river and lake sediments were associated with four main types: rock weathering, fertilizer and pesticide use, mining and manufacturing, and domestic waste discharge (including domestic solid waste, domestic wastewater, and automobile exhaust). The potential sources associated with heavy metals were provided in Table 6. The source apportionment of metal pollution in river and lake sediments was conducted using the SPSS V17.0 software (IBM Corp., Armonk, NY, USA).

3 Results

3.1 Trends of Heavy Metal Pollution in River Sediments

The concentrations of heavy metals in global river sediments differed between the five decades (Table 2, Fig. 1). For example, the Cd concentration was 3.4 g kg^{-1}, 4.0 g kg^{-1}, 1.7 g kg^{-1}, 3.1 g kg^{-1}, and 1.5 g kg^{-1} in the 1970s, 1980s, 1990s, 2000s,

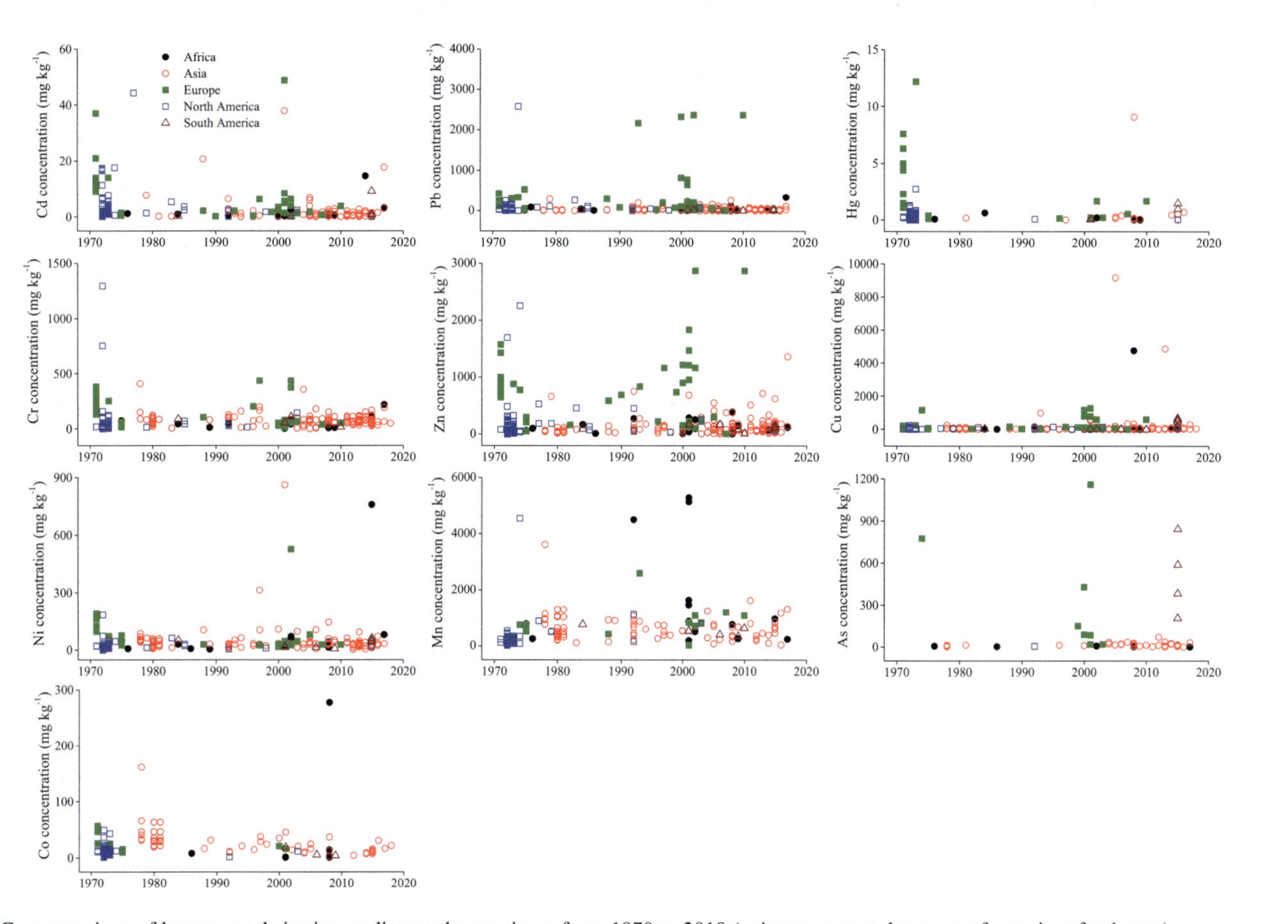

Fig. 1 Concentrations of heavy metals in river sediments by continent from 1970 to 2018 (points represent the mean of one river for 1 year)

and 2010s, respectively. From 1970 to 2018, river sediments showed increasing trends for Cd, Cr, Ni, Mn, and Co, and decreasing trends for Hg.

Heavy metal pollution in river sediments varied greatly between continents (Table 3). Most heavy metal species had the highest concentrations in Europe and the lowest concentrations in Africa or North America. For example, the descending order for Cd concentration is Europe > North America > South America > Asia > Africa, and for Cr concentration is Europe > Asia > South America > North America > Africa.

3.2 *Trends of Heavy Metal Pollution in Lake Sediments*

The concentrations of heavy metals in global lake sediments also differed between the five time groups (Table 4, Fig. 2). For example, the Pb concentration was 42.7 g kg^{-1} in the 1970s, 167.3 g kg^{-1} in the 1980s, 253.6 g kg^{-1} in the 1990s, 36.2 g kg^{-1} in the 2000s, and 41.8 g kg^{-1} in the 2010s. From 1970 to 2018, lake sediment showed increasing trends for Pb, Hg, Cr, and Mn, and decreasing trends for Cd, Zn, and As.

Heavy metal pollution in lake sediments also varied greatly between continents (Table 5). Most heavy metal species had the highest concentrations in North America and the lowest concentrations in Africa or Asia. For example, the descending order for Cd concentrations is North America > Africa > Europe > Asia, and for Pb concentration is North America > Europe > Asia > Africa.

3.3 *Sources of Sediment Heavy Metal Pollution Over Time*

The main pollution sources in combined river and lake sediments changed significantly over time (Table 7). In the 1970s, the first three principal component factors explained 82.7% of the heavy metal pollution in sediments. The main metal sources were mining and manufacturing activities, with a contribution of 58.0%. In the 1980s, the first two principal component factors explained 63.4% of the heavy metal pollution in sediments, and mining and manufacturing was main metal source, with a contribution of 77.5%. In the 1990s, the first three principal component factors explained 59.7% of the heavy metal pollution in sediments; mining and manufacturing activities were the dominant metal source, with a contribution of 92.7%. In the 2000s, the first three principal component factors explained 59.2% of the heavy metal pollution in sediments; domestic waste discharge was the dominant source, with a contribution of 48.3%. In the 2010s, the first two principal component factors explained 51.0% of the heavy metal pollution in sediments; domestic waste discharge was the dominant metal source, with a contribution of 90.3%.

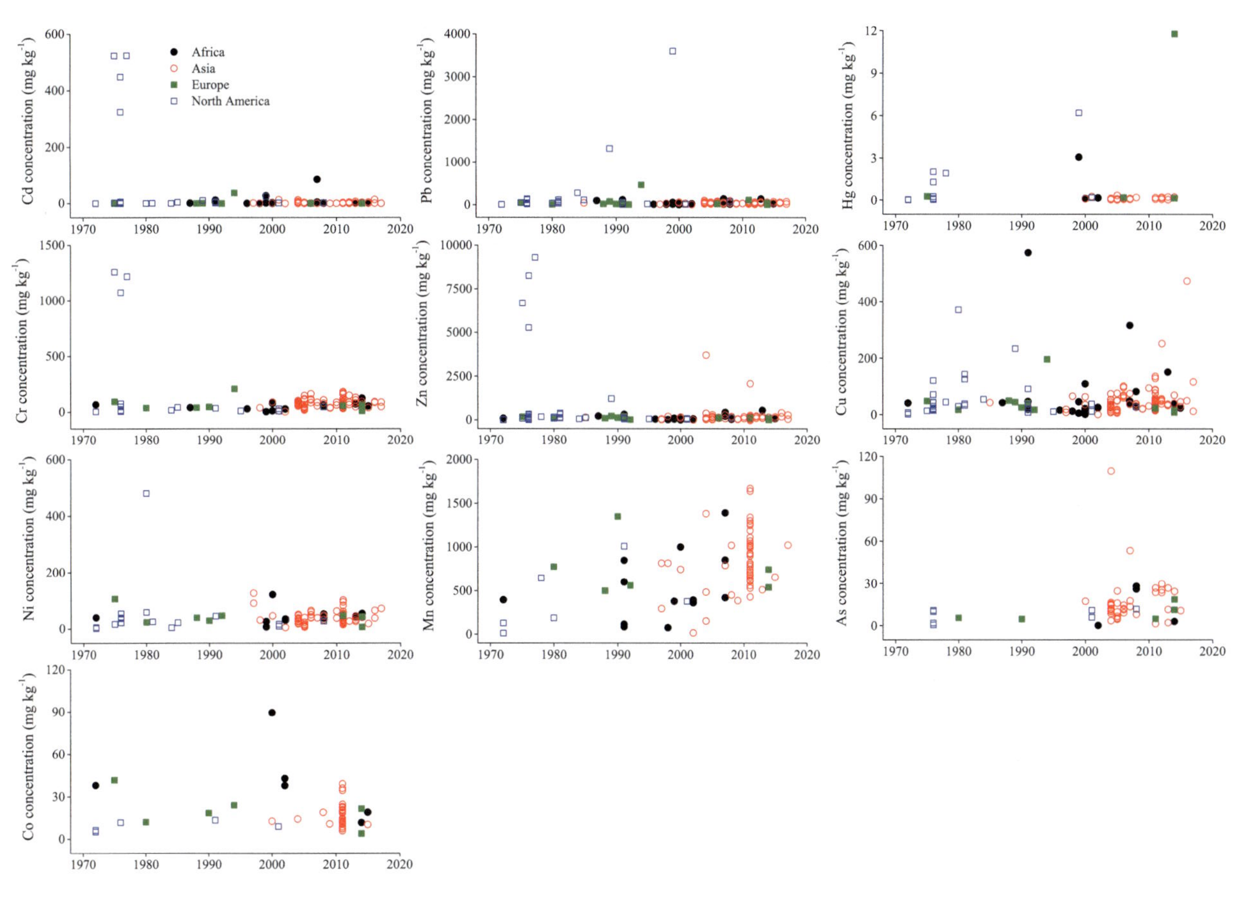

Fig. 2 Concentrations of heavy metals in lake sediments by continent from 1970 to 2018 (points represent the mean of one lake for 1 year)

Table 7 Rotated component matrix of selected metal concentrations in global combined river and lake sediments by decade

	1970s			1980s		1990s			2000s			2010s	
Metals	PC1	PC2	PC3	PC1	PC2	PC1	PC2	PC3	PC1	PC2	PC3	PC1	PC2
Cd	0.98	−0.01	−0.02	0.11	0.70	−0.26	0.01	0.77	0.21	−0.06	0.70	0.14	0.17
Pb	0.04	0.95	0.10	0.92	0.03	0.31	0.04	0.73	0.73	−0.06	0.16	0.81	−0.08
Cr	0.93	0.01	0.13	0.08	0.87	−0.05	0.82	−0.11	0.02	0.84	0.01	0.01	0.77
Zn	0.97	0.12	0.05	0.88	0.03	0.74	0.05	0.21	0.78	−0.02	0.31	0.86	−0.06
Cu	0.02	0.09	0.74	0.79	0.22	0.57	0.58	0.01	0.04	0.05	0.60	0.75	0.21
Ni	0.08	0.08	0.76	0.01	0.60	−0.02	0.64	0.14	−0.06	0.84	−0.02	−0.08	0.74
Mn	0.04	0.94	0.12	0.10	0.72	0.69	−0.11	−0.18	0.61	0.05	−0.44	0.44	0.41
Eigenvalue	2.88	1.89	1.02	2.67	1.77	1.75	1.23	1.20	1.72	1.39	1.05	2.23	1.34
Cumulative percentage	41.10	68.09	82.66	38.16	63.43	24.96	42.57	59.73	24.58	44.46	59.52	31.88	51.04
Possible source[a]	MM	RW	FP	MM	RW	MM	RW	DW	DW	MM	FP	DW	MM
Contribution percentage	58.03	30.22	11.75	77.45	22.55	92.67	2.19	5.14	48.31	4.86	46.83	90.27	9.73

[a] *RW* rock weathering, *FP* fertilizer and pesticide use, *MM* mining and manufacturing, *DW* domestic waste discharge

3.4 Sources of Sediment Heavy Metal Pollution by Continent

The main pollution sources in sediments also differed by continent (Table 8). In Africa, the first three principal component factors explained 57.2% of the heavy metal pollution in sediments; the main metal source was rock weathering, with a contribution of 55.2%. In Asia, the first two principal component factors explained 49.0% of the heavy metal pollution in sediments; the main source was domestic waste discharge, with a contribution of 87.2%. In Europe, the first three principal component factors explained 70.0% of the heavy metal pollution in sediments; domestic waste discharge was the dominant source, with a contribution of 86.9%. In North America, the first three principal component factors explained 83.8% of the heavy metal pollution in sediments; mining and manufacturing activities were the main source, with a contribution of 58.2%.

4 Discussion

4.1 Trends in Heavy Metal Pollution of River and Lake Sediments

From 1970 to 2018, global river sediments showed increasing trends for Cd, Cr, Ni, Mn, and Co, and decreasing trends for Hg, indicating that rivers acted as a sink for the former and a source for the latter. For example, the Cd concentrations were 3.4 mg kg^{-1}, 4.0 mg kg^{-1}, 1.7 mg kg^{-1}, 3.1 mg kg^{-1}, and 1.5 mg kg^{-1} in 1970s, 1980s, 1990s, 2000s, and 2010s and Hg concentrations were 0.6 mg kg^{-1}, 0.4 mg kg^{-1}, 0.1 mg kg^{-1}, 0.7 mg kg^{-1}, and 1.0 mg kg^{-1} in the five decades, respectively.

Meanwhile, global lake sediments showed increasing trends for Pb, Hg, Cr, and Mn, and decreasing trends for Cd, Zn, and As, indicating that lakes acted as a sink for the former and a source for the latter. For example, the Pb concentrations were 42.7 mg kg^{-1}, 167.3 mg kg^{-1}, 253.6 mg kg^{-1}, 36.2 mg kg^{-1}, and 41.8 mg kg^{-1} in 1970s, 1980s, 1990s, 2000s, and 2010s and Cd concentrations were 122.8 mg kg^{-1}, 2.3 mg kg^{-1}, 8.7 mg kg^{-1}, 2.9 mg kg^{-1}, and 1.2 mg kg^{-1} in the five decades, respectively.

4.2 Heavy Metal Concentrations in Sediments by Continents

Heavy metal pollution in river sediments varied greatly between continents. Compared with Africa and South America, Europe had relatively high metal concentrations in river sediments. For example, in Africa's Nile River, Pb concentration in sediment was 10.4 mg kg^{-1} in 1975, 13.4 mg kg^{-1} in 1992, and 12.6 mg kg^{-1} in

Table 8 Rotated component matrix of selected metal concentrations in combined river and lake sediments by continent

	Africa			Asia		Europe			North America		
Metals	PC1	PC2	PC3	PC1	PC 2	PC 1	PC 2	PC 3	PC 1	PC 2	PC 3
Cd	0.18	−0.69	−0.19	0.61	0.05	0.45	0.15	0.72	0.98	0.01	0.02
Pb	0.78	−0.16	0.01	0.87	0.04	0.74	−0.14	−0.13	0.05	0.88	0.19
Cr	0.49	−0.40	0.49	−0.03	0.78	0.02	0.86	0.18	0.94	0.01	0.07
Zn	0.73	0.41	−0.11	0.50	0.20	0.86	−0.02	0.12	0.96	0.17	0.06
Cu	0.19	0.03	−0.46	0.69	−0.08	0.75	0.06	0.04	0.18	0.47	0.69
Ni	0.17	0.11	0.81	0.10	0.56	−0.08	0.87	−0.11	−0.01	−0.06	0.90
Mn	0.17	0.70	−0.13	0.06	0.78	0.49	0.09	−0.69	0.03	0.83	−0.05
Eigenvalue	1.59	1.37	1.05	1.96	1.47	2.30	1.57	1.04	2.97	1.81	1.08
Cumulative percentage	22.67	42.22	57.18	28.01	49.03	32.82	55.22	70.03	42.47	68.30	83.77
Possible source[a]	FP	RW	MM	DW	MM	DW	MM	RW	DW	RW	MM
Contribution percentage	28.41	55.19	16.40	87.22	12.78	86.86	0.70	11.44	58.17	34.31	7.52

[a] *RW* rock weathering, *FP* fertilizer and pesticide use, *MM* mining and manufacturing, *DW* domestic waste discharge

2009 (Guerzoni et al. 1984; Birch et al. 1996; Taher and Soliman 1999). In Europe's Po River, the Pb concentration in sediment was 58.0 mg kg^{-1} in 1975, 27.0 mg kg^{-1} in 1996, and 52.7 mg kg^{-1} in 2005 (Zayed et al. 1994; Rifaat 2005; Osman and Kloas 2010).

In addition, heavy metal pollution in lake sediments also varied greatly between continents. Compared with Africa and Asia, North America had relatively higher metal concentrations in lake sediments. For example, In Asia's Poyang Lake, Pb concentration in sediment was 40.1 mg kg^{-1} in 1985 and 49.3 mg kg^{-1} in 2005 (Chen et al. 1989; Yuan et al. 2011). In North America's Erie Lake, the Pb concentration in sediment was 113.7 mg kg^{-1} in 1976 and 75.5 mg kg^{-1} in 2008 (Nriagu et al. 1979; Opfer et al. 2011). The low and high polluted sediment of rivers and lakes in studied continents were showed in Table 9.

4.3 Source of Heavy Metal Pollution in River and Lake Sediments

From the 1970s to 1990s, the main metal source was mining and manufacturing. A study in 1976 indicated that Hg in the sediments of the Wabigoon River, Ontario, Canada, originated from mining and manufacturing activities (Jackson 1979). Abdel-Moati and El-Sammak (1997) found that industrial discharge was the main sources for heavy metals in sediments of the Manzala Lake. However, from the 2000s to 2010s the main metal source was domestic waste discharge. Li et al. found that Zn, Pb, Cd, and As in sediments of China's Dongting Lake mainly originated from wastewater (Li et al. 2013). This suggests that metal sources changed from initial production activities to the secondary discharge of metal-contaminated waste over time.

Additionally, metal sources in sediments differed greatly by continent, with rock weathering dominant in Africa, mining and manufacturing dominant in North America, and domestic waste discharge dominant in Asia and Europe. A study on America's River Arkansas indicated that As, Cd, Cu, Mn, Pb, and Zn in sediments originated from mine dumps and tailings piles (Kimball et al. 1995). Yin et al. found that Hg, Cu, Cr, Cd, and Pb in sediments of China's Taihu Lake mainly originated from sewage and wastewater (Yin et al. 2011). Therefore, pollution-control measures should concentrate on these continent-specific sources.

4.4 Control Measures for Heavy Metal Pollution in Sediments

Decreasing trends in Hg in global river sediments and in Cd, Zn, and As in global lake sediments illustrate that pollution-control measures produced meaningful effects on these substances. In Europe, the maximum Cd concentration permitted

Table 9 Low and high polluted sediment in river sediments of five selected continents and lake sediments of four selected continents

Water body	Africa		Asia		Europe		North America		South America	
	LPS[a]	HPS[b]	LPS	HPS	LPS	HPS	LPS	HPS	LPS	HPS
River	Kuja River, Nile River, Zambezi River,	Kafue River, Nakivubo Channelized Stream	Bendimahi River, Palaru River, Yellow River	Wardha River, Shing Mun River, Xiang River, Pasig River	Danube River, Minor River, Sicily River, Sodo River	Tinto River, Odiel River, Lot River, Asopos River	Dead River, Tahquamenon River, Thessalon River, Indian River	Coeur d'Alene River, Harvey Canal, Milwaukee River, South Platte River, Tippecanoe River	Parnaíba River, Suquía River	Kaw River, Villa River
Lake	Lake Edku, Qarun Lake, Lake Tanganyika, Lake Victoria	Lake Manzala, Lake Mariout, Lake Wadi El Natrun	Lake Uluabat, Hongze Lake, Kovada Lake	Lake Beyşehir, Lake Dongting, Vembanad Lake	Cruhlig Lake, Venice Lagoon	Lake Constance, Malter Reservoir	Lake Erie, Lake Monona	Bruère Lake, Fontana Lake, Patroon Lake, Lake Palestine	–	–

[a] *LPS* Low polluted sediment
[b] *HPS* High polluted sediment

in fertilizers has been regulated since the last century. For instance, the Dutch government proposed regulations limiting the maximum Cd concentration in P fertilizers to 35 mg kg^{-1} in the mid-1980s (Anon 1989). Switzerland proposed changing the highest Cd concentration permitted to add to P fertilizers to 50 mg kg^{-1}. To alleviate metal pollution in the Rhine River, the European Community (now called the European Union [EU]) limited Cd concentrations in industrial effluents discharged into rivers (Mortvedt 1996). In 1998, the United Nations Economic Commission for Europe signed the Aarhus Protocol on Heavy Metals to control harmful levels of Cd and Hg (Duan and Tan 2013). In addition, European legislation prescribes the following order of priority to be applied to waste: waste treatment (including prevention, reuse, and recycling), other recovery (e.g., energy recovery), and disposal (Kelessidis and Stasinakis 2012). In the US, from 1994, the US EPA's Part 503 rule restricted the ceiling concentration of Zn (2,800 mg kg^{-1}) in biosolids applied to land (Agency 1994). After an industrial pretreatment program was implemented in 1976 to control the discharge of heavy metals into publicly owned treatment facilities, the metal load to municipal wastewater treatment facilities and metal concentrations in biosolids decreased over time (Mortvedt 1996).

Since the Chinese government enacted the Law of Environmental Protection in 1986, China gradually establishes a comprehensive legal system for environmental protection. In 1997, leaded gasoline was first prohibited in Beijing, and then the production and use of leaded gasoline were prohibited nationwide in 2000 (Duan and Tan 2013). After this implementation, the ratios of blood lead levels exceeding 100 μg L^{-1} for children of 1–6 years in Shanghai dropped from 37.8% to 3.9% (Yuan and Shen 2006). In 2011, the "12th Five-Year Plan on Prevention and Control of Heavy Metal Pollution" was approved by the State Council of China (Järup 2003). The Ministry of Environmental Protection switched the fuel used in boilers for electrical generation and providing hot water from coal to gas, and issued strict emission standards for coal-burning boilers (Duan and Tan 2013). After implementing this fuel replacement initiative, consumption of gas increased from 0.75 billion cubic meters in 1999 to 6.90 billion cubic meters in 2010 in Beijing, whereas the consumption of coal decreased slightly from 26.5 million tons in 1999 to 26.4 million tons in 2010 (Duan and Tan 2013). These measures, encompassing the implementation of rigorous standards on metal emissions, limitations to metals added in industrial products, and pretreatment of metal-contaminated waste, were effective at controlling heavy metal pollution in Chinese rivers and lakes.

In addition to such environmental measures, ecological approaches should be considered for heavy metal treatment. For example, after landscape reclamation in catchments, heavy metal inputs to the United Kingdom's Fendrod Lake decreased (Blake et al. 2007). Phytoremediation, the use of plants to extract, sequester, or detoxify pollutants, has shown great promise in removing heavy metals from sediments and water (Peng et al. 2009). In Dongting Lake, China, an area highly polluted by heavy metals, it was estimated that 700 kg of Cd, 22,900 kg of Cu, 3,100 kg of Pb, and 95,900 kg of Zn per year could be potentially accumulated by aboveground organs of *Miscanthus sacchariflorus* and removed by harvesting this plant for paper manufacture(Yao et al. 2018). Therefore, the ecological measures could be considered for pollutant interception and removal in remediation of heavy metal-polluted rivers and lakes.

5 Conclusions

This study clearly showed that global river sediments had increasing trends in Cd, Cr, Ni, Mn, and Co and decreasing trends in Hg over the past period from 1970 to 2018. Meanwhile, global lake sediments showed increasing trends in Pb, Hg, Cr, and Mn and decreasing trends in Cd, Zn, and As. Heavy metal pollution in river and lake sediments varied greatly between continents. Collectively, mean metal concentrations were generally higher in Europe and North America than in Africa, Asia, and South America.

The sources of metal pollution in sediment differed greatly by time and space. From the 1970s to 1990s, the main metal source was mining and manufacturing, changing to domestic waste discharge from the 2000s to 2010s. Over the past period from 1970 to 2018, dominant metal sources were rock weathering in Africa, mining and manufacturing in North America, and domestic waste discharge in Asia and Europe. Therefore, regional pollution-control measures should concentrate on such region-specific sources.

Decreasing trends in Hg in global river sediments and in Cd, Zn, and As in global lake sediments illustrate that pollution-control measures produced meaningful effects on these substances. These measures, such as the implementation of rigorous standards on metal emissions, limitations on metal concentrations in manufactured products, and the pretreatment of metal-contaminated waste, should be extended globally.

Acknowledgments This study was financially supported by the Chenzhou National Sustainable Development Agenda Innovation Demonstration Zone Construction Project (2019sfq21); Key Projects in Social Development Field of Hunan Innovative Province Construction (2019NK2011); the Research Foundation of Education Bureau of Hunan Province, China (19A225).

Conflict of Interest The authors declare that they have no conflict of interest.

References

Abdel-Moati M, El-Sammak A (1997) Man-made impact on the geochemistry of the Nile Delta Lakes. A study of metals concentrations in sediments. Water Air Soil Pollut 97:413

Abernathy AR, Larson GL, Mathews RC Jr (1984) Heavy metals in the surficial sediments of Fontana Lake, North Carolina. Water Res 18:351–354

Adams TG, Atchison GJ, Vetter RJ (1980) The impact of an industrially contaminated lake on heavy metal levels in its effluent stream. Hydrobiologia 69:187–193

Addo-Bediako A, Matlou K, Makushu E (2018) Heavy metal concentrations in water and sediment of the Steelpoort River, Olifants River System, South Africa. Afr J Aquat Sci 43:413–416

Agency USEP (1994) A plain English guide to the EPA part 503 biosolids rule. Management OOW, Washington, DC

Ajmal M, Uddin R, Khan AU (1988) Heavy metals in water, sediments, plants and fish of Kali Nadi UP (India). Environ Int 14:515–523

Akbulut A, Akbulut NE (2010) The study of heavy metal pollution and accumulation in water, sediment, and fish tissue in Kızılırmak River Basin in Turkey. Environ Monit Assess 167:521–526

Akcay H, Oguz A, Karapire C (2003) Study of heavy metal pollution and speciation in Buyak Menderes and Gediz river sediments. Water Res 37:813–822

Ali MM, Ali ML, Islam MS, Rahman MZ (2016) Preliminary assessment of heavy metals in water and sediment of Karnaphuli River, Bangladesh. Environ Nanotechnol Monitoring Manag 5:27–35

Al-Saadi HA, Al-Lami AA, Hassan FA, Al-Dulymi AA (2002) Heavy metals in water, suspended particles, sediments and aquatic plants of Habbaniya Lake, Iraq. Int J Environ Stud 59:589–598

Altındağ A, Yiğit S (2005) Assessment of heavy metal concentrations in the food web of lake Beyşehir, Turkey. Chemosphere 60:552–556

An YJ, Kampbell DH (2003) Total, dissolved, and bioavailable metals at Lake Texoma marinas. Environ Pollut 122:253–259

Anon (1989) Cadmium in phosphates: one part of a wider environmental problem. Phosphorus Potassium 162:23–30

Arnason JG, Fletcher BA (2003) A 40+ year record of Cd, Hg, Pb, and U deposition in sediments of Patroon Reservoir, Albany County, NY, USA. Environ Pollut 123:383–391

Ashayeri NY, Keshavarzi B, Moore F, Kersten M, Yazdi M, Lahijanzadeh AR (2018) Presence of polycyclic aromatic hydrocarbons in sediments and surface water from Shadegan wetland-Iran: a focus on source apportionment, human and ecological risk assessment and sediment-water exchange. Ecotoxicol Environ Saf 148:1054–1066

Aucoin J, Blanchard R, Billiot C, Partridge C, Schultz D, Mandhare K, Beck MJ, Beck JN (1999) Trace metals in fish and sediments from Lake Boeuf, Southeastern Louisiana. Microchem J 62:299–307

Audry S, Schäfer J, Blanc G, Jouanneau JM (2004) Fifty-year sedimentary record of heavy metal pollution (Cd, Zn, Cu, Pb) in the Lot River reservoirs (France). Environ Pollut 132:413–426

Babich H, Stotzky G (1985) Heavy metal toxicity to microbe-mediated ecologic processes: a review and potential application to regulatory policies. Environ Res 36:111–137

Bai J, Cui B, Chen B, Zhang K, Deng W, Gao H, Xiao R (2011a) Spatial distribution and ecological risk assessment of heavy metals in surface sediments from a typical plateau lake wetland, China. Ecol Model 222:301–306

Bai J, Xiao R, Cui B, Zhang K, Wang Q, Liu X, Gao H, Huang L (2011b) Assessment of heavy metal pollution in wetland soils from the young and old reclaimed regions in the Pearl River Estuary, South China. Environ Pollut 159:817–824

Bai J, Xiao R, Zhang K, Gao H (2012) Arsenic and heavy metal pollution in wetland soils from tidal freshwater and salt marshes before and after the flow-sediment regulation regime in the Yellow River Delta, China. J Hydrol 450:244–253

Barlas N, Akbulut N, Aydoğan M (2005) Assessment of heavy metal residues in the sediment and water samples of Uluabat Lake, Turkey. Bull Environ Contam Toxicol 74:286–293

Begy RC, Preoteasa L, Timar-Gabor A, Mihăiescu R, Tănăselia C, Kelemen S, Simon H (2016) Sediment dynamics and heavy metal pollution history of the Cruhlig Lake (Danube Delta, Romania). J Environ Radioact 153:167–175

Beltrán R, De La Rosa J, Santos J, Beltrán M, Gomez-Ariza J (2010) Heavy metal mobility assessment in sediments from the Odiel River (Iberian Pyritic Belt) using sequential extraction. Environ Earth Sci 61:1493–1503

Bermejo JS, Beltrán R, Ariza JG (2003) Spatial variations of heavy metals contamination in sediments from Odiel river (Southwest Spain). Environ Int 29:69–77

Biksham G, Subramanian V (1988) Elemental composition of Godavari sediments (central and southern Indian subcontinent). Chem Geol 70:275–286

Bing H, Wu Y, Sun Z, Yao S (2011) Historical trends of heavy metal contamination and their sources in lacustrine sediment from Xijiu Lake, Taihu Lake Catchment, China. J Environ Sci 23:1671–1678

Bing H, Wu Y, Zhou J, Sun H, Wang X, Zhu H (2019) Spatial variation of heavy metal contamination in the riparian sediments after two-year flow regulation in the Three Gorges Reservoir, China. Sci Total Environ 649:1004–1016

Birch L, Hanselmann KW, Bachofen R (1996) Heavy metal conservation in Lake Cadagno sediments: Historical records of anthropogenic emissions in a meromictic alpine lake. Water Res 30:679–687

Blake WH, Walsh RP, Reed JM, Barnsley MJ, Smith J (2007) Impacts of landscape remediation on the heavy metal pollution dynamics of a lake surrounded by non-ferrous smelter waste. Environ Pollut 148:268–280

Bonanno G, Giudice RL (2010) Heavy metal bioaccumulation by the organs of *Phragmites australis* (common reed) and their potential use as contamination indicators. Ecol Indic 10:639–645

Botsou F, Karageorgis A, Dassenakis E, Scoullos M (2011) Assessment of heavy metal contamination and mineral magnetic characterization of the Asopos River sediments (Central Greece). Mar Pollut Bull 62:547–563

Bradley S, Woods WL (1997) Cd, Cr, Cu, Ni and Pb in the water column and sediments of the Ob-Irtysh Rivers, Russia. Mar Pollut Bull 35:270–279

Bryan G, Hummerstone L (1977) Indicators of heavy-metal contamination in the Looe Estuary (Cornwall) with particular regard to silver and lead. J Mar Biol Assoc U K 57:75–92

CEC (Council of the European Communities) (1991) Council Directive of 21 May 1991 concerning urban waste water treatment (91/271/EEC). Off J Eur Communities 135:40–52

Chale F (2002) Trace metal concentrations in water, sediments and fish tissue from Lake Tanganyika. Sci Total Environ 299:115–121

Chaparro MAE, Sinito AM, Ramasamy V, Marinelli C, Chaparro MAE, Mullainathan S, Murugesan S (2008) Magnetic measurements and pollutants of sediments from Cauvery and Palaru River, India. Environ Geol 56:425–437

Chaparro MA, Chaparro MA, Rajkumar P, Ramasamy V, Sinito AM (2011) Magnetic parameters, trace elements, and multivariate statistical studies of river sediments from southeastern India: a case study from the Vellar River. Environ Earth Sci 63:297–310

Che Y, He Q, Lin WQ (2003) The distributions of particulate heavy metals and its indication to the transfer of sediments in the Changjiang Estuary and Hangzhou Bay, China. Mar Pollut Bull 46:123–131

Chen MH, Wu HT (1995) Copper, cadmium and lead in sediments from the Kaohsiung River and its harbour area, Taiwan. Mar Pollut Bull 30:879–884

Chen J, Dong L, Deng B (1989) A study on heavy metal partitioning in sediments from Poyang Lake in China. Hydrobiologia 176:159–170

Chen Y, Jiang Y, Huang H, Mou L, Ru J, Zhao J, Xiao S (2018) Long-term and high-concentration heavy-metal contamination strongly influences the microbiome and functional genes in Yellow River sediments. Sci Total Environ 637:1400–1412

Cheng H, Li M, Zhao C, Yang K, Li K, Peng M, Yang Z, Liu F, Liu Y, Bai R, Cui Y, Huang Z, Li L, Liao Q, Luo J, Jia S, Pang X, Yang J, Yin G (2015a) Concentrations of toxic metals and ecological risk assessment for sediments of major freshwater lakes in China. J Geochem Explor 157:15–26

Cheng Q, Wang R, Huang W, Wang W, Li X (2015b) Assessment of heavy metal contamination in the sediments from the Yellow River Wetland National Nature Reserve (the Sanmenxia section), China. Environ Sci Pollut Res 22:8586–8593

Chenhall B, Yassini I, Jones B (1992) Heavy metal concentrations in lagoonal saltmarsh species, Illawarra region, southeastern Australia. Sci Total Environ 125:203–225

Cui B, Zhang Q, Zhang K, Liu X, Zhang H (2011) Analyzing trophic transfer of heavy metals for food webs in the newly-formed wetlands of the Yellow River Delta, China. Environ Pollut 159:1297–1306

Dalai TK, Rengarajan R, Patel PP (2004) Sediment geochemistry of the Yamuna River System in the Himalaya: Implications to weathering and transport. Geochem J 38:441–453

Datta DK, Subramanian V (1998) Distribution and fractionation of heavy metals in the surface sediments of the Ganges-Brahmaputra-Meghna river system in the Bengal basin. Environ Geol 36:93–101

Dauvalter V, Rognerud S (2001) Heavy metal pollution in sediments of the Pasvik River drainage. Chemosphere 42:9–18

Davies CA, Tomlinson K, Stephenson T (1991) Heavy metals in River Tees estuary sediments. Environ Technol 12:961–972

de Paula Filho FJ, de Lacerda LD, Marins RV, Aguiar JE, Peres TF (2015) Background values for evaluation of heavy metal contamination in sediments in the Parnaíba River Delta estuary, NE/Brazil. Mar Pollut Bull 91:424–428

De Vevey E, Bitton G, Rossel D, Ramos L, Guerrero LM, Tarradellas J (1993) Concentration and bioavailability of heavy metals in sediments in Lake Yojoa (Honduras). Bull Environ Contam Toxicol 50:253–259

Demirak A, Yilmaz F, Tuna AL, Ozdemir N (2006) Heavy metals in water, sediment and tissues of *Leuciscus cephalus* from a stream in southwestern Turkey. Chemosphere 63:1451–1458

Dhanakumar S, Solaraj G, Mohanraj R (2015) Heavy metal partitioning in sediments and bioaccumulation in commercial fish species of three major reservoirs of river Cauvery delta region, India. Ecotoxicol Environ Saf 113:145–151

Duan J, Tan J (2013) Atmospheric heavy metals and arsenic in China: situation, sources and control policies. Atmos Environ 74:93–101

Duncan AE, De Vries N, Nyarko KB (2018) Assessment of heavy metal pollution in the sediments of the River Pra and its tributaries. Water Air Soil Pollut 229:272

Dundar MS, Altundag H (2007) Investigation of heavy metal contaminations in the lower Sakarya river water and sediments. Environ Monit Assess 128:177–181

Duodu GO, Goonetilleke A, Ayoko GA (2016) Comparison of pollution indices for the assessment of heavy metal in Brisbane River sediment. Environ Pollut 219:1077–1091

Dupré B, Gaillardet J, Rousseau D, Allègre CJ (1996) Major and trace elements of river-borne material: the Congo Basin. Geochim Cosmochim Acta 60:1301–1321

El-Amier YA, Elnaggar AA, El-Alfy MA (2017) Evaluation and mapping spatial distribution of bottom sediment heavy metal contamination in Burullus Lake, Egypt. Egypt J Basic Appl Sci 4:55–66

Elghobashy HA, Zaghloul KH, Metwally M (2001) Effect of some water pollutants on the Nile tilapia *Oreochromis niloticus* collected from the River Nile and some Egyptian Lakes. Egypt J Aquatic Biol Fish 5:251–279

El-Sayed S, Moussa E, El-Sabagh M (2015) Evaluation of heavy metal content in Qaroun Lake, El-Fayoum, Egypt. Part I: bottom sediments. J Radiation Res Appl Sci 8:276–285

Facchinelli A, Sacchi E, Mallen L (2001) Multivariate statistical and GIS-based approach to identify heavy metal sources in soils. Environ Pollut 114:313–324

Farkas A, Erratico C, Vigano L (2007) Assessment of the environmental significance of heavy metal pollution in surficial sediments of the River Po. Chemosphere 68:761–768

Feng H, Han X, Zhang W, Yu L (2004) A preliminary study of heavy metal contamination in Yangtze River intertidal zone due to urbanization. Mar Pollut Bull 49:910–915

Fitchko J, Hutchinson T (1975) A comparative study of heavy metal concentrations in river mouth sediments around the Great Lakes. J Great Lakes Res 1:46–78

Förstner U (1976) Lake sediments as indicators of heavy-metal pollution. Naturwissenschaften 63:465–470

Förstner U, Müller G (1973) Heavy metal accumulation in river sediments: a response to environmental pollution. Geoforum 4:53–61

Fu J, Zhao C, Luo Y, Liu C, Kyzas GZ, Luo Y, Zhao D, An S, Zhu H (2014) Heavy metals in surface sediments of the Jialu River, China: their relations to environmental factors. J Hazard Mater 270:102–109

Gagneten A, Paggi J (2009) Effects of heavy metal contamination (Cr, Cu, Pb, Cd) and eutrophication on zooplankton in the lower basin of the Salado River (Argentina). Water Air Soil Pollut 198:317–334

Galán E, Gómez-Ariza J, González I, Fernández-Caliani J, Morales E, Giráldez I (2003) Heavy metal partitioning in river sediments severely polluted by acid mine drainage in the Iberian Pyrite Belt. Appl Geochem 18:409–421

Gao Q, Li Y, Cheng Q, Yu M, Hu B, Wang Z, Yu Z (2016) Analysis and assessment of the nutrients, biochemical indexes and heavy metals in the Three Gorges Reservoir, China, from 2008 to 2013. Water Res 92:262–274

García-Ordiales E, Esbrí JM, Covelli S, López-Berdonces MA, Higueras PL, Loredo J (2016) Heavy metal contamination in sediments of an artificial reservoir impacted by long-term mining activity in the Almadén mercury district (Spain). Environ Sci Pollut Res 23:6024–6038

Gewurtz SB, Helm PA, Waltho J, Stern GA, Reiner EJ, Painter S, Marvin CH (2007) Spatial distributions and temporal trends in sediment contamination in Lake St. Clair. J Great Lakes Res 33:668–685

Glooschenko WA (1982) Sediment trace metal chemistry of second marsh, Oshawa, Ontario. Wetlands 2:207–215

Glooschenko W, Capocianco J, Coburn J, Glooschenko V (1981) Geochemical distribution of trace metals and organochlorine contaminants of a Lake Ontario shoreline marsh. Water Air Soil Pollut 15:197–213

Gorenc S, Kostaschuk R, Chen Z (2004) Spatial variations in heavy metals on tidal flats in the Yangtze Estuary, China. Environ Geol 45:1101–1108

Grabowski LA, Houpis JL, Woods WI, Johnson KA (2001) Seasonal bioavailability of sediment-associated heavy metals along the Mississippi river floodplain. Chemosphere 45:643–651

Grieve D, Fletcher W (1976) Heavy metals in deltaic sediments of the Fraser River, British Columbia. Can J Earth Sci 13:1683–1693

Guerzoni S, Frignani M, Giordani P, Frascari F (1984) Heavy metals in sediments from different environments of a northern Adriatic Sea area, Italy. Environ Geol Water Sci 6:111–119

Guo W, Huo S, Xi B, Zhang J, Wu F (2015) Heavy metal contamination in sediments from typical lakes in the five geographic regions of China: distribution, bioavailability, and risk. Ecol Eng 81:243–255

Gupta A, Rai DK, Pandey RS, Sharma B (2009) Analysis of some heavy metals in the riverine water, sediments and fish from river Ganges at Allahabad. Environ Monit Assess 157:449–458

Hahladakis J, Smaragdaki E, Vasilaki G, Gidarakos E (2013) Use of sediment quality guidelines and pollution indicators for the assessment of heavy metal and PAH contamination in Greek surficial sea and lake sediments. Environ Monit Assess 185:2843–2853

Harguinteguy CA, Cirelli AF, Pignata ML (2014) Heavy metal accumulation in leaves of aquatic plant Stuckenia filiformis and its relationship with sediment and water in the Suquía river (Argentina). Microchem J 114:111–118

Heiny JS, Tate C (1997) Concentration, distribution, and comparison of selected trace elements in bed sediment and fish tissue in the South Platte River Basin, USA, 1992–1993. Arch Environ Contam Toxicol 32:246–259

Hu H, Jin Q, Kavan P (2014) A Study of Heavy Metal Pollution in China: Current Status, Pollution-Control Policies and Countermeasures. Sustainability 6:5820–5838

Hu C, Deng ZM, Xie YH, Chen XS, Li F (2015) The risk assessment of sediment heavy metal pollution in the East Dongting Lake Wetland. J Chem 2015:835487
Hu C, Yang X, Dong J, Zhang X (2018) Heavy metal concentrations and chemical fractions in sediment from Swan Lagoon, China: Their relation to the physiochemical properties of sediment. Chemosphere 209:848–856
Huang X, Hu J, Li C, Deng J, Long J, Qin F (2009) Heavy-metal pollution and potential ecological risk assessment of sediments from Baihua Lake, Guizhou, PR China. Int J Environ Health Res 19:405–419
Huang L, Pu X, Pan JF, Wang B (2013) Heavy metal pollution status in surface sediments of Swan Lake lagoon and Rongcheng Bay in the northern Yellow Sea. Chemosphere 93:1957–1964
Huang Y, Li T, Wu C, He Z, Japenga J, Deng M, Yang X (2015) An integrated approach to assess heavy metal source apportionment in peri-urban agricultural soils. J Hazard Mater 299:540–549
Hudson-Edwards K, Macklinb M, Taylorb M (1997) Historic metal mining inputs to Tees river sediment. Sci Total Environ 194–195:437–445
Ikenaka Y, Nakayama SMM, Muzandu K, Choongo K, Teraoka H, Mizuno N, Ishizuka M (2010) Heavy metal contamination of soil and sediment in Zambia. Afr J Environ Sci Technol 4:729–739
Ip CC, Li XD, Zhang G, Wai OW, Li YS (2007) Trace metal distribution in sediments of the Pearl River Estuary and the surrounding coastal area, South China. Environ Pollut 147:311–323
Islam MS, Ahmed MK, Habibullah-Al-Mamun M, Hoque MF (2015a) Preliminary assessment of heavy metal contamination in surface sediments from a river in Bangladesh. Environ Earth Sci 73:1837–1848
Islam MS, Ahmed MK, Raknuzzaman M, Habibullah-Al-Mamun M, Islam MK (2015b) Heavy metal pollution in surface water and sediment: a preliminary assessment of an urban river in a developing country. Ecol Indic 48:282–291
Islam S, Bhuiyan MAH, Rume T, Mohinuzzaman M (2016) Assessing heavy metal contamination in the bottom sediments of Shitalakhya River, Bangladesh; using pollution evaluation indices and geo-spatial analysis. Pollution 2:299–312
Islam MS, Hossain MB, Matin A, Sarker MSI (2018) Assessment of heavy metal pollution, distribution and source apportionment in the sediment from Feni River estuary, Bangladesh. Chemosphere 202:25–32
Jackson TA (1979) Sources of heavy metal contamination in a river-lake system. Environ Pollut 18:131–138
Järup L (2003) Hazards of heavy metal contamination. Hazards Heavy Metal Contam 68:167–182
Jiang Q, Liu M, Wang J, Liu F (2018) Feasibility of using visible and near-infrared reflectance spectroscopy to monitor heavy metal contaminants in urban lake sediment. Catena 162:72–79
Kang M, Tian Y, Peng S, Wang M (2019) Effect of dissolved oxygen and nutrient levels on heavy metal contents and fractions in river surface sediments. Sci Total Environ 648:861–870
Karadede H, Ünlü E (2000) Concentrations of some heavy metals in water, sediment and fish species from the Atatürk Dam Lake (Euphrates), Turkey. Chemosphere 41:1371–1376
Karadede-Akin H, Ünlü E (2007) Heavy metal concentrations in water, sediment, fish and some benthic organisms from Tigris River, Turkey. Environ Monit Assess 131:323–337
Karbassi A, Monavari S, Bidhendi GRN, Nouri J, Nematpour K (2008) Metal pollution assessment of sediment and water in the Shur River. Environ Monit Assess 147:107–116
Ke X, Gui S, Huang H, Zhang H, Wang C, Guo W (2017) Ecological risk assessment and source identification for heavy metals in surface sediment from the Liaohe River protected area, China. Chemosphere 175:473–481
Kelessidis A, Stasinakis AS (2012) Comparative study of the methods used for treatment and final disposal of sewage sludge in European countries. Waste Manag 32:1186–1195
Kendall MG (1975) Rank correlation methods, 4th edn. Charles Griffin, London
Kim Y, Kim BK, Kim K (2010) Distribution and speciation of heavy metals and their sources in Kumho River sediment, Korea. Environ Earth Sci 60:943–952

Kimball BA, Callender E, Axtmann EV (1995) Effects of colloids on metal transport in a river receiving acid mine drainage, upper Arkansas River, Colorado, USA. Appl Geochem 10:285–306

Kishe M, Machiwa J (2003) Distribution of heavy metals in sediments of Mwanza Gulf of Lake Victoria, Tanzania. Environ Int 28:619–625

Kisi O, Ay M (2014) Comparison of Mann-Kendall and innovative trend method for water quality parameters of the Kizilirmak River, Turkey. J Hydrol 513:362–375

Kumari P, Chowdhury A, Maiti SK (2018) Assessment of heavy metal in the water, sediment, and two edible fish species of Jamshedpur Urban Agglomeration, India with special emphasis on human health risk. Hum Ecol Risk Assess 24:1477–1500

Lacal J, Da Silva MP, Garcıa R, Sevilla MT, Procopio JR, Hernandez L (2003) Study of fractionation and potential mobility of metal in sludge from pyrite mining and affected river sediments: changes in mobility over time and use of artificial ageing as a tool in environmental impact assessment. Environ Pollut 124:291–305

Lalah J, Ochieng E, Wandiga S (2008) Sources of heavy metal input into Winam Gulf, Kenya. Bull Environ Contam Toxicol 81:277–284

Larsen RK, Baker JE (2003) Source apportionment of polycyclic aromatic hydrocarbons in the urban atmosphere: a comparison of three methods. Environ Sci Technol 37:1873–1881

Lasheen M, Ammar N (2009) Speciation of some heavy metals in River Nile sediments, Cairo, Egypt. Environmentalist 29:8–16

Li Q, Wu Z, Chu B, Zhang N, Cai S, Fang J (2007) Heavy metals in coastal wetland sediments of the Pearl River Estuary, China. Environ Pollut 149:158–164

Li F, Huang J, Zeng G, Yuan X, Li X, Liang J, Wang X, Tang X, Bai B (2013) Spatial risk assessment and sources identification of heavy metals in surface sediments from the Dongting Lake, Middle China. J Geochem Explor 132:75–83

Li P, Qian H, Howard KW, Wu J (2015) Heavy metal contamination of Yellow River alluvial sediments, northwest China. Environ Earth Sci 73:3403–3415

Li N, Tian Y, Zhang J, Zuo W, Zhan W, Zhang J (2017) Heavy metal contamination status and source apportionment in sediments of Songhua River Harbin region, Northeast China. Environ Sci Pollut Res 24:3214–3225

Li Y, Zhou Q, Ren B, Luo J, Yuan J, Ding X, Bian H, Yao X (2020) Trends and health risks of dissolved heavy metal pollution in global river and lake water from 1970 to 2017. Rev Environ Contam Toxicol 251:1–24

Liang J, Liu J, Yuan X, Zeng G, Lai X, Li X, Wu H, Yuan Y, Li F (2015) Spatial and temporal variation of heavy metal risk and source in sediments of Dongting Lake wetland, mid-south China. J Environ Sci Health Part A Toxic/Hazardous Subst Environ Eng 50:100–108

Liang P, Wu SC, Zhang J, Cao Y, Yu S, Wong MH (2016) The effects of mariculture on heavy metal distribution in sediments and cultured fish around the Pearl River Delta region, south China. Chemosphere 148:171–717

Lintern A, Leahy PJ, Heijnis H, Zawadzki A, Gadd P, Jacobsen G, Deletic A, Mccarthy DT (2016) Identifying heavy metal levels in historical flood water deposits using sediment cores. Water Res 105:34–46

Liu W, Li X, Shen Z, Wang D, Wai O, Li Y (2003) Multivariate statistical study of heavy metal enrichment in sediments of the Pearl River Estuary. Environ Pollut 121:377–388

Liu C, Xu J, Liu C, Zhang P, Dai M (2009a) Heavy metals in the surface sediments in Lanzhou Reach of Yellow River, China. Bull Environ Contam Toxicol 82:26–30

Liu J, Li Y, Zhang B, Cao J, Cao Z, Domagalski J (2009b) Ecological risk of heavy metals in sediments of the Luan River source water. Ecotoxicology 18:748–758

Liu J, Liang J, Yuan X, Yuan X, Zeng G, Yuan Y, Wu H, Huang X, Liu J, Hua S, Li F, Li X (2015a) An integrated model for assessing heavy metal exposure risk to migratory birds in wetland ecosystem: a case study in Dongting Lake Wetland, China. Chemosphere 135:14–19

Liu JL, Xu XR, Ding ZH, Peng JX, Jin MH, Wang YS, Hong YG, Yue WZ (2015b) Heavy metals in wild marine fish from South China Sea: levels, tissue- and species-specific accumulation and potential risk to humans. Ecotoxicology 24:1583–1592

Ma X, Zuo H, Tian M, Zhang L, Meng J, Zhou X, Min N, Chang X, Liu Y (2016) Assessment of heavy metals contamination in sediments from three adjacent regions of the Yellow River using metal chemical fractions and multivariate analysis techniques. Chemosphere 144:264–272

Maanan M, Saddik M, Maanan M, Chaibi M, Assobhei O, Zourarah B (2015) Environmental and ecological risk assessment of heavy metals in sediments of Nador lagoon, Morocco. Ecol Indic 48:616–626

Malm O, Pfeiffer WC, Fiszman M, Azcue J (1989) Heavy metal concentrations and availability in the bottom sediments of the Paraiba do Sul-Guandu river system, RJ, Brazil. Environ Technol 10:675–680

Malvandi H (2017) Preliminary evaluation of heavy metal contamination in the Zarrin-Gol River sediments, Iran. Mar Pollut Bull 117:547–553

Mamat Z, Haximu S, Yong Zhang Z, Aji R (2016) An ecological risk assessment of heavy metal contamination in the surface sediments of Bosten Lake, northwest China. Environ Sci Pollut Res 23:7255–7265

Mann HB (1945) Nonparametric tests against trend. Econometrica 13:245–259

Marchand C, Lallier-Vergès E, Baltzer F, Albéric P, Cossa D, Baillif P (2006) Heavy metals distribution in mangrove sediments along the mobile coastline of French Guiana. Mar Chem 98:1–17

Mariyanto M, Amir MF, Utama W, Hamdan AM, Bijaksana S, Pratama A, Yunginger R, Sudarningsih S (2019) Heavy metal contents and magnetic properties of surface sediments in volcanic and tropical environment from Brantas River, Jawa Timur Province, Indonesia. Sci Total Environ 675:632–641

Marrugo-Negrete J, Pinedo-Hernández J, Díez S (2017) Assessment of heavy metal pollution, spatial distribution and origin in agricultural soils along the Sinú River Basin, Colombia. Environ Res 154:380–388

Martin CW (2004) Heavy metal storage in near channel sediments of the Lahn River, Germany. Geomorphology 61:275–285

Martin JM, Huang WW, Yoon YY (1994) Level and fate of trace metals in the lagoon of Venice (Italy). Mar Chem 46:371–386

Mavura W, Wangila P (2003) The pollution status of Lake Nakuru, Kenya: heavy metals and pesticide residues, 1999/2000. Afr J Aquat Sci 28:13–18

Mazej Z, Al Sayegh-Petkovšek S, Pokorny B (2010) Heavy metal concentrations in food chain of Lake Velenjsko jezero, Slovenia: an artificial lake from mining. Arch Environ Contam Toxicol 58:998–1007

Mendil D, Ünal ÖF, Tüzen M, Soylak M (2010) Determination of trace metals in different fish species and sediments from the River Yeşilırmak in Tokat, Turkey. Food Chem Toxicol 48:1383–1392

Milenkovic N, Damjanovic M, Ristic M (2005) Study of heavy metal pollution in sediments from the Iron Gate (Danube River), Serbia and Montenegro. Pol J Environ Stud 14:781–787

Mohiuddin K, Zakir H, Otomo K, Sharmin S, Shikazono N (2010) Geochemical distribution of trace metal pollutants in water and sediments of downstream of an urban river. Int J Environ Sci Technol 7:17–28

Morillo J, Usero J, Gracia I (2002a) Heavy metal fractionation in sediments from the Tinto River (Spain). Int J Environ Anal Chem 82:245–257

Morillo J, Usero J, Gracia I (2002b) Partitioning of metals in sediments from the Odiel River (Spain). Environ Int 28:263–271

Mortvedt JJ (1996) Heavy metal contaminants in inorganic and organic fertilizers. Fertilizer Res 43:55–61

Mothersill JS (1976) The mineralogy and geochemistry of the sediments of northwestern Lake Victoria. Sedimentology 23:553–565

Muhammad S, Shah MT, Khan S (2011) Health risk assessment of heavy metals and their source apportionment in drinking water of Kohistan region, northern Pakistan. Microchem J 98:334–343

Müller G, Förstner U (1975) Heavy metals in sediments of the Rhine and Elbe estuaries: mobilization or mixing effect? Environ Geol 1:33–39

Müller G, Grimmer G, Böhnke H (1977) Sedimentary record of heavy metals and polycyclic aromatic hydrocarbons in Lake Constance. Naturwissenschaften 64:427–431

Müller J, Ruppert H, Muramatsu Y, Schneider J (2000) Reservoir sediments–a witness of mining and industrial development (Malter Reservoir, eastern Erzgebirge, Germany). Environ Geol 39:1341–1351

Murray KS (1996) Statistical comparisons of heavy-metal concentrations in river sediments. Environ Geol 27:54–58

Nakayama SM, Ikenaka Y, Muzandu K, Choongo K, Oroszlany B, Teraoka H, Mizuno N, Ishizuka M (2010) Heavy metal accumulation in lake sediments, fish (*Oreochromis niloticus* and *Serranochromis thumbergi*), and crayfish (*Cherax quadricarinatus*) in Lake Itezhi-tezhi and Lake Kariba, Zambia. Arch Environ Contam Toxicol 59:291–300

Ndiokwere C (1984) An investigation of the heavy metal content of sediments and algae from the River Niger and Nigerian Atlantic coastal waters. Environ Pollut B 7:247–254

Neumann T, Leipe T, Shimmield G (1998) Heavy-metal enrichment in surficial sediments in the Oder River discharge area: source or sink for heavy metals? Appl Geochem 13:329–337

Nguyen TTH, Zhang W, Li Z, Li J, Liu J, Bai X, Feng H, Yu L (2016) Assessment of heavy metal pollution in Red River surface sediments, Vietnam. Mar Pollut Bull 113:513–519

Niu H, Deng W, Wu Q, Chen X (2009) Potential toxic risk of heavy metals from sediment of the Pearl River in South China. J Environ Sci 21:1053–1058

Nriagu JO (1986) Chemistry of the river Niger II. Trace metals. Sci Total Environ 58:89–92

Nriagu J, Kemp A, Wong H, Harper N (1979) Sedimentary record of heavy metal pollution in Lake Erie. Geochim Cosmochim Acta 43:247–258

Olivares-Rieumont S, De La Rosa D, Lima L, Graham DW, Alessandro KD, Borroto J, Martínez F, Sánchze J (2005) Assessment of heavy metal levels in Almendares River sediments – Havana City, Cuba. Water Res 39:3945–3953

Oliver BG (1973) Heavy metal levels of Ottawa and Rideau River sediments. Environ Sci Technol 7:135–137

Onyari JM, Wandiga SO (1989) Distribution of Cr, Pb, Cd, Zn, Fe and Mn in Lake Victoria sediments, East Africa. Bull Environ Contam Toxicol 42:807–813

Opfer SE, Farver JR, Miner JG, Krieger K (2011) Heavy metals in sediments and uptake by burrowing mayflies in western Lake Erie basin. J Great Lakes Res 37:1–8

Osman AG, Kloas W (2010) Water quality and heavy metal monitoring in water, sediments, and tissues of the African catfish *Clarias gariepinus* (Burchell, 1822) from the river Nile, Egypt. J Environ Prot 1:389–400

Oyewale A, Musa I (2006) Pollution assessment of the lower basin of Lakes Kainji/Jebba, Nigeria: heavy metal status of the waters, sediments and fishes. Environ Geochem Health 28:273–381

Pandey J, Singh R (2017) Heavy metals in sediments of Ganga River: up-and downstream urban influences. Appl Water Sci 7:1669–1678

Paramasivam K, Ramasamy V, Suresh G (2015) Impact of sediment characteristics on the heavy metal concentration and their ecological risk level of surface sediments of Vaigai river, Tamil Nadu, India. Spectrochim Acta A Mol Biomol Spectrosc 137:397–407

Patel P, Raju NJ, Reddy BSR, Suresh U, Sankar D, Reddy T (2018) Heavy metal contamination in river water and sediments of the Swarnamukhi River Basin, India: risk assessment and environmental implications. Environ Geochem Health 40:609–623

Pathiratne A, Chandrasekera L, Pathiratne K (2009) Use of biomarkers in Nile tilapia (*Oreochromis niloticus*) to assess the impacts of pollution in Bolgoda Lake, an urban water body in Sri Lanka. Environ Monit Assess 156:361–375

Pekey H, Karaka D, Bakoğlu M (2004) Source apportionment of trace metals in surface waters of a polluted stream using multivariate statistical analyses. Mar Pollut Bull 49:809–818

Peng JF, Song YH, Yuan P, Cui XY, Qiu GL (2009) The remediation of heavy metals contaminated sediment. J Hazard Mater 161:633–640

Pesantes A, Carpio AP, Vitvar E, López TMM, Juan MMA (2019) A multi-Index analysis approach to heavy metal pollution assessment in river sediments in the Ponce Enríquez Area, Ecuador. Water 11:590

Phillips D, Human L, Adams J (2015) Wetland plants as indicators of heavy metal contamination. Mar Pollut Bull 92:227–232

Prasad MBK, Ramanathan A, Shrivastav SK, Saxena R (2006) Metal fractionation studies in surfacial and core sediments in the Achankovil river basin in India. Environ Monit Assess 121:77–102

Prudente MS, Ichihashi H, Tatsukawa R (1994) Heavy metal concentrations in sediments from Manila Bay, Philippines and inflowing rivers. Environ Pollut 86:83–88

Qian Y, Zheng MH, Gao L, Zhang B, Liu W, Jiao W, Zhao X, Xiao K (2005) Heavy metal contamination and its environmental risk assessment in surface sediments from Lake Dongting, People's Republic of China. Bull Environ Contam Toxicol 75:204–210

Rahman MS, Saha N, Molla AH (2014) Potential ecological risk assessment of heavy metal contamination in sediment and water body around Dhaka export processing zone, Bangladesh. Environ Earth Sci 71:2293–2308

Raju KV, Somashekar R, Prakash K (2012) Heavy metal status of sediment in river Cauvery, Karnataka. Environ Monit Assess 184:361–373

Ramachandra T, Sudarshan P, Mahesh M, Vinay S (2018) Spatial patterns of heavy metal accumulation in sediments and macrophytes of Bellandur wetland, Bangalore. J Environ Manag 206:1204–1210

Ramesh R, Subramanian V, Van Grieken R, Van't Dack L (1989) The elemental chemistry of sediments in the Krishna river basin, India. Chem Geol 74:331–341

Ranjbar GA (1998) Heavy metal concentration in surficial sediments from Anzali wetland, Iran. Water Air Soil Pollut 104:305–312

Rashed M (2001) Monitoring of environmental heavy metals in fish from Nasser Lake. Environ Int 27:27–33

Reboredo FHS, Ribeiro CG (1984) Vertical distribution of Al, Cu, Fe and Zn in the soil salt marshes of the Sado estuary, Portugal. Int J Environ Stud 23:249–253

Reece D, Felkey JR, Wai C (1978) Heavy metal pollution in the sediments of the Coeur d'Alene River, Idaho. Environ Geol 2:289–293

Ren J, Williams PN, Luo J, Ma H, Wang X (2015) Sediment metal bioavailability in Lake Taihu, China: evaluation of sequential extraction, DGT, and PBET techniques. Environ Sci Pollut Res 22:12919–11228

Rifaat AE (2005) Major controls of some metals' distribution in sediments off the Nile Delta, Egypt. Egypt J Aquat Res 31:16–28

Rowan D, Kalff J (1993) Predicting sediment metal concentrations in lakes without point sources. Water Air Soil Pollut 66:145–161

Ruilian Y, Xing Y, Yuanhui Z, Gongren H, Xianglin T (2008) Heavy metal pollution in intertidal sediments from Quanzhou Bay, China. J Environ Sci 20:664–669

Sabri AW, Khalid AR, Thaer IK (1993) Heavy metals in the water, suspended solids and sediment of the river Tigris impoundment at Samarra. Water Res 27:1099–10103

Saeed SM, Shaker IM (2008) Assessment of heavy metals pollution in water and sediments and their effect on *Oreochromis niloticus* in the northern delta lakes, Egypt. In: Proceedings of the 8th International Symposium on Tilapia in Aquaculture. Central Laboratory for Aquaculture Research, Agricultural Research Center. Limnology Department, Cairo, pp 475–490

Saeki K, Okazaki M, Kubota M (1993) Heavy metal accumulations in a semi-enclosed hypereutrophic system: Lake Teganuma, Japan. Water Air Soil Pollut 69:79–91

Saha P, Hossain M (2011) Assessment of heavy metal contamination and sediment quality in the Buriganga River, Bangladesh. In: Proceedings of the 2nd International Conference on Environmental Science and Technology. IPCBEE, Singapore, pp 26–28
Sakan SM, Đorđević DS, Manojlović DD, Predrag PS (2009) Assessment of heavy metal pollutants accumulation in the Tisza river sediments. J Environ Manag 90:3382–3390
Salati S, Moore F (2010) Assessment of heavy metal concentration in the Khoshk River water and sediment, Shiraz, Southwest Iran. Environ Monit Assess 164:677–689
Sekabira K, Origa HO, Basamba T, Mutumba G, Kakudidi E (2010) Assessment of heavy metal pollution in the urban stream sediments and its tributaries. Int J Environ Sci Technol 7:435–446
Selvam AP, Priya SL, Banerjee K, Hariharan G, Purvaja R, Ramesh R (2012) Heavy metal assessment using geochemical and statistical tools in the surface sediments of Vembanad Lake, Southwest Coast of India. Environ Monit Assess 184:5899–5915
Shafie NA, Aris AZ, Zakaria MP, Haris H, Lim WY, Isa NM (2013) Application of geoaccumulation index and enrichment factors on the assessment of heavy metal pollution in the sediments. J Environ Sci Health A 48:182–190
Sharley DJ, Sharp SM, Bourgues S, Pettigrove VJ (2016) Detecting long-term temporal trends in sediment-bound trace metals from urbanised catchments. Environ Pollut 219:705–713
Shen Z, Dong W, Zhang L, Chen X (1989) Geochemical characteristics of heavy metals in the Xiangjiang River, China. Hydrobiologia 176:253–262
Siddiqui E, Pandey J (2019) Assessment of heavy metal pollution in water and surface sediment and evaluation of ecological risks associated with sediment contamination in the Ganga River: a basin-scale study. Environ Sci Pollut Res 26:10926–10940
Simpson RL, Good RE, Walker R, Frasco BR (1983) The role of Delaware River freshwater tidal wetlands in the retention of nutrients and heavy metals. J Environ Qual 12:41–48
Sin S, Chua H, Lo W, Ng L (2001) Assessment of heavy metal cations in sediments of Shing Mun River, Hong Kong. Environ Int 26:297–301
Singh M, Ansari A, Müller G, Singh I (1997) Heavy metals in freshly deposited sediments of the Gomati River (a tributary of the Ganga River): effects of human activities. Environ Geol 29:246–252
Singh O, Labana S, Pandey G, Budhiraja R, Jain R (2003) Phytoremediation: an overview of metallic ion decontamination from soil. Appl Microbiol Biotechnol 61:405–412
Singh KP, Malik A, Sinha S, Singh VK, Murthy RC (2005) Estimation of source of heavy metal contamination in sediments of Gomti River (India) using principal component analysis. Water Air Soil Pollut 166:321–341
Subramanian V, Van't Dack L, Van Grieken R (1985) Chemical composition of river sediments from the Indian sub-continent. Chem Geol 48:271–279
Subramanian V, Van Grieken R, Van't Dack L (1987) Heavy metals distribution in the sediments of Ganges and Brahmaputra rivers. Environ Geol Water Sci 9:93–103
Suresh G, Ramasamy V, Meenakshisundaram V, Venkatachalapathy R, Ponnusamy V (2011) Influence of mineralogical and heavy metal composition on natural radionuclide concentrations in the river sediments. Appl Radiat Isot 69:1466–1474
Suresh G, Sutharsan P, Ramasamy V, Venkatachalapathy R (2012) Assessment of spatial distribution and potential ecological risk of the heavy metals in relation to granulometric contents of Veeranam lake sediments, India. Ecotoxicol Environ Saf 84:117–124
Swarnalatha K, Letha J, Ayoob S, Nair AG (2015) Risk assessment of heavy metal contamination in sediments of a tropical lake. Environ Monit Assess 187:322
Taher AG, Soliman AA (1999) Heavy metal concentrations in surficial sediments from Wadi El Natrun saline lakes, Egypt. Int J Salt Lake Res 8:75–92
Tao Y, Yuan Z, Xiaona H, Wei M (2012) Distribution and bioaccumulation of heavy metals in aquatic organisms of different trophic levels and potential health risk assessment from Taihu lake, China. Ecotoxicol Environ Saf 81:55–64
Taylor GJ, Crowder A (1983) Accumulation of atmospherically deposited metals in wetland soils of Sudbury, Ontario. Water Air Soil Pollut 19:29–42

Tekin-Zan S (2008) Determination of heavy metal levels in water, sediment and tissues of tench (*Tinca tinca* L., 1758) from Beyşehir Lake (Turkey). Environ Monit Assess 145:295–302

Tekin-Zan S, Kir İ (2005) Comparative study on the accumulation of heavy metals in different organs of tench (*Tinca tinca* L. 1758) and plerocercoids of its endoparasite *Ligula intestinalis*. Parasitol Res 97:156–159

Tenhola M (1993) Use of lake sediments in sulphur and heavy metal pollution monitoring: example from Finland. Appl Geochem 8:171–173

Tessier A, Campbell P, Auclair J, Bisson M (1984) Relationships between the partitioning of trace metals in sediments and their accumulation in the tissues of the freshwater mollusc *Elliptio complanata* in a mining area. Can J Fish Aquat Sci 41:1463–1472

Trocine RP, Trefry JH (1996) Metal concentrations in sediment, water and clams from the Indian River Lagoon, Florida. Mar Pollut Bull 10:754–759

Vandecasteele B, De Vos B, Tack F (2003) Temporal-spatial trends in heavy metal contents in sediment-derived soils along the Sea Scheldt river (Belgium). Environ Pollut 122:7–18

Varol M (2011) Assessment of heavy metal contamination in sediments of the Tigris River (Turkey) using pollution indices and multivariate statistical techniques. J Hazard Mater 195:355–364

Viganò L, Arillo A, Buffagni A, Camusso M, Ciannarella R, Crosa G, Falugi C, Galassi S, Guzzella L, Lopez A, Mingazzini M, Pagnotta R, Patrolecco L, Tartari G, Valsecchi S (2003) Quality assessment of bed sediments of the Po River (Italy). Water Res 37:501–518

Von Gunten H, Sturm M, Moser R (1997) 200-year record of metals in lake sediments and natural background concentrations. Environ Sci Technol 31:2193–2197

Walker J, Colwell R (1974) Mercury-resistant bacteria and petroleum degradation. Appl Microbiol 27:285–287

Wang S, Cao Z, Lan D, Zheng Z, Li G (2008) Concentration distribution and assessment of several heavy metals in sediments of west-four Pearl River Estuary. Environ Geol 55:963–975

Wang Y, Yang L, Kong L, Liu E, Wang L, Zhu J (2015) Spatial distribution, ecological risk assessment and source identification for heavy metals in surface sediments from Dongping Lake, Shandong, East China. Catena 125:200–205

Wang H, Liu T, Tsang DC, Feng S (2017) Transformation of heavy metal fraction distribution in contaminated river sediment treated by chemical-enhanced washing. J Soils Sediments 17:1208–1218

Wei B, Yang L (2010) A review of heavy metal contaminations in urban soils, urban road dusts and agricultural soils from China. Microchem J 94:99–107

Wen J, Yi Y, Zeng G (2016) Effects of modified zeolite on the removal and stabilization of heavy metals in contaminated lake sediment using BCR sequential extraction. J Environ Manag 178:63–69

Wenchuan Q, Dickman M, Sumin W (2001) Multivariate analysis of heavy metal and nutrient concentrations in sediments of Taihu Lake, China. Hydrobiologia 450:83–89

Wentsel R, Mcintosh A, Anderson V (1977a) Sediment contamination and benthic macroinvertebrate distribution in a metal-impacted lake. Environ Pollut 14:187–193

Wentsel R, Mcintosh A, Atchison G (1977b) Sublethal effects of heavy metal contaminated sediment on midge larvae (*Chironomus tentans*). Hydrobiologia 56:153–156

Wentsel R, Mcintosh A, Mccafferty WP, Atchison G, Anderson V (1977c) Avoidance response of midge larvae (*Chironomus tentans*) to sediments containing heavy metals. Hydrobiologia 55:171–175

White KD, Tittlebaum ME (1985) Metal distribution and contamination in sediments. J Environ Eng 111:161–175

Wilber WG, Hunter JV (1979) The impact of urbanization on the distribution of heavy metals in bottom sediments of the Saddle River. J Am Water Resour Assoc 15:790–800

Williams S, Simpson H, Olsen C, Bopp R (1978) Sources of heavy metals in sediments of the Hudson River estuary. Mar Chem 6:195–213

Woitke P, Wellmitz J, Helm D, Kube P, Lepom P, Litheraty P (2003) Analysis and assessment of heavy metal pollution in suspended solids and sediments of the river Danube. Chemosphere 51:633–642

Wong KW, Yap CK, Nulit R, Hamzah MS, Chen SK, Cheng WK, Karami A, Al-Shami SA (2017) Effects of anthropogenic activities on the heavy metal levels in the clams and sediments in a tropical river. Environ Sci Pollut Res 24:116–134

Xia F, Qu L, Wang T, Luo L, Chen H, Dahlgren RA, Zhang M, Mei K, Huang H (2018) Distribution and source analysis of heavy metal pollutants in sediments of a rapid developing urban river system. Chemosphere 207:218–228

Xiong C, Wang D, Tam NF, Dai Y, Zhang X, Tang X, Yang Y (2018) Enhancement of active thin-layer capping with natural zeolite to simultaneously inhibit nutrient and heavy metal release from sediments. Ecol Eng 119:64–72

Xu F, Liu Z, Cao Y, Qiu L, Feng J, Xu F, Tian X (2017) Assessment of heavy metal contamination in urban river sediments in the Jiaozhou Bay catchment, Qingdao, China. Catena 150:9–16

Yan N, Liu W, Xie H, Gao L, Han Y, Wang M, Li H (2016) Distribution and assessment of heavy metals in the surface sediment of Yellow River, China. J Environ Sci 39:45–51

Yan X, Liu M, Zhong J, Guo J, Wu W (2018) How human activities affect heavy metal contamination of soil and sediment in a long-term reclaimed area of the Liaohe River Delta, North China. Sustainability 10:338

Yang Z, Li B, Li G, Wang W (2007) Nutrient elements and heavy metals in the sediment of Baiyangdian and Taihu Lakes: A comparative analysis of pollution trends. Front Agric China 1:203–209

Yang Z, Wang Y, Shen Z, Niu J, Tang Z (2009) Distribution and speciation of heavy metals in sediments from the mainstream, tributaries, and lakes of the Yangtze River catchment of Wuhan, China. J Hazard Mater 166:1186–1194

Yang Y, Chen F, Zhang L, Liu J, Wu S, Kang M (2012) Comprehensive assessment of heavy metal contamination in sediment of the Pearl River Estuary and adjacent shelf. Mar Pollut Bull 64:1947–1955

Yang Y, Christakos G, Guo M, Xiao L, Huang W (2017) Space-time quantitative source apportionment of soil heavy metal concentration increments. Environ Pollut 223:560–566

Yao X, Niu Y, Li Y, Zou D, Ding X, Bian H (2018) Heavy metal bioaccumulation by *Miscanthus sacchariflorus* and its potential for removing metals from the Dongting Lake wetlands, China. Environ Sci Pollut Res 25:20003–20011

Yim WWS (1976) Heavy metal accumulation in estuarine sediments in a historical mining of Cornwall. Mar Pollut Bull 7:147–150

Yin H, Gao Y, Fan C (2011) Distribution, sources and ecological risk assessment of heavy metals in surface sediments from Lake Taihu, China. Environ Res Lett 6:044012

Yousef Y, Wanielista M, Hvitved-Jacobsen T, Harper H (1984) Fate of heavy metals in stormwater runoff from highway bridges. Sci Total Environ 33:233–244

Yu T, Zhang Y, Hu X, Meng W (2012) Distribution and bioaccumulation of heavy metals in aquatic organisms of different trophic levels and potential health risk assessment from Taihu lake, China. Ecotoxicol Environ Saf 81:55–64

Yuan CH, Shen XM (2006) Present state and mission of the prevention and treatment for lead poisoning in children in China. Chin J Pract Pediat 21:161–163

Yuan GL, Liu C, Chen L, Yang Z (2011) Inputting history of heavy metals into the inland lake recorded in sediment profiles: Poyang Lake in China. J Hazard Mater 185:336–345

Yuan X, Zhang L, Li J, Wang C, Ji J (2014) Sediment properties and heavy metal pollution assessment in the river, estuary and lake environments of a fluvial plain, China. Catena 119:52–60

Zan F, Huo S, Xi B, Su J, Li X, Zhang J, Yeager KM (2011) A 100 year sedimentary record of heavy metal pollution in a shallow eutrophic lake, Lake Chaohu, China. J Environ Monit 13:2788–2797

Zayed M, Eldien FN, Rabie K (1994) Comparative study of seasonal variation in metal concentrations in River Nile sediment, fish, and water by atomic absorption spectrometry. Microchem J 49:27–35

Zhang J (1999) Heavy metal compositions of suspended sediments in the Changjiang (Yangtze River) estuary: significance of riverine transport to the ocean. Cont Shelf Res 19:1521–1543

Zhang W, Yu L, Hutchinson S, Xu S, Chen Z, Gao X (2001) China's Yangtze Estuary: I. Geomorphic influence on heavy metal accumulation in intertidal sediments. Geomorphology 41:195–205

Zhang W, Feng H, Chang J, Qu J, Xie H, Yu L (2009) Heavy metal contamination in surface sediments of Yangtze River intertidal zone: an assessment from different indexes. Environ Pollut 157:1533–1543

Zhang L, Liao Q, Shao S, Zhang N, Shen Q, Liu C (2015) Heavy metal pollution, fractionation, and potential ecological risks in sediments from Lake Chaohu (Eastern China) and the surrounding rivers. Int J Environ Res Public Health 12:14115–14131

Zhang Z, Juying L, Mamat Z, Qingfu Y (2016) Sources identification and pollution evaluation of heavy metals in the surface sediments of Bortala River, Northwest China. Ecotoxicol Environ Saf 126:94–101

Zhang C, Shan B, Tang W, Dong L, Zhang W, Pei Y (2017a) Heavy metal concentrations and speciation in riverine sediments and the risks posed in three urban belts in the Haihe Basin. Ecotoxicol Environ Saf 139:263–271

Zhang G, Bai J, Xiao R, Zhao Q, Jia J, Cui B, Liu X (2017b) Heavy metal fractions and ecological risk assessment in sediments from urban, rural and reclamation-affected rivers of the Pearl River Estuary, China. Chemosphere 184:278–288

Zhang H, Jiang Y, Ding M, Xie Z (2017c) Level, source identification, and risk analysis of heavy metal in surface sediments from river-lake ecosystems in the Poyang Lake, China. Environ Sci Pollut Res 24:21902–21916

Zhang Z, Lu Y, Li H, Tu Y, Liu B, Yang Z (2018) Assessment of heavy metal contamination, distribution and source identification in the sediments from the Zijiang River, China. Sci Total Environ 645:235–243

Zhao G, Ye S, Yuan H, Ding X, Wang J (2017) Surface sediment properties and heavy metal pollution assessment in the Pearl River Estuary, China. Environ Sci Pollut Res 24:2966–2979

Zheng LG, Liu GJ, Kang Y, Yang RK (2010) Some potential hazardous trace elements contamination and their ecological risk in sediments of western Chaohu Lake, China. Environ Monit Assess 166:379–386

Zhuang W, Liu Y, Chen Q, Wang Q, Zhou F (2016) A new index for assessing heavy metal contamination in sediments of the Beijing-Hangzhou Grand Canal (Zaozhuang Segment): a case study. Ecol Indic 69:252–260

Zingde M, Rokade M, Mandalia A (1988) Heavy metals in Mindhola river estuary, India. Mar Pollut Bull 19:538–540

Zorer S, Ceylan H, Doğru M (2009) Determination of heavy metals and comparison to gross radioactivity concentration in soil and sediment samples of the Bendimahi River Basin (Van, Turkey). Water Air Soil Pollut 196:75–87

Zoumis T, Schmidt A, Grigorova L, Calmano W (2001) Contaminants in sediments: remobilisation and demobilisation. Sci Total Environ 266:195–202

A Review on Prediction Models for Pesticide Use, Transmission, and Its Impacts

Edwin Prem Kumar Gilbert and Lydia Edwin

Contents

1 Introduction 38
2 Mathematical Models for Prediction 40
  2.1 Prediction Based on Regression Models 40
  2.2 Prediction Using Non-linear Models 41
  2.3 Performance Metrics for Prediction Models 42
3 Prediction Models for Agriculture 43
  3.1 Prediction Models for Pesticide Usage 43
  3.2 Prediction Models for Pesticide Residue in the Environment 49
  3.3 Prediction Models for Human Health Disorders Due to Pesticide Exposure 52
  3.4 Prediction Models for Antimicrobial Resistance Due to Pesticides 54
4 Modeling Strategies 57
  4.1 Pesticide Simulation Models 57
  4.2 Machine Learning-Based Models 57
  4.3 Experimental Study 59
5 Mission Ahead 59
6 Conclusion 60
References 61

Abstract The lure of increased productivity and crop yield has caused the imprudent use of pesticides in great quantity that has unfavorably affected environmental health. Pesticides are chemicals intended for avoiding, eliminating, and mitigating any pests that affect the crop. Lack of awareness, improper management, and negligent disposal of pesticide containers have led to the permeation of pesticide residues into the food chain and other environmental pathways, leading to

E. P. K. Gilbert (✉)
Department of Information Technology, Sri Krishna College of Engineering and Technology, Coimbatore, Tamil Nadu, India
e-mail: edwinpremkumar@gmail.com

L. Edwin
Department of Mechatronics Engineering, Sri Krishna College of Engineering and Technology, Coimbatore, Tamil Nadu, India
e-mail: lydiaedwin.05@gmail.com

P. de Voogt (ed.), *Reviews of Environmental Contamination and Toxicology Volume 257*,
Reviews of Environmental Contamination and Toxicology 257,
https://doi.org/10.1007/398_2020_64

environmental degradation. Sufficient steps must be undertaken at various levels to monitor and ensure judicious use of pesticides. Development of prediction models for optimum use of pesticides, pesticide management, and their impact would be of great help in monitoring and controlling the ill effects of excessive use of pesticides. This paper aims to present an exhaustive review of the prediction models developed and modeling strategies used to optimize the use of pesticides.

Keywords Machine learning · Modeling strategy · Pesticides · Prediction models · Residues

1 Introduction

The advancement of science and technology has revolutionized and redefined our lives in many amazing ways. Unfortunately, the adverse side effects have crippled us in many ways. Mankind as a whole is staring at the grim reality of the ill effects that have led to human health hazards and environmental degradation. According to the United Nations (UN) estimate, the population of the world is expected to reach 9.7 billion in 2050, and the food demand is likely to go up by 59–98%. Such statistics lay down the need for increased food production (UN/FAO 2017).

The need for increased productivity, lack of availability of agricultural land, and various other reasons has led to the rampant use of pesticides all over the world. Though the use of pesticides fostered the green revolution and has enhanced food security and productivity, it has also had several disastrous effects on the environment. It is doubted that these sustained years of increased productivity have been powered by the disproportionate use of pesticides and fertilizers.

Pesticides come in the form of insecticides, rodenticides, bactericides/disinfectants, fungicides, and herbicides. The World Health Organization (WHO) classifies pesticides based on their toxicity levels: Class I-a as extremely hazardous, Class I-b as highly hazardous, Class II moderately hazardous, and Class III slightly hazardous. Highly Hazardous Pesticides (HHP) cause very severe or irreversible harm to health and the environment (WHO/FAO 2014). Pesticide management involves the control of all aspects of the pesticide life cycle, including its formulation, import, distribution, transportation, sale, storage, application, and disposal of pesticide containers to minimize the risk of pesticide exposure. As pesticide risk is a function of both hazard and exposure, reduction in risk can be attained either by a reduction in hazard or reduction in exposure (FAO/WHO 2016). Reduced dependence on pesticides, usage of pesticides with lower risk, and ensuring appropriate use of selected products are the important steps to be adhered to reduce pesticide risk.

Pesticide lifecycle management (PLM) is a very critical global issue as it incorporates the legal matters, regulation, its manufacturing, distribution, usage, risk mitigation, monitoring, and disposal of pesticide wastage. In a survey conducted to review the global status of PLM in 2017–2018, it was observed that the gaps between pesticide efficiency and safety were highest in low-income countries (Berg et al. 2020). Three important shortcomings were observed in PLM in terms of deficiencies in pesticide legislation, registration and protection of humans, and the environment from pesticide residues. Lack of regulatory policy and research and development efforts in the areas of integrated pest management (IPM) and integrated vector management (IVM) were also major areas of concern.

Pesticides continue to be a threat not only for the farmers but to all living beings at large directly or indirectly. Human exposure occurs both through direct means and dietary means. We are exposed to pesticides through the air we breathe, the food we eat, the water we drink, and other subtle means. Pesticides that are air-sprayed, get carried away along with the wind, are washed off into the nearby water bodies, or seep into the groundwater. Pesticide residues are likely to be found in agricultural products like fruits and vegetables. Contaminated feed and veterinary pesticide applications have resulted in pesticides found in animal products too. In accordance with Good Agricultural Practice, the maximum residue limit (MRL) is the highest level of pesticide residue that is legally tolerated in or on food or feed after the application of pesticides (FAO/WHO 2019a). The WHO guidelines for ensuring food quality (FAO/WHO 2019b) and water quality (WHO 2017) serve as a monitoring mechanism in this regard.

Young children seem to be the worst affected lot (UNICEF 2018). The agricultural sector accounts for 71% of child labor and around 108 million children are involved in agricultural work. Children are exposed to pesticides from their mothers through blood, placental tissue, and breast milk. They are exposed at home through stored chemicals or by ingestion of contaminated fruits and vegetables. When exposed to aerially sprayed pesticides, children tend to inhale double the amount of an average adult and the inhaled chemicals are likely to be ten times more toxic than of an adult, due to the small size of the liver and kidneys (UNICEF 2018). Playing children get exposed to pesticide containers, intake of contaminated soil, and due to residues on surfaces (WHO 2019). Continued monitoring and evaluation and implementation of robust government legislations are vital to ensure that children are protected from these hazards. The majority of children exposed to pesticides scored badly in cognitive tasks (Kumar and Reddy 2017).

There are several health impacts associated with pesticide exposure. It is estimated that nearly 300,000 deaths worldwide occur every year due to pesticide poisonings (Sabarwal et al. 2018). Pesticide exposure includes long-term high-level occupational exposure and low-level non-occupational exposure. According to Sabarwal et al. (2018), the cancers associated with pesticides include cancers of the prostrate, breast, and colorectal regions. The role of pesticides in Parkinson's disease, Alzheimer's disease, and various other disorders in respiratory and reproductive tracts has also been reported. Pesticides have been found to have a detrimental impact on the cellular mechanism of the human body, resulting in endocrine disruption and other genetic damages. Kim et al. (2017) reviewed the health effects associated with exposure to pesticides. According to, Bolognesi and Merlo (2019), the delayed health impacts in agricultural workers, farmers, and sprayers include leukemia, sarcomas, and cancers. The association between parental exposure and childhood cancers has also been reported.

2 Mathematical Models for Prediction

Predictive analytics is a challenging area of research that incorporates a wide variety of statistical techniques like machine learning (ML), regression models, predictive modeling, data mining, and various other areas to analyze the current and historical data and make predictions for the future. Predictive analytics in agriculture is poised to revolutionize farming practices and improve crop performance leveraging the latest technological innovations and smart decision making. Predictive modeling and data analytics have the immense potential to reshape agriculture and all related industries, market boundaries, and relations (Pham and Stack 2018).

2.1 Prediction Based on Regression Models

Quantitative techniques for forecasting use historical data and a prediction model. A set of observations $y(t)$ recorded at a specific time t comprise a time series. Data pertaining to weather conditions, climate indices, crop yield, etc. that are usually recorded with time stamp are time series data. Time series analysis comprises formulating time series models that are used to provide a compact description of the data, in terms of the statistical relationship between current and historical data (Montgomery et al. 2008). The smoothing model provides the forecast by employing a simple function of earlier observations. Based on the number of variables used, prediction models can be classified as univariate and multivariate models. Model-identification, parameter estimation, and diagnostic checking are the three important steps involved in building a time series model. It is observed that better performing accurate models can be obtained by combining different models and also by using parallel-series hybrid structures (Hajirahimi and Khashei 2019). Use of combined forecasting models improved forecast accuracy, ensured the appropriate selection of models, and also simplified the model selection procedure.

2.1.1 Univariate Models

Univariate models are built on only one variable. In an AutoRegressive (AR) model, the variable of interest is forecast based on the linear combination of its past variables. The AR model structure is given by the following equation:

$$A(q)y(t) = e(t) \tag{1}$$

where $y(t)$ is the variable of interest, $A(q)y(t)$ is the autoregressive part, $e(t)$ is the white noise, and q is the backshift operator. $A(q) = 1 + a_1q^{-1} + \ldots + a_{n_a}q^{-n_a}$ where $a_1, \ldots, a_{n_a}$ are the AR parameters and n_a is the AR order. The AR model parameters can be estimated using variants of the least squares method. The

AutoRegressive Moving Average (ARMA) model for a univariate model is given by the following equation:

$$A(q)y(t) = C(q)e(t) \tag{2}$$

where $C(q)e(t)$ is the moving average (MA) part and $C(q) = 1 + c_1q^{-1} + \ldots + c_{n_c}q^{-n_c}$ where $c_1, \ldots, c_{n_c}$ are the MA parameters and n_c is the MA order (Montgomery et al. 2008). The ARMA models can be more generalized for nonstationary series using differenced data, which in turn results in AutoRegressive Integrated Moving Average (ARIMA) models. The seasonal component of period p can be eliminated easily, by differencing the series at lag p. By fitting this differenced series in ARMA, a special model known as seasonal ARIMA (SARIMA) is obtained (Brockwell and Davis 2002).

2.1.2 Multivariate Models

Multivariate models are built when more than one parameter is available and the variation in the parameter of interest can be realized as a statistical function involving the current and historic values of all the known variables. The structure of a linear AutoRegressive model with eXogenous inputs (ARX) for multiple inputs and a single output is given by the following equation:

$$A(q)y(t) = B(q)u(t) + e(t) \tag{3}$$

where $B(q)u(t)$ corresponds to the exogenous inputs and $\mathrm{B}(q) = b_1q^{-1} + \ldots + b_{n_b}q^{-n_b}$ where $b_1, \ldots, b_{n_b}$ are the parameters of the exogenous inputs and n_b is the input order. A linear time-invariant ARMAX model can be written as follows:

$$A(q)y(t) = B(q)u(t) + C(q)e(t) \tag{4}$$

Vector ARIMA, Vector AR, AutoRegressive Conditional Heteroskedasticity (ARCH) and Generalized ARCH (GARCH), state-space models, direct forecasting of percentiles, hybrid prediction models are various other models that can be used for developing prediction models (Montgomery et al. 2008).

2.2 *Prediction Using Non-linear Models*

A system can be described in a better way using a non-linear relationship between the input and output variables (Ljung 1999). Unlike the linear models, the predictor function is realized as a non-linear function of the past observations in a non-linear model. The non-linear form of AR, known as NAR, the non-linear realization of ARX, known as NLARX and NL-ARMAX can be written, respectively, as given below:

$$y(t) = f(y(t-1), y(t-2), \ldots y(t-n_a)) \tag{5}$$

$$y(t) = f(y(t-1), y(t-2), \ldots y(t-n_a), u(t-1), u(t-2) \ldots u(t-n_b)) \tag{6}$$

$$y(t) = f(y(t-1), \ldots y(t-n_a), u(t-1), \ldots u(t-n_b), e(t-1), \ldots e(t-n_c) \tag{7}$$

where f is a non-linear function.

It can be observed that the next value of the output signal is expressed as a non-linear function of the past values of input values and exogenous parameters. The non-linear models can be realized using estimators like Artificial Neural Network (ANN), wavelet network, sigmoid network, decision tree, and various other algorithms.

2.3 Performance Metrics for Prediction Models

The selection of an appropriate forecasting model depends on the accurate performance of the model. The performance of the prediction model is measured in terms of prediction error. The standard measures of forecast accuracy include Average or Mean Error (ME), Mean Absolute Deviation (MAD), Mean Squared Error (MSE), Relative forecast Error (RE), Mean Percent forecast Error (MPE), and Mean Absolute Percent forecast Error (MAPE).

For n number of observations, if y_t is the actual observation and $\widehat{y}_t$ is the estimated value, error s_t is given by $s_t = y_t - \widehat{y}_t$.

$$ME = \frac{1}{n}\sum\nolimits_{t=1}^{n} s_t \tag{8}$$

$$MAD = \frac{1}{n}\sum\nolimits_{t=1}^{n} |s_t| \tag{9}$$

$$MSE = \frac{1}{n}\sum\nolimits_{t=1}^{n} [s_t]^2 \tag{10}$$

$$RE = \frac{s_t}{y_t} * 100 \tag{11}$$

$$MPE = \frac{1}{n}\sum\nolimits_{t=1}^{n} RE \tag{12}$$

$$MAPE = \frac{1}{n}\sum\nolimits_{t=1}^{n} |rs_t| \tag{13}$$

Performance measures can also be computed from the confusion matrix, in terms of accuracy, recall, precision, and F-measure.

3 Prediction Models for Agriculture

Prediction models based on these techniques can be used to find the optimal quantity of pesticide usage and to estimate pesticide residues in the environment. These models can also help in identifying the human disorders that would result due to the rampant usage of pesticides and can also aid in assessing the antimicrobial resistance that would be developed in the living organisms through the food chain. Elavarasan et al. (2018) have reviewed the application of various analytical models like decision trees, regression models, support vector machines (SVM), and other latest techniques including ML algorithms for agriculture purposes. These techniques can help in forecasting, crop analysis using image processing, pest and weed identification, etc.

3.1 Prediction Models for Pesticide Usage

The optimal usage of pesticides will depend on the knowledge of the time of occurrence of pests, its intensity, and spread. Prediction models can be used to predict the occurrence of pests and to ensure optimum usage of pesticides. Subash et al. (2017) analyzed the trends in the usage of pesticides in the Indian agricultural market and defined certain policy issues. They concluded that the use of economical and ecologically safe pesticides is the need of the hour. A critical review of the nature and degree of pesticide contamination in the urban environment has been presented by Meftaul et al. (2020). Usage of pesticides with additives of indefinite concentrations with variable toxicities is a highly worrying cause of concern of national and international relevance.

The environmental risk caused by agricultural pesticides in China was evaluated by Jiao et al. (2020b), for the period from 2004 to 2017. They analyzed the spatio-temporal variability of pesticide distribution and also quantified the pesticide risk hotspots. Malaj et al. (2020) evaluated the spatial distribution of agricultural pesticides in the Canadian Prairie region. They combined the impact of pesticide use density, wetland density, precipitation patterns, and physicochemical properties and developed an index of predicted pesticide occurrence in wetlands. Devi et al. (2017) presented a spatio-temporal study of pesticide usage across various states of India over the period 1992–2012. They emphasized the need for regulated policy and governance in the policy sector. Mohanty et al. (2013) studied the correlation between knowledge and practice of pesticide usage among farmers in Puducherry. A questionnaire, based on WHO's 'Exposure to Pesticides Standard Protocol' was adapted and around 100 agricultural workers were interviewed. They concluded that the overall level of awareness of farmers on pesticide application and disposal was inadequate. Specialized training and guiding services are required for farmers, to increase their awareness of prudent use of pesticides (Houbraken et al. 2017). The authors studied the farmers' behavior towards pesticides and also assessed the ecological risk of pesticide residues in the lakes of Rwanda.

3.1.1 Prediction of Climatic Indices

Climate change and its related factors have been proven to have an impact on the occurrence of pests. The climate change model can be realized using global circulation models that portray the climate using a 3D-grid and representative concentration pathways (RCP). These RCPs represent the greenhouse gas concentration trajectory and other seasonal and climatic variables that can be obtained from the meteorological department. As pollution is at the highest during winter, the increased level of precipitation of pesticides during this season needs to be considered. Spatial variability should be ascertained based on the spatial data obtained from geographic information systems (GIS). The data obtained should be preprocessed and a suitable prediction model for climate change (Fig. 1) can be developed for the best performance.

Caffara et al. (2012) have modeled the effect of climate change on the interaction between grapevine and its pests and pathogens. The need to understand the impact of climate change on various combinations of host–pest/pathogen systems was also emphasized. The possible spreading of an insect pest, spotted lanternfly particularly due to climate change was modeled by Jung et al. (2017). These prediction models could be used to devise strategies for preventing the occurrence of future invasions. Osawa et al. (2018) have proved that climate-related factors play a vital role in the population dynamics of pests. The consequences of a varying climate especially due to temperature deviations on biological pest control mechanisms have been analyzed by Guzman et al. (2016). The significance of knowing the implications of tritrophic relations due to climate change on pest management was emphasized by Castex et al. (2018). Thomson et al. (2010) predicted the effects of climate change on biological control of agricultural pests. The effect of varying climate on the occurrence of crop diseases was modeled by Newbery et al. (2016). According to them, the emergence of new pathogens and uncertainty in prediction models are some of the major challenges in this. Xu et al. (2020a) used the maximum entropy (MaxEnt) model to comprehend the role of climatic factors on the spread of the *H. vitessoides* pest.

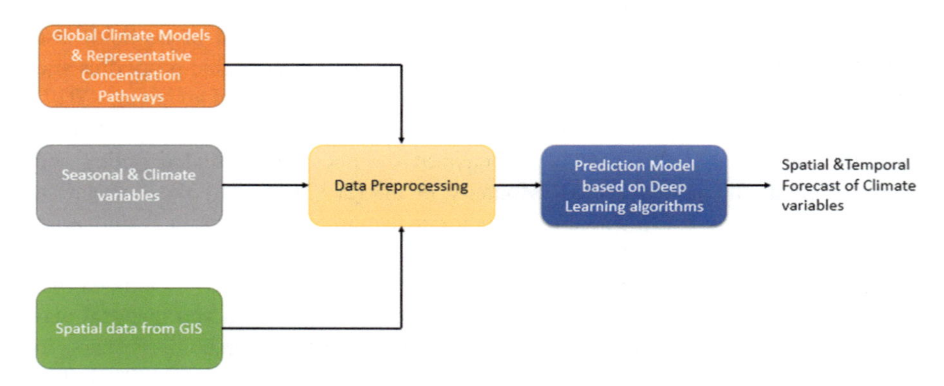

Fig. 1 Prediction model for climatic factors

Table 1 Review on prediction models for climate-induced pest occurrence

Authors	Objective	Model	Study area	Software
Xu et al. (2020a)	Predict the risk of *H. vitessoides*	Maximum Entropy (MaxEnt) modeling	China	ArcGIS
Caffara et al. (2012)	To model the impact of climate change on grapevine pests and pathogens	Linear regression model, FE NOVITIS model Climate Model: Hadley Centre's Atmosphere-Ocean General Circulation Model (AOGCM) HADCM3	Central-Eastern Italian Alps	SPSS (v15)
Jung et al. (2017)	Prediction of potential spread of invasive insect pest	Model based determination of ecoclimatic index	South Korea	CLIMEX
Osawa et al. (2018)	To predict the population dynamics of rice pest due to climate change	Generalized Linear Models (GLMs) with Poisson distribution and Wald's test	Japan	R (v3.3.1)
Castex et al. (2018)	Pest management under climate change	Phenological models, demographic models	–	CLIMEX
Mol et al. (2018)	Weather dependent occurrence of wheat stem base disease	Random forest, Correlation, Generalized Linear Models	German federal states	R

It was proved that precipitation of warmest and wettest periods, mean temperature of the warmest period, and minimum temperature of wettest periods were the most significant variables responsible for the distribution of the pest. The MaxEnt model was based on statistical analysis and ML. Mol et al. (2018) proved that the weather played an important role in the occurrence of wheat stem base diseases. Table 1 presents the overview of prediction models used for climate-induced pest occurrence.

3.1.2 Prediction of Occurrence of Pests and Crop Diseases

Pest occurrences are generally driven by ecological processes at different spatial and temporal scales. With increasing globalization, transboundary pests and diseases are posing a great threat to agriculture (UN/FAO 2017). Modeling the incidence of pests can aid in the planning and distribution of appropriate pesticides in various regions. Pest prediction models will be based on the mathematical relationship between the occurrence of pest and its life cycle which would invariably depend on environmental parameters like humidity, temperature, rainfall, etc. (Collier 2017). Figure 2 shows the methodology that can be used for predicting the occurrence of pests and crop diseases. Since the occurrence of pest is very much correlated to the climatic variables, the forecast output of climatic conditions can be considered as one of the inputs. The current climatic variables and their spatial variations will also be

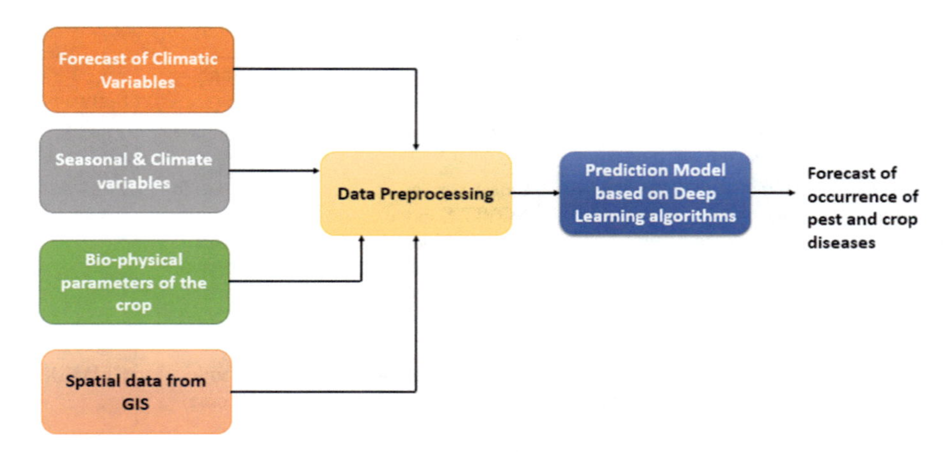

Fig. 2 Prediction model for occurrence of pest and crop disease

considered. The occurrence of different types of insects with changing weather conditions will also be predicted. The climatic conditions responsible for good insects useful for cross-pollination will be predicted to ensure that farmers would abstain from using pesticides that would kill them. Climatic conditions responsible for the spread of pests will also be determined. The biophysical parameters of the crop, like the leaf area index, the height of the crop, any noticeable change in color, shape, or texture of the plant, etc. will be considered to forecast the intensity of the disease.

Wildemeersch et al. (2019) modeled the outbreak of pests at different scales, taking into consideration host–pest interactions, the topology of landscape, and connectivity. They used spatio-temporal datasets for network construction, small-scale and large-scale modeling. They implemented the susceptible-infested-susceptible (SIS) model for Geometrid moth outbreaks in birch forests and the susceptible-infested-removed (SIR) model for European spruce bark beetle outbreaks in temperate forests. The node probabilities for the outbreaks and trajectories of the expected infected area were predicted using these landscape-scale models.

Weather-based forecasting models for the occurrence of pests and crop diseases can aid in prudent usage of pesticides. A neural network-based prediction model was proposed by Yang et al. (2009) for forecasting the population of paddy stem borer. Congdon et al. (2017) developed an empirical forecasting model for the occurrence of pea seed-borne mosaic virus in pea crops. The multidimensional data pertaining to the number of pests, soil, environment, climatic and meteorological factors were used to build a forecasting model for vegetable pests by Cai et al. (2019). The early warning model used environmental sensor data pertaining to vegetable soil, environment, ecology climate and meteorology, and pest occurrence data, and was field-tested in South China. The data preprocessing was done using Lagrangian interpolation and k-means algorithm, the feature selection was done based on Pearson's correlation coefficient, and the degree of gray correlation was calculated using the gray correlation algorithm. The degree of pest occurrence classified as mild, moderate, moderately severe, and severe was predicted using a neural network model. It was proved that it could warn about the incidence and intensity of the pest outbreak up to an accuracy of 96.7%. Groot and Ogris developed two prediction models for bark beetle outbreak in coniferous forests (2019).

Handling pest images are complicated during the varying size of pest targets, instances, and species. To overcome this challenge, a convolutional neural network (CNN) for agricultural pest detection was proposed based on feature fusion maps (Jiao et al. 2020a). They proposed the anchor-free region convolutional neural network (AF-RCNN) for exactly detecting and classifying 24-classes of pests. The anchor-free region proposals generation network (AFRPN) used for obtaining the pest region information and the CNN-based approach used for extraction of pest features are combined into a single network in the AF-RCNN framework. The proposed model was proved to perform well in real-time intelligent pest detection.

Alves et al. (2020) proposed a novel model based on deep residual networks for the classification of cotton pests using field-based images. They investigated two different classification problems involving a set of six pest images and another dataset of all 15 pests with no pest class. Models based on Linear Binary Features (LBF), Support Vector Machine (SVM), and CNN models like AlexNet, ResNet34, and ResNet 50 were used for comparison. The proposed ResNet34* outperformed all these models.

The population density distribution (PDD) of invasive pests was investigated and a prediction model for the same was proposed based on discrete-time model by Gao et al. (2020). They concluded that in a given two-dimensional ecosystem, uniform distribution, target distribution, irregular speckle distribution, and turbulence are the four different types of PDDs. A foreknowledge of population density of the pests can well be handy in ascertaining the quantity of pesticides needed to be used in a particular field at any given point of time. A short review of the prediction models used for pest outbreaks and management has been tabulated in Table 2.

3.1.3 Prediction of Quantity and Methodology of Pesticide Application

The proper estimation of incidence and intensity of pest outbreaks can be used to determine the optimum quantity of pesticide application. The methodology that can be used for predicting the quantity of pesticides to be used is shown in Fig. 3. This prediction model is useful to avoid the overuse of pesticides and to prevent wastage during heavy rainfall or winds, this prediction model is developed. This would aid the farmer in determining the exact quantity of pesticides to be used in proportion to the intensity of the disease. The farmer can refrain from the application of pesticides if he is notified about a windy or rainy-day forecast.

Meite et al. (2018) studied the influence of rainfall frequency, duration, and volume that would cause the spread of pesticides and other pollutants from agricultural soils. Accurate forecasting models for rainfall can curb the occurrence of such export. Asaei et al. (2019) developed an orchard sprayer integrated with machine vision for site-specific application of pesticides exactly on diseased trees. A software

Table 2 Review on prediction models for pest outbreaks and management

Authors	Objective	Model	Study area	Software
Wildemeersch et al. (2019)	To model pest outbreaks	Multi-scale model incorporating SIS/SIR epidemic models	Norway, Sweden, Finland	–
Cai et al. (2019)	Early warning model for vegetable pests	Backpropagation neural network	China	–
Jiao et al. (2020a, b)	Pest detection model	AF-RCNN	China	Ubuntu 14.04 + Python 2.7
Alves et al. (2020)	Classification of cotton pests	ResNet34* (Deep residual learning model)	–	Fastai deep learning library with Pytorch
Gao et al. (2020)	Prediction of distribution of pests	Discrete-time model		Numerical simulation
Yang et al. (2009)	Prediction model for occurrence of paddy stem borer	Backpropagation ANN	JianShui County, Yunnan	MATLAB
Congdon et al. (2017)	Forecasting model for pea seed-borne mosaic virus epidemic	Linear models with Leave One Out Cross Validation (LOOCV)	Mediterranean-type environment	R (v3.0.2)

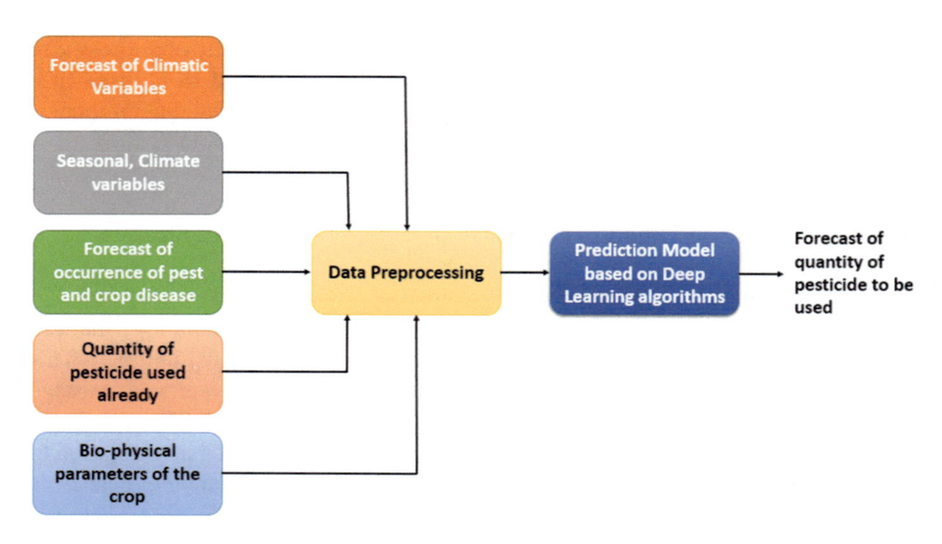

Fig. 3 Prediction model for quantity of pesticide to be used

Table 3 Review on prediction models for pesticide usage

Authors	Objective	Model	Study area	Software
Bagheri (2019)	To investigate farmer's intention to use pesticides in agriculture	Expanded version of Theory of Planned Behavior	Moghan region, Iran	Statistical Package for Social Sciences (SPSS) AMOS
Basir et al. (2019)	Pest management	Integrated pest management based mathematical model	–	Numerical simulation
Abadi (2018)	To establish the determinants of pesticide use behavior (PUB) among cucumber farmers	Theory of planned behavior, Innovation diffusion model, Technology acceptance model, Social cognitive model	Central Iran	SPSS 15.0, AMOS v20
Vaz et al. (2020)	Prediction of applicability of potential pesticide	Combination of FeatMorgan fingerprints with Random Forest algorithm	–	Python v3.6
Mubushar et al. (2019)	Modeling farmers' knowledge regarding safe pesticide usage	Logit regression model	Pakistan	SPSS (v21)

application named SAAS was developed by Hong et al. (2018) for approximating the effectiveness of pesticide spray and assessing the drift of air-borne pesticides.

The farmers' intention to use pesticides was modeled extensively based on the theory of planned behavior (Bagheri et al. 2019). Since beliefs and attitudes form the basis of human behavior, these can be modeled to mitigate the risks due to pesticide usage. Crop pest management and awareness in farming techniques are directly related to each other. The greater the level of awareness, the greater is the density of pests in the field (Basir et al. 2019). The authors have proposed a mathematical model for the same and have concluded that increasing the level of awareness among farmers with a tolerable time delay will augur well for prudent use of pesticides and aid in integrated pest management in a big way. The proposed mathematical model was validated using numerical simulation. A brief review on the prediction models used for pesticide usage is presented in Table 3.

3.2 Prediction Models for Pesticide Residue in the Environment

Prediction models for pesticide pollution are of great importance as it ensures protection of human health, flora, fauna, and the environment (Pan et al. 2019).

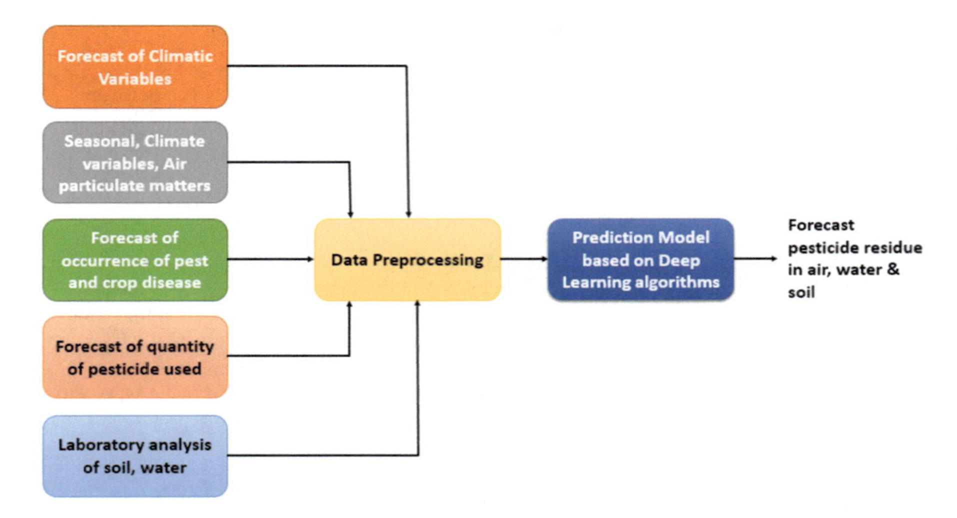

Fig. 4 Prediction model for pesticide residues in air, water, and soil

When a total of 76 pesticide residues were investigated in 317 European topsoil samples, around 83% of the agricultural soil contained pesticide residues and 58% contained multiple residues (Silva et al. 2019). Michael et al. (2020) suggested a two-tier model for environmental risk assessment in Ghana. The first tier included the use of Pesticides Risk in the Tropics to Man, Environment and Trade (PRIMET) model, and the application of the Species Sensitivity Distribution (SSD) concept as the second tier. They expected the PRIMET model to help in identifying the pesticides that threaten the aquatic and terrestrial environment and SSD to determine the threshold levels for safe use of pesticides.

The spatial and temporal spread of pesticide residues in air, water, and soil can be predicted using the methodology depicted in Fig. 4. The samples of groundwater, nearby water bodies, and soil near the farmlands can be collected and analyzed to determine the presence of pesticide residues used in the farm. The air particulate matter can also be measured to find out the pesticide-induced pollutants.

McGrath et al. (2019) developed a stochastic forecast model for estimating the concentration of pesticides in the soil. They used the Bayesian model calibration for the forecasting model. The influence of uncertainties related to climate, agricultural practices, soil and pesticide qualities was used to model pesticide leaching by Lammoglia et al. (2018). The destiny and transmission pathways of pesticides applied to farmlands are complex, involving evaporation into the atmosphere, aerial drift, overflow into local waterbodies, and leaching into groundwater basins. Pesticide fate modeling was performed by Wang et al. (2019a), Queyrel et al. (2016), and Lammoglia et al. (2017). The temporal and spatial distribution of three different pesticides, namely atrazine, isoprothiolane, and oxadiazon in an agricultural watershed was carried out by Ouyang et al. (2017). Lee et al. (2019) examined the incidence of pharmaceutical products and six different pesticides present in the groundwater in agricultural regions in Korea. The effect of pesticide deposits in pollen and nectar was analyzed by Gierer et al. (2019). Shein et al. (2017) developed a prediction model for the transmission of pesticides in soils. To dynamically estimate the dispersal of pesticide residues in soil, Li developed a Bayesian generalized log-normal model (Li 2018a). Qu et al. (2019) analyzed the health impacts of spatial distribution and air–soil exchange of pesticides present in the soils in certain regions of Italy. Jin et al. (2018) suggested a simple system to collect empty pesticide packages to alleviate land pollution. The statistics pertaining to the number of empty pesticide containers found in agronomic areas were analyzed by Marnasidis et al. (2018). They calibrated the following indices pertaining to empty pesticide containers (EPC) and empty plastic pesticide containers (EPPC): EPC_{pieces} per hectare and per crop of the studied area, $EPPC_{weight}$ per hectare, and $EPPC_{weight}$/farmer/year. Meticulous data collection in this regard can help in building prediction models to estimate the pesticide plastic waste amounts and to thereafter carefully plan for their disposal. Braun et al. (2018) analyzed the cause for spatial distribution of pesticides in agronomic soils in a certain region in Vietnam.

Yadav et al. (2015) presented an exhaustive review on pesticide residues in the air, water, and soil and its probable impact on nearby countries. Mishra et al. (2012) evaluated the pollution levels and spatial spread of organochlorine pesticides in Indian soils. The presence of organochlorine pesticides in seven major Indian cities, including New Delhi, Agra, Kolkata, Mumbai, Goa, Chennai, and Bangalore were analyzed by Chakraborty et al. (2015). The seasonal variation and accumulation of organochlorine pesticides were carried out in several regions of Tamil Nadu, in south India (Chakraborty et al. 2019; Arisekar et al. 2019). Mondal et al. (2018) assessed the environmental risk in the Hooghly river basin in West Bengal. The health impacts caused by pesticide residues in water bodies in China were analyzed by Lai (2017). It was observed that a 10% rise in the use of rice pesticides affected public health significantly. Li (2018b) assessed the international pesticide groundwater regulations incorporating both drinking water and soil standards and developed a health-based supervisory framework. Zdravkovic et al. (2018) developed a prediction model based on the Monte Carlo method to estimate the retention time of pesticide residues. Utami et al. (2020) proposed a method to prioritize 31 different pesticides based on their measured and predicted concentrations in the aquatic environment. Estimation of emission, fate prediction, and comparison of modeled water concentrations were also carried out. The methodology developed aimed at the judicious use of pesticides and water quality management. Norman et al. (2020) analyzed the temporal variation of pesticide residues in daily and weekly samples of surface water. They estimated the pesticide toxicity index (PTI) and also its impact on aquatic life.

Detection of pesticide residue in the environment can be done using the application of fluorescence spectroscopy (Lin et al. 2020). A neural network with multiple hidden layers was proposed to estimate the content of pesticide residues. The effectiveness of the model was proved using four traditional pesticides, namely zhongshengmycin, paclobutrazol, boscalid, and pyridaben, dissolved in drinking water. Excitation emission matrix fluorescence spectra were quantitatively analyzed by Yuan et al. (2020) for the determination of multi-pesticide residues. They also employed a multivariate calibration algorithm for predicting the pesticidal residues in water samples.

Management of pesticide contamination in surface water catchments was carefully analyzed by Villamizar et al. (2020). They compared the efficiency of technical and system-based approaches in mitigating this type of contamination. The most sensitive model parameters included threshold depth of water in the shallow aquifer, percolation fraction of deep aquifer, the time lag between water exiting soil profile and shallow aquifer, travel times for lateral flow within the catchment, etc. A summary of prediction models used for determining pesticide residue in the environment is presented in Table 4.

3.3 Prediction Models for Human Health Disorders Due to Pesticide Exposure

A geographic information system and remote sensing-based estimation of human health impact due to pesticide exposure were carried out by VoPham et al. (2015). Li (2018c) developed the DALY (disability-adjusted life-year) metric to assess the effect of long-term pesticide exposure on human health. Hussein et al. (2019) developed an early prediction strategy to predict the occurrence of liver cancers in workers who are exposed to pesticides.

Bhandari et al. (2019) investigated on the pesticide residues in Nepalese vegetables and their likely health hazards. An artificial intelligence-based methodology was proposed by Elahi et al. (2019) to determine the relationship between agrochemicals and human health. Yadav et al. (2015) have reviewed the health influences of organic pesticides, the monitoring measures, and India's commitment towards their elimination. Sharma and Peshin (2016) evaluated the influence of integrated pest management of vegetables on the use of pesticides in a few regions in Jammu, India. The pesticide residues in fruits and vegetables grown in the Western Indian Himalayan region were analyzed and the corresponding health impacts were studied (Kumari and John 2019).

The equivalent biological exposure limits (EBEL) was established using real-life data and acceptable operator exposure level (AOEL) using mancozeb as reference (Rajcevic et al. 2020). The study involved the vineyard pesticide applicators and used a linear regression model for modeling the EBEL. A brief review of prediction models used to determine pesticide residues in food items is presented in Table 5.

Prediction models for human health hazards due to pesticide exposure would be of great help in devising preventive and prescriptive methodologies.

Table 4 Review on prediction models for determining pesticide residue in the environment

Authors	Objective	Model	Study area	Software
Utami et al. (2020)	Prediction of pesticide concentrations in aquatic environments	Model based on empirical relationship between rainfall vs. runoff volume	Indonesia	–
Lin et al. (2020)	Fluorescence detection of pesticides	Neural network	–	–
Yuan et al. (2020)	Detection of pesticide residues	Multivariate calibration algorithms	–	MATLAB
Villamizar et al. (2020)	Management of pesticide contamination in surface water catchments	Multiple regression system with Latin hypercube sampling	–	ArcSWAT, ArcGIS
Norman et al. (2020)	Estimation of pesticide toxicity index (PTI)	Chemical analysis and mathematical models	Midwest Stream Quality Assessment (MSQA) and South East Stream Quality Assessment (SESQA) study sites, US	–
Quaglia et al. (2019)	Identification of priority areas for pesticide pollution mitigation	Geospatial emission modeling	–	GIS-based tool
Pose-Juan et al. (2015)	Pesticide residue in vineyard soils	Statistical techniques applied to spatial and temporal distribution of pesticide residues	Spain	IBM SPSS
Bhandari et al. (2020)	Environmental concentrations of pesticides	Statistical analysis and mathematical models	Nepal	Canoco 5, SPSS 23
Shoda et al. (2016)	Prediction of pesticide toxicity	Watershed Regressions for Pesticides-Pesticide Toxicity Index (WARP-PTI) model	MSQA sites, US	–
Sybertz et al. (2020)	Prediction of mixture risk for different pesticide combinations	Mixture toxicity of application spray series (MITAS) model	Germany	–
Santos et al. (2020)	Prediction of air-borne	Linear interpolation, Thiessen polygons, Inverse distance	Colombia	R

(continued)

Table 4 (continued)

Authors	Objective	Model	Study area	Software
	pesticide drift deposits on soil	weighting, Universal Kriging, Co-Kriging, Spatial vine copulas, Karhunen–Loeve expansion, Integrated nested Laplace approximations		
Ccanccapa et al. (2016)	Detection of pesticide residues using spatio-temporal patterns	Multiple step-wise linear regression models	Spain	IBM SPSS
Kumar et al. (2010)	Prediction of chronic lethality of selected pesticides	Regression model, Log-inverse time extrapolation, Log-log method	Australia	–
Dashtbozorgi et al. (2013)	Prediction of retention time of pesticide residues	Quantitative structure property relationship (QSPR), Genetic algorithm-partial least squares (GA-PLS), Artificial Neural Network (ANN), Support Vector Machine (SVM)	–	Dragon (v3)
Wang et al. (2020a)	Prediction model for detecting pesticide mixture	Missing Data Recovery -Parallel Factor Analysis (MDR-PARAFAC)	–	MATLAB 2014b
Wang et al. (2020b)	Prediction model for detecting carbamate pesticide mixture	Genetic Algorithm (GA)-Backpropagation (BP) network	–	MATLAB 2014b
Islam et al. (2020)	Prediction of pesticide soil-air partitioning coefficient (Ksa)	Multiple Linear Regression-Quantitative Structural Property Relationship (MLR-QSPR)	–	R

3.4 *Prediction Models for Antimicrobial Resistance Due to Pesticides*

Antimicrobial resistance has become an alarmingly concern globally. Ramakrishnan et al. (2019) have analyzed how the local application of pesticides has resulted in global consequences. The inappropriate storage facility, indiscriminate use of

Table 5 Review on prediction models for pesticide residue in food items

Authors	Objective	Model	Software
Rajcevic et al. (2020)	Determination of EBEL	Univariate Linear regression	–
Zhan-qi et al. (2018)	Identification of different concentrations of pesticide residues on spinach leaves	Support vectors classification, K nearest neighbor, Random forest algorithm, Linear discriminant analysis	MATLAB 2016b, IBM SPSS, Python 3.6
Pagani and Ibanez (2019)	Prediction of pesticide residues in fruits and vegetables	Unfolded Partial Least Squares coupled to Residual Trilinearization (U-PLS/RTL)	MATLAB Graphical interface of MVC3 (Multivariate Calibration for third-order) toolbox
Sousa et al. (2020)	Quantification of seven pesticide residues in vegetables	Multivariate Curve Resolution – Alternating Least Squares (MCR-ALS) algorithm	MATLAB
Wang et al. (2019b)	To model the transfer of pesticide residue during tea brewing	Linear regression	SPSS 19.0
Qian et al. (2019)	To determine the effect of pesticides on the minerals used to classify rice based on geographic origin	Fischer multivariant discrimination analysis model	SPSS 18.0
Huan et al. (2016)	Assessment of pesticide residues in cowpea	Probabilistic assessment using Monte Carlo simulation	Oracle Crystal Ball 11.1.2.1.000
Wang et al. (2017)	To model the impact of farmers' knowledge and risk perception on their practices to eliminate pesticide residues	Structural equation modeling	Lisrel 8.8
Schreinemachers et al. (2020)	Estimation of pesticide overuse in vegetable production	Linear regression model	STATA
Drabova et al. (2019)	Estimation of pesticide overuse in oregano	Orthogonal Partial Least Squares Discriminant Analysis (OPLS-DA)	Simca 13.0
Galimberti et al. (2020)	Pesticide risk assessment	Multiple Linear Regression (MLR) Quantitative Structure–Activity Relationship (QSAR) models by Ordinary Least Squares (OLS) and Genetic Algorithm (GA)	QSARINS

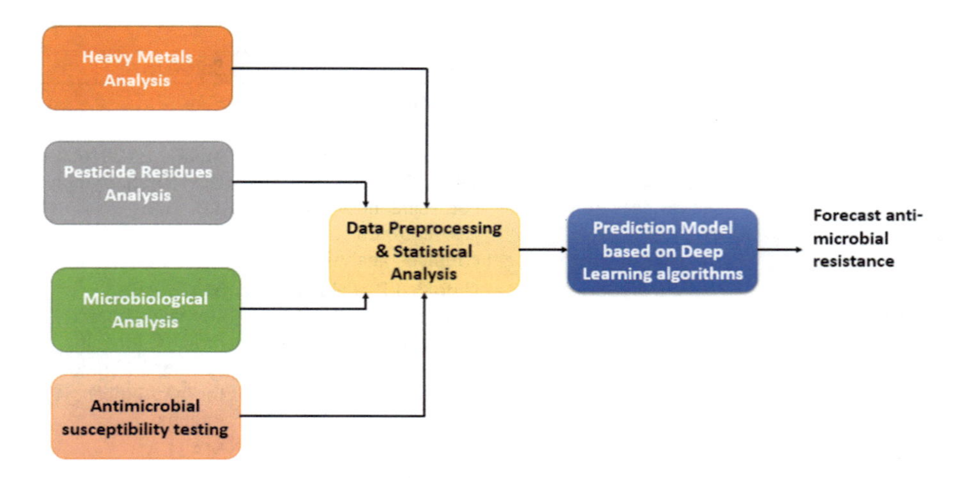

Fig. 5 Prediction model for pesticide-induced AMR

pesticides has impaired the global food chain. Microbes try to tolerate, resist, or degrade these chemicals. The pesticide degraders have a greater risk of possessing antimicrobial resistance. Ahmed et al. (2019) investigated the presence of pesticides and antibiotics resistance pathogens in vegetable salads in certain regions in Bangladesh. Curutiu et al. (2017) have pointed out clearly that the rampant use of pesticides has led to the selection and occurrence of multiple-antibiotic resistance microbes. Though a life without antibiotics or pesticides is nearly unimaginable, a delicate balance needs to be maintained and judicious usage of both is required. Braun et al. (2019) analyzed the presence of antibiotics and pesticides in rice, rice-shrimp, and permanent shrimp systems in the coastal regions of Vietnam. The occurrence and distribution of antibiotics in coastal regions (Xie et al. 2019), antibiotic resistance in urban runoffs (Almakki et al. 2019), groundwater (Szekeres et al. 2018), drinking water (Sanganyado and Gwenzi 2019), and aquatic environments (Gao et al. 2018) have been investigated.

The AMR developed due to the use of pesticides in plants can be predicted based on the methodology depicted in Fig. 5.

Prolonged exposure and accumulation of pesticide residue lead to the development of AMR in the flora and fauna available in land, water, and air. The effects of pesticide exposure on soil bacteria and the development of antibiotic resistance have been investigated by Rangasamy et al. (2017, 2018).

4 Modeling Strategies

The whole life cycle of pesticides, since its production to disposal, can be modeled using several strategies including pesticide simulation models, ML-based models, and models based on experimental study. This section presents a review of these modeling strategies.

4.1 Pesticide Simulation Models

The Soil and Water Assessment Tool (SWAT) is an ecohydrological model used for mimicking pesticide fate and transport. Wang et al. (2019a, b) presented a review of the current status and research concerning the simulation of pesticide fate and transport. They improved the algorithm for the SWAT model in simulation of pesticide fate, representation of pathway, and control of pollution and transport. Agatz and Brown (2017) introduced a 2-DROPS (2-Dimensional Roots and Pesticide Simulation) model for two-dimensional simulation of pesticide within the soil-root zone. Lammoglia et al. (2018) used the STICS (Simulateur muiTIdisciplinaire pour les Cultures Standard)-MACRO to model the impact of uncertainties in climate, soil, and pesticide properties. The HYDRUS-1D is an accomplished model to simulate the pesticide and water transport in soil. Anlauf et al. (2018) combined HYDRUS-1D with ArcGIS to assess the risk of leaching and accumulation of pesticides at regional levels in Thailand. Villamizar and Brown (2017) used the pesticide fate model MACRO and catchment scale model SPIDER to analyze the pesticide concentrations in river flow. The pesticide runoff from agricultural fields was evaluated using three models by Zhang and Goh (2015). The models they used were PRZM (Pesticide Root Zone Model), RZWQM (Root Zone Water Quality Model), OpusCZ. Table 6 presents an overview of various pesticide attenuation models.

4.2 Machine Learning-Based Models

Artificial Intelligence (AI) based models and ML algorithms have revolutionized the development of prediction models for agricultural uses. ML algorithms can be classified into supervised, unsupervised, and reinforcement learning methods. Prediction models are generally in the supervised category and they include classification and regression algorithms. Unsupervised learning methods generally include dimension reduction algorithms like principal component analysis, clustering algorithms, etc. Techniques like decision trees, random forests, support vector machines (SVM), Bayesian networks, and Artificial Neural Networks (ANN) were used to predict crop yield (Elavarasan et al. 2018). Deep learning networks like AlexNet –

Table 6 Review on pesticide fate and transport models

Authors	Objective	Model	Software
Agatz et al. (2020)	To estimate the development and spatial distribution of corn roots and corn rootworm pests; fate distribution and toxic action of insecticide	COMPASS-Rootworm	SigmaPlot (v12.5)
Baan (2020)	Calculation of national risk indicators for valuation of pesticide risk potential	SYNOPS Least Absolute Shrinkage and Selection Operator (LASSO) and Random Forest regression algorithms	R (v3.5.1)
Farlin et al. (2017)	Assessment of pesticide attenuation from water dating	Lumped parameter model, Exponential piston flow model	–
Zhang et al. (2020)	Mathematical estimation of emission and fate modeling of pesticides	SWAT	–
Boulange et al. (2017)	Pesticide fate and transport modeling	Bayesian methodology using Markov Chain Monte Carlo technique	–
Farlin et al. (2013)	Estimation of field degradation half-life due to pesticides	PEARL coupled with SWAP	PEST software
Farlin et al. (2012)	Pesticide attenuation in fractured aquifer	Lumped parameter model, Exponential piston flow model	FLOWPC
Guardo and Finizio (2016)	Methodology to identify vulnerability areas to pesticides	Pesticide Leaching Model (PELMO)	SAGA-GIS VULPES software (VULnerability to PESticide)

CNNs (Convolution Neural Networks) were used by Jiang et al. (2018) along with machine vision technology to identify pesticide residues in postharvest apples. ML-based models based on SVM were developed by He et al. (2019) to ascertain the toxicity of pesticides towards aquatic organisms. SVM based models were used to evaluate chemical and biological indicators of the effect of pesticides on beehives used for biomonitoring by Niell et al. (2018). Random forest-based ML algorithm was developed by Wrzesien et al. (2019) to predict the occurrence of apple scab. Zhan-qi et al. (2018) developed an ML-based model to detect the presence of dimethoate on spinach leaves. ML algorithms based on neural networks and tree-based learners were used to predict bioconcentration factors of pesticides in fishes and other invertebrates (Miller et al. 2019). The performance of these models was proven based on the error metrics of the forecast model. The commonly used metrics include mean absolute error (MAE), root mean square error (RMSE), relative root mean square error (RRMSE), coefficient of determination (R^2), etc.

4.3 Experimental Study

Prediction models for pesticide occurrence and its impact in the environment can also be determined using chemical analysis and experimental evidence obtained in laboratories. Pesticide residue analysis for the presence of heavy metals can be ascertained using microbiological analysis and antimicrobial susceptibility testing.

Carrao et al. (2019) predicted the pesticide–drug interactions using chemical analysis. Alves et al. (2018) carried out chemical toxicity prediction of different pesticides and other chemicals based on QSAR (Quantitative structure–activity relationship) models. The QSAR model has been used to predict the toxicity of pesticides by several researchers (Yang et al. 2020a, b; Xu et al. 2020b; Benigni et al. 2020). The ecotoxicity of pesticides was predicted by Stergiopoulos et al. (2019) using immobilized artificial membrane chromatography. Martin et al. (2017) used two-dimensional chemical descriptors and target species classification for prediction of acute toxicity of pesticides. An experimental study using PEARL 4.4.4 model was used to predict the migration of pesticides in soil by Shein et al. (2017).

5 Mission Ahead

Sir Albert Howard, the Father of Modern Organic Health, said "The health of soil, plant, animal and man is one and indivisible." Yes, to ensure a healthy environment, we need to aim at safer and more sustainable modes of food production. Environment-friendly integrated pest management techniques and enforcement mechanisms should be in place. The challenge of food production, food storage, food loss, food distribution should be critically analyzed and carefully addressed. Prediction models can go a long way in assessing the need for pesticides, the right quantity, quality and medium of application, and its impacts in the food chain.

Varjani et al. (2019) have proposed the addition of biochar as a measure for pesticide remediation and have evaluated its performance on various agricultural fields. According to them, an increase in soil water holding capacity, improved soil aeration, rise in the number of microorganisms that result in enhanced metabolic activity and pesticide degradation are some of the benefits that can be reaped with the addition of biochar in a pesticide-contaminated soil. Steingrimsdottir et al. (2018) proposed a framework to find alternatives for chemical pesticides that would be more sustainable and environmentally friendly. They defined pesticide substitution scenarios for every crop of interest based on the knowledge of one or more pest that can potentially affect the crop, available pesticides, its registration status, the quantity of application, market price, and toxicity related details. They proved that their substitution scenarios led to more judicious use of pesticides.

Kumar et al. (2019) pointed out the efficacy of pesticide nanoformulation and nanoencapsulation media in targeted delivery of pesticides on the infected area. The probable application of nanotechnology in integrated pest management would include assessing the intensity of pest infestation, pest identification, targeted delivery of pesticide nanoformulations, and monitoring its impact on pests. Mfarrej and Rara (2019) suggested the use of natural organic pesticides as a substitute for synthetic pesticides. More research and advancement in the area of sustainable and environmentally friendly methods for remediation of pesticide-contaminated agricultural soil is the need of the hour (Meftaul et al. 2020). All these technologies can be suitably assisted by advanced machine learning-based prediction models and result in synergistic outcomes, both for the betterment of human health and the environment.

The Government can also in turn adopt policies to phase out highly hazardous pesticides and initiate a support system for eco-friendly measures. The pesticide industry can ensure the safe disposal of pesticides from the production to its disposal. They can also take necessary steps to educate the people who handle pesticides from production to disposal on their harmful effects and the precautionary measures that have to be taken. The food industry can also apprise the farmers about available eco-friendlier alternatives.

Conservation agriculture with limited natural resources and climate-smart agriculture is the need of the hour (UN/FAO 2017). IPM, IVM, and well-coordinated pesticide management should be carefully planned. Strategies for pesticide use-reduction, targeted interventions, and meticulous mobilization of resources will lead to breakthroughs in the structural improvement of PLM (Berg et al. 2020). Alternative cropping systems with reduced dependence on pesticides, novel pesticide formulations like microcapsule suspensions, use of appropriate equipment for pesticide spraying will make a great difference by reducing the adverse risk of pesticides (Michael et al. 2020).

6 Conclusion

The use of AI and ML-based prediction models can mitigate the indiscriminate use of pesticides. This paper presented an exhausted review of the significance of prediction models in integrated pest management that are aimed at mitigating the impact of pesticide exposure. Prediction models can predict the optimum quantity and quality of pesticides and prevent unnecessary leaching into the soil by predicting the climatic factors as well. The human health impacts and antimicrobial resistance due to pesticide residues can also be predicted. Accurate prediction models can aid in proper planning, management, and prevention of loss of lives and preservation of the environment. Modeling strategies used for developing these models have also been reviewed. The alternate path of effective prediction, monitoring, and use of regulatory mechanisms for pesticide usage and attenuation model, to preserve our earth with all its variety and vigor and hand it over safely to the future generations is the greatest need of the hour.

Declarations *Funding* (information that explains whether and by whom the research was supported): The research received no funding.

Conflicts of interest/Competing interests (include appropriate disclosures): The authors declare that they no known competing financial interests or personal relationships that could have appeared to influence the work reported in this paper.

Availability of data and material (data transparency): Not applicable.

Code availability (software application or custom code): Not applicable.

Authors' contributions (optional: please review the submission guidelines from the journal whether statements are mandatory): Not applicable.

References

Abadi B (2018) The determinants of cucumber farmers' pesticide use behaviour in Central Iran: implications for the pesticide use management. J Clean Prod 205:1069–1081

Agatz A, Brown CD (2017) Introducing the 2-DROPS model for two-dimensional simulation of crop roots and pesticide within the soil-root zone. Sci Total Environ 586:966–975

Agatz A, Ashauer R, Sweeney P, Brown CD (2020) A knowledge-based approach to designing control strategies for agricultural pests. Agric Syst 183:102865

Ahmed S, Siddique MA, Rahman M, Bari ML, Ferdousi S (2019) A study on the prevalence of heavy metals, pesticides and microbial contaminants and antibiotics resistance pathogens in raw salad vegetables sold in Dhaka, Bangladesh. Heliyon 5(2):e01205

Almakki A, Bilak EJ, Marchandin H, Fajardo PL (2019) Antibiotic resistance in urban runoff. Sci Total Environ 667:64–76

Alves VM, Muratov EN, Zakharov A, Muratov NN, Andrade CH, Tropsha A (2018) Chemical toxicity prediction for major classes of industrial chemicals: Is it possible to develop universal models covering cosmetics, drugs and pesticides? Food Chem Toxicol 112:526–534

Alves AN, Souza WSR, Borges DL (2020) Cotton pests classification in field-based images using deep residual networks. Comput Electron Agric 174:105488

Anlauf R, Schaefer J, Kajitvichyanukul P (2018) Coupling HYDRUS-1D with ArcGIS to estimate pesticide accumulation and leaching risk on a regional basis. J Environ Manag 217:980–990

Arisekar U, Shakila RJ, Jeyasekaran G, Shalin R, Kumar P, Malani AH, Rani V (2019) Accumulation of organochlorine pesticide residues in fish, water and sediments in the Thamirabarani river system of southern pensinsular India. Environ Nanotechnol Monitoring Manag 11:100194

Asaei H, Jafari A, Loghavi M (2019) Site-specific orchard sprayer equipped with machine vision for chemical usage management. Comput Electron Agric 162:431–439

Baan L (2020) Sensitivity analysis of the aquatic pesticide fate models in SYNOPS and their parametrization for Switzerland. Sci Total Environ 715:136881

Bagheri A, Bondori A, Allahyari MS, Damalas CA (2019) Modeling farmers' intention to use pesticides: an expanded version of the theory of planned behaviour. J Environ Manag 248:109291

Basir FA, Banerjee A, Ray S (2019) Role of farming awareness in crop pest management – a mathematical model. J Theor Biol 461:59–67

Benigni R, Serafimova R, Morte JMP, Battistelli CL, Bossa C, Giuliani A, Fioravanzo E, Bassan A, Gatnik MF, Rathman J, Yang C, Szlichtyng AM, Sacher O, Tcheremenskaia O (2020) Evaluation of the applicability of existing (Q)SAR models for predicting the genotoxicity of pesticides and similarity analysis related with genotoxicity of pesticides for facilitating of grouping and read across: An EFSA funded project. Regul Toxicol Pharmacol 114:104658

Berg HVD, Gu B, Grenier B, Kohlschmid E, Al-Eryani S, Bezerra HSS, Nagpal BN, Chanda E, Gasimov E, Velayudhan R, Yadav RS (2020) Pesticide lifecycle management in agriculture and public health: where are the gaps? Sci Total Environ 742:140598

Bhandari G, Zomer P, Atreya K, Mol HGJ, Yang X, Geissen V (2019) Pesticide residues in Nepalese vegetables and potential health risks. Environ Res 172:511–521

Bhandari G, Atreya K, Scheepers PTJ, Geissen V (2020) Concentration and distribution of pesticide residues in soil: non-dietary human health risk assessment. Chemosphere 253:126594

Bolognesi C, Merlo FD (2019) Pesticides: human health effects. In: Encyclopedia of environmental health, 2nd edn. Elsevier, Burlington, pp 118–132

Boulange J, Watanabe H, Akai S (2017) A Markov Chain Monte Carlo technique for parameter estimation and inference in pesticide fate and transport modeling. Ecol Model 360:270–278

Braun G, Sebesvari Z, Braun M, Kruse J, Amelung W, An NT, Renaud FG (2018) Does sea-dyke construction affect the spatial distribution of pesticides in agricultural soils? – a case study from the Red River Delta, Vietnam. Environ Pollut 243(Part B):890–899

Braun G, Braun M, Kruse J, Amelung W, Renaud FG, Khoi CM, Duong MV, Sebesvari Z (2019) Pesticides and antibiotics in permanent rice, alternating rice-shrimp and permanent shrimp systems of the coastal Mekong Delta, Vietnam. Environ Int 127:442–451

Brockwell PJ, Davis RA (2002) Introduction to time series and forecasting, 2nd edn. Springer, New York

Caffara A, Rinaldi M, Eccel E, Rossi V, Pertot I (2012) Modelling the impact of climate change on the interaction between grapevine and its pests and pathogens: European grapevine moth and powdery mildew. Agric Ecosyst Environ 148:89–101

Cai J, Xiao D, Lv L, Ye Y (2019) An early warning model for vegetable pests based on multidimensional data. Comput Electron Agric 156:217–226

Carrao DB, Habenchus MD, Albuquerque NCP, Silva RM, Lopes NP, Oliveria ARM (2019) In vitro inhibition of human CYP2D6 by the chiral pesticide fipronil and its metabolite fipronil sulfone: prediction of pesticide-drug interactions. Toxicol Lett 313:196–204

Castex V, Beniston M, Calanca P, Fleury D, Moreau J (2018) Pest management under climate change: the importance of understanding tritrophic relations. Sci Total Environ 616–617:397–407

Ccanccapa A, Masia A, Andreu V, Pico Y (2016) Spatio-temporal patterns of pesticide residues in the Turia and Jucar rivers (Spain). Sci Total Environ 540:200–210

Chakraborty P, Zhang G, Li J, Sivakumar A, Jones KC (2015) Occurrence and sources of selected organochlorine pesticides in the soil of seven major Indian cities: assessment of air-soil exchange. Environ Pollut 204:74–80

Chakraborty P, Zhang G, Li J, Sampathkumar P, Balasubramanian T, Kathiresan K, Takahashi S, Subramanian A, Tanabe S, Jones KC (2019) Seasonal variation of atmospheric organochlorine pesticides and polybrominated diphenyl ethers in Parangipettai, Tamil Nadu, India: Implication for atmospheric transport. Sci Total Environ 649:1653–1660

Collier RH (2017) Pest and disease prediction models. In: Encyclopedia of applied plant sciences, vol 3, 2nd edn. Academic Press, Waltham, pp 120–123

Congdon BS, Coutts BA, Jones RAC, Renton M (2017) Forecasting model for Pea seed-borne mosaic virus epidemics in field pea crops in a Mediterranean-type environment. Virus Res 241:163–171

Curutiu C, Lazar V, Chifiriuc MC (2017) Pesticides and antimicrobial resistance: from environmental compartments to animal and human infections. In: New pesticides and soil sensors. Academic Press, Waltham, pp 373–392

Dashtbozorgi Z, Golmohammadi H, Konoz E (2013) Support vector regression based QSPR for the prediction of retention time of pesticide residues in gas chromatography–mass spectroscopy. Microchem J 106:51–60

Devi PI, Thomas J, Raju RK (2017) Pesticide consumption in India: a spatiotemporal analysis. Agric Econ Res Rev 30(1):163–172

Drabova L, Rivera GA, Suchanova M, Schusterova D, Pulkrabova J, Tomaniova M, Kocourek V, Chevallier O, Elliott C, Hajslova J (2019) Food fraud in oregano: pesticide residues as adulteration markers. Food Chem 276:726–734

Elahi E, Weijun C, Zhang H, Nazeer M (2019) Agricultural intensification and damages to human health in relation to agrochemicals: application of artificial intelligence. Land Use Policy 83:461–474

Elavarasan D, Vincent DR, Sharma V, Zomaya AY, Srinivasan K (2018) Forecasting yield by integrating agrarian factors and machine learning models: a survey. Comp Elect Agric 155:257–282

FAO/WHO (2016) The international code of conduct on pesticide management – guidelines on highly hazardous pesticides. Food and Agriculture Organization of the United Nations/World Health Organization, Rome/Geneva

FAO/WHO (2019a) Maximum residue limits (MRLs). Codex Alimentarius. International Food Standards. Food and Agriculture Organization of the United Nations/World Health Organization, Rome/Geneva

FAO/WHO (2019b) Reports of the joint meeting on pesticide residues. FAO plant production and protection paper series. Food and Agriculture Organization of the United Nations/World Health Organization, Rome/Geneva

Farlin J, Galle T, Bayerle M, Pittois D, Braun C, Khabbaz HE, Elsner M, Maloszewski P (2012) Predicting pesticide attenuation in a fractured aquifer using lumped-parameter models. Groundwater 51:276–285. 1–10

Farlin J, Galle T, Bayerle M, Pittois D, Braun C, Khabbaz HE, Lallement C, Leopold U, Vanderborght J, Weihermueller L (2013) Using the long-term memory effect of pesticide and metabolite soil residues to estimate field degradation half-life and test leaching predictions. Geoderma 207–208:15–24

Farlin J, Bayerle M, Pittois D, Galle T (2017) Estimating pesticide attenuation from water dating and the ratio of metabolite to parent compound. Groundwater 55:550–557. 1–8

Galimberti F, Moretto A, Papa E (2020) Application of chemometric methods and QSAR models to support pesticide risk assessment starting from ecotoxicological datasets. Water Res 174:115583

Gao H, Zhang L, Lu Z, He C, Li Q, Na G (2018) Complex migration of antibiotic resistance in natural aquatic environments. Environ Pollut 232:1–9

Gao J, Gu C, Yang H, Weng T (2020) Prediction of spatial distribution of invasive alien pests in two-dimensional systems based on a discrete time model. Ecol Eng 143:105673

Gierer F, Vaughan S, Slater M, Thompson HM, Elmore JS, Girling RD (2019) A review of the factors that influence pesticide residues in pollen and nectar: future research requirements for optimizing the estimation of pollinator exposure. Environ Pollut 249:236–247

Groot M, Ogris N (2019) Short-term forecasting of bark beetle outbreaks on two economically important conifer tree species. For Ecol Manag 450:117495

Guardo AD, Finizio A (2016) A moni-modelling approach to manage groundwater risk to pesticide leaching at regional scale. Sci Total Environ 545–546:200–209

Guzman C, Fenollosa EA, Sahun RM, Boyero JR, Vela JM, Wong E, Jaques JA, Montserrat M (2016) Temperature-specific competition in predatory mites: implications for biological pest control in a changing climate. Agric Ecosyst Environ 216:89–97

Hajirahimi Z, Khashei M (2019) Hybrid structures in time series modeling and forecasting: a review. Eng App Artif Intell 86:83–106

He L, Xiao K, Zhou C, Li G, Yang H, Li Z, Cheng J (2019) Insights into pesticide toxicity against aquatic organism: QSTR models on *Daphnia Magna*. Ecotoxicol Environ Saf 173:285–292

Hong SW, Zhao L, Zhu H (2018) SAAS, a computer program for estimating pesticide spray efficiency and drift of air-assisted pesticide applications. Comput Electron Agric 155:58–68

Houbraken M, Habimana V, Senaeve D, Davila EL, Spanoghe P (2017) Multi-residue determination and ecological risk assessment of pesticides in the lakes of Rwanda. Sci Total Environ 576:888–894

Huan Z, Xu Z, Luo J, Xie D (2016) Monitoring and exposure assessment of pesticide residues in cowpea (*Vigna unguiculata* L.Walp) from five provinces of southern China. Regul Toxicol Pharmacol 81:260–267

Hussein AS, Beshir S, Taha MM, Shahy EM, Shaheen W, Shafy EAA, Thabet E (2019) Early prediction of liver carcinogenicity due to occupational exposure to pesticides. Mutat Res Genet Toxicol Environ Mutagen 838:46–53

Islam MN, Huang L, Siciliano SD (2020) Inclusion of molecular descriptors in predictive models improves pesticide soil-air partitioning estimates. Chemosphere 248:126031

Jiang B, He J, Yang S, Fu H, Li T, Song H, He D (2018) Fusion of machine vision technology and AlexNet-CNNs deep learning network for the detection of postharvest apple pesticide residues. Artif Intell Agric 1:1–8

Jiao L, Dong S, Zhang S, Xie C, Wang H (2020a) AF-RCNN: an anchor-free convolutional neural network for multi-categories agricultural pest detection. Comp Elect Agric 174:1–9

Jiao C, Chen L, Sun C, Jiang Y, Zhai L, Liu H, Shen Z (2020b) Evaluating national ecological risk of agricultural pesticides from 2004 to 2017 in China. Environ Pollut 259:113778

Jin S, Bluemling B, Mol APJ (2018) Mitigating land pollution through pesticide packages – the case of a collection scheme in Rural China. Sci Total Environ 622–623:502–509

Jung JM, Jung S, Byeon DH, Lee WH (2017) Model-based prediction of potential distribution of the invasive insect pest, spotted lanternfly *Lycorma delicatula* (Hemiptera: Fulgoridae), by using CLIMEX. J Asia-Pacific Biodiv 10(4):532–538

Kim KH, Kahir E, Jahan SA (2017) Exposure to pesticides and the associated human health effects. Sci Total Environ 575:525–535

Kumar ADD, Reddy DN (2017) High pesticide use in India: health implications. Health Action 1:7–12

Kumar A, Correll R, Grocke S, Bajet C (2010) Toxicity of selected pesticides to freshwater shrimp, *Paratya australiensis* (Decapoda: Atyidae): use of time series acute toxicity data to predict chronic lethality. Ecotoxicol Environ Saf 73:360–369

Kumar S, Nehra M, Dilbaghi N, Marrazza G, Hassan AA, Kim KH (2019) Nano-based smart pesticide formulations: emerging opportunities for agriculture. J Control Release 294:131–153

Kumari D, John S (2019) Health risk assessment of pesticide residues in fruits and vegetables from farms and markets of Western Indian Himalayan region. Chemosphere 224:162–167

Lai W (2017) Pesticide use and health outcomes: evidence from agricultural water pollution in China. J Environ Econ Manag 86:93–120

Lammoglia SK, Makowski D, Moeys J, Justes E, Barrisuo E, Mamy L (2017) Sensitivity analysis of the STICS-MACRO model to identify cropping practices reducing pesticide losses. Sci Total Environ 580:117–129

Lammoglia SK, Brun F, Quemar T, Moeys J, Barrisuo E, Gabrielle B, Mamy L (2018) Modelling pesticides leaching in cropping systems: effect of uncertainties in climate, agricultural practices, soil and pesticide properties. Environ Model Softw 109:342–352

Lee HJ, Kim KY, Hamm SY, Kim M, Kim HK, Oh JE (2019) Occurrence and distribution of pharmaceutical and personal care products, artificial sweeteners and pesticides in groundwater from an agricultural area in Korea. Sci Total Environ 659:168–176

Li Z (2018a) A Bayesian generalized log-normal model to dynamically evaluate the distribution of pesticide residues in soil associated with population health risks. Environ Int 121:620–634

Li Z (2018b) A health-based regulatory chain framework to evaluate international pesticide groundwater regulations integrating soil and drinking water standards. Environ Int 121:1253–1278

Li Z (2018c) The use of a disability-adjusted life-year (DALY) metric to measure human health damage resulting from pesticide maximum legal exposures. Sci Total Environ 639:438–456

Lin G, Ji R, Yao H, Chen R, Yu Y, Wang X, Yang X, Zhu T, Bian H (2020) Fluorescence detection of multiple kinds of pesticides with multi hidden layers neural network algorithm. Optik- Int J Light Electron Optics 211:164632

Ljung L (1999) System identification: theory for the user, 2nd edn. Prentice Hall, Upper Saddle River

Malaj E, Liber K, Morrissey CA (2020) Spatial distribution of agricultural pesticide use and predicted wetland exposure in the Canadian Prairie Pothole Region. Sci Total Environ 718:134765

Marnasidis S, Stamatelatou K, Verikouki E, Kazantzis K (2018) Assessment of the generation of empty pesticide containers in agricultural areas. J Environ Manag 224:37–48

Martin TM, Lilavois CR, Barron MG (2017) Prediction of pesticide acute toxicity using two dimensional chemical descriptors and target species classification. SAR QSAR Environ Res 28(6):525–539

McGrath G, Rao PSC, Mellander PE, Kennedy I, Rose M, Zwieten LV (2019) Real-time forecasting of pesticide concentrations in soil. Sci Total Environ 663:709–717

Meftaul IM, Venkateswarlu K, Dharmarajan R, Annamalai P, Megharaj M (2020) Pesticides in the urban environment: a potential threat that knocks at the door. Sci Total Environ 711:134612

Meite F, Zaldivar PA, Crochet A, Wiegert C, Payraudeau S, Imfeld G (2018) Impact of rainfall patterns and frequency on the export of pesticides and heavy metals from agricultural soils. Sci Total Environ 616–617:500–509

Mfarrej MFB, Rara FM (2019) Competitive, sustainable natural pesticides. Acta Ecol Sin 39 (2):145–151

Michael OK, Hogarh JN, Brink PJV (2020) Environmental risk assessment of pesticides currently applied in Ghana. Chemosphere 254:126845

Miller TH, Gallidabino MD, MacRae JI, Owen SF, Bury NR, Barron LP (2019) Prediction of bioconcentration factors in fish and invertebrates using machine learning. Sci Total Environ 648:80–89

Mishra K, Sharma RC, Kumar S (2012) Contamination levels and spatial distribution of organochlorine pesticides in soils from India. Ecotoxicol Environ Saf 76:215–225

Mohanty MK, Behera BK, Jena SK, Srikanth S, Mogane C, Samal S, Behera AA (2013) Knowledge attitude and practice of pesticide use among agricultural workers in Puducherry. South India J Forensic Legal Med 20:1028–1031

Mol FD, Winter M, Gerowitt B (2018) Weather determines the occurrence of wheat stem base diseases in biogas cropping systems. Crop Prot 114:1–11

Mondal R, Mukherjee A, Biswas S, Kole RK (2018) GC-MS/MS determination and ecological risk assessment of pesticides in aquatic system: a case study in Hooghly river basin in West Bengal, India. Chemosphere 206:217–230

Montgomery DC, Jennings CL, Kulahci M (2008) Introduction to time series analysis and forecasting. Wiley, Hoboken

Mubushar M, Aldosari FO, Baig MB, Alotaibi BM, Khan AQ (2019) Assessment of farmers on their knowledge regarding pesticide usage and biosafety. Saudi J Biol Sci 26(7):1903–1910

Newbery F, Qi A, Fitt BDL (2016) Modelling impacts of climate change on arable crop diseases: progress, challenges and applications. Curr Opin Plant Biol 32:101–109

Niell S, Jesus F, Diaz R, Mendoza Y, Notte G, Santos E, Gerez N, Cesio V, Cancela H, Heinzein H (2018) Beehives biomonitor pesticides in agroecosystems: Simple chemical and biological indicators evaluation using Support Vector Machines (SVM). Ecol Indic 91:149–154

Norman JE, Mahler BJ, Nowell LH, Metre PCV, Sandstrom MW, Corbin MA, Qian Y, Pankow JF, Luo W, Fitzgerald NB, Asher WE, McWhirter KJ (2020) Daily stream samples reveal highly complex pesticide occurrence and potential toxicity to aquatic life. Sci Total Environ 715:136795

Osawa T, Yamasaki K, Tabuchi K, Yoshioka A, Ishigooka Y, Sudo S, Takada MB (2018) Climate-mediated population dynamics enhance distribution range expansion in a rice pest insect. Basic Appl Ecol 30:41–51

Ouyang W, Cai G, Tysklind M, Yang W, Hao F, Liu H (2017) Temporal-spatial patterns of three types of pesticide loadings in a middle-high latitude agricultural watershed. Water Res 122:377–386

Pagani AP, Ibanez GA (2019) Pesticide residues in fruits and vegetables: high-order calibration based on spectrofluorimetric/pH data. Microchem J 149:104042

Pan X, Dong F, Wu X, Xu J, Liu X, Zheng Y (2019) Progress of the discovery, application and control technologies of chemical pesticides in China. J Integr Agric 18(4):840–853

Pham X, Stack M (2018) How data analytics is transforming agriculture. Bus Horiz 61(1):125–133

Pose-Juan E, Sanchez Martin MJ, Andrades MS, Rodriguez-Cruz MS, Hernandez EH (2015) Pesticide residues in vineyard soils in Spain: spatial and temporal distributions. Sci Total Environ 514:351–358

Qian L, Zhang C, Zuo F, Zheng L, Li D, Zhang A, Zhang D (2019) Effects of fertilizers and pesticides on the mineral elements used for the geographical origin traceability of rice. J Food Compos Anal 83:103276

Qu C, Albanese S, Li J, Cicchella D, Zuzolo D, Hope D, Cerino P, Pizzolante A, Doherty AL, Lima A, Vivo BD (2019) Organochlorine pesticides in the soils from Benevento provincial territory, Southern Italy: spatial distribution, air-soil exchange and implications for environmental health. Sci Total Environ 674:159–170

Quaglia G, Joris I, Broekx S, Desmet N, Koopmans K, Vandaele K, Seuntjens P (2019) A spatial approach to identify priority areas for pesticide pollution mitigation. J Environ Manag 246:583–593

Queyrel W, Habets F, Blanchoud H, Ripoche D, Launay M (2016) Pesticide fate modelling in soils with the crop model STICS: feasibility for assessment of agricultural practices. Sci Total Environ 542:782–802

Rajcevic SM, Rubino FM, Colosio C (2020) Establishing health-based biological exposure limits for pesticides: a proof of principle study using mancozeb. Regul Toxicol Pharmocol 115:104689

Ramakrishnan B, Venkateswarlu K, Sethunathan N, Megharaj M (2019) Local applications but global implications: can pesticides drive microorganisms to develop antimicrobial resistance? Sci Total Environ 654:177–189

Rangasamy K, Athiappan M, Devarajan N, Parray JA (2017) Emergence of multi drug resistance among soil bacteria exposing to insecticides. Microb Pathog 105:153–165

Rangasamy K, Athiappan M, Devarajan N, Samykannu G, Parray JA, Aruljothi KN, Shameem N, Alqarawi AA, Hashem A, Abd Allah EF (2018) Pesticide degrading natural multidrug resistance bacterial flora. Microb Pathog 114:304–310

Sabarwal A, Kumar K, Singh RP (2018) Hazardous effects of chemical pesticides on human health – cancer and other associated disorders. Environ Toxicol Phar 63:103–114

Sanganyado E, Gwenzi W (2019) Antibiotic resistance in drinking water systems: occurrence, removal and human health risks. Sci Total Environ 669:785–797

Santos GG, Scheiber M, Pilz J (2020) Spatial interpolation methods to predict airborne pesticide drift deposits on soils using knapsack sprayers. Chemosphere 258:127231

Schreinemachers P, Grovermann C, Praneetvatakul S, Heng P, Nguyen TTC, Buntong B, Le NT, Pinn T (2020) How much is too much? Quantifying pesticide overuse in vegetable production in Southeast Asia. J Clean Prod 244:118738

Sharma R, Peshin R (2016) Impact of integrated pest management of vegetables on pesticide use in subtropical Jammu, India. Crop Prot 84:105–112

Shein EV, Belik AA, Kokoreva AA, Kolupaeva VN, Pletenev PA (2017) Prediction of pesticide migration in soils: the role of experimental soil control. Moscow Univ Soil Sci Bull 72 (4):185–190

Shoda ME, Stone WW, Nowell LH (2016) Prediction of pesticide toxicity in Midwest streams. J Environ Qual 45:1856–1864

Silva V, Mol HGJ, Zomer P, Tienstra M, Ritsema CJ, Geissen V (2019) Pesticide residues in European agricultural soils – a hidden reality unfolded. Sci Total Environ 653:1532–1545

Sousa ES, Schneider MP, Pinto L, Araujo MCU, Gomes AA (2020) Chromatographic quantification of seven pesticide residues in vegetables: univariate and multiway calibration comparison. Microchem J 152:104301

Steingrimsdottir MM, Peterson A, Fantke P (2018) A screening framework for pesticide substitution in agriculture. J Clean Prod 192:306–315

Stergiopoulos C, Makarouni D, Kakoulidou AT, Petropoulou MO, Tsopelas F (2019) Immobilized artificial membrane chromatography as a tool for the prediction of ecotoxicity of pesticides. Chemosphere 224:128–139

Subash SP, Chand P, Pavithra S, Balaji SJ, Pal S (2017) Pesticide use in Indian agriculture: trends, market structure and policy issues. Policy in brief. Indian Council of Agricultural Research, New Delhi, pp 1–5

Sybertz A, Ottermanns R, Schaffer A, Starke BS, Daniels B, Frische T, Bar S, Ullrich C, Nickoll MR (2020) Simulating spray series of pesticides in agricultural practice reveals evidence for accumulation of environmental risk in soil. Sci Total Environ 710:135004

Szekeres E, Chiriac CM, Baricz A, Nagy TS, Lung I, Soran ML, Rudi K, Dragos N, Coman C (2018) Investigating antibiotics, antibiotic resistance genes, and microbial contaminants in groundwater in relation to the proximity of urban areas. Environ Pollut 236:734–744

Thomson LJ, Macfadyen S, Hoffmann AA (2010) Predicting the effects of climate change on natural enemies of agricultural pests. Biol Control 52(3):296–306

UN/FAO (2017) The future of food and agriculture: trends and challenges. Summary Version: 1–47. www.fao.org/publications

UNICEF (2018) Understanding the impacts of pesticides on children: a discussion paper. UNICEF, New York

Utami RR, Geerling GW, Salami IRS, Notodarmojo S, Ragas AMJ (2020) Environmental prioritization of pesticide in the Upper Citarum River Basin, Indonesia, using predicted and measured concentrations. Sci Total Environ 738:140130

Varjani S, Kumar G, Rene ER (2019) Developments in biochar application for pesticide remediation: current knowledge and future research directions. J Environ Manag 232:505–513

Vaz WF, D'Oliveira GDC, Perez CN, Neves BJ, Napolitano HB (2020) Machine learning prediction of the potential pesticide applicability of three dihydroquinoline derivatives: syntheses, crystal structures and physical properties. J Mol Struct 1206:127732

Villamizar ML, Brown CD (2017) A modelling framework to simulate river flow and pesticide loss via preferential flow at the catchment scale. Catena 149(1):120–130

Villamizar ML, Stoate C, Biggs J, Morris C, Szczur J, Brown CD (2020) Comparison of technical and systems-based approaches to managing pesticide contamination in surface water catchments. J Environ Manag 260:110027

VoPham T, Wilson JP, Ruddell D, Rashed T, Brooks MM, Yuan JM, Talbott EO, Chang CCH, Weissfeld JL (2015) Linking pesticides and human health: a geographic information system (GIS) and Landsat remote sensing method to estimate agricultural pesticide exposure. Appl Geogr 62:171–181

Wang J, Tao J, Yang C, Chu M, Lam H (2017) A general framework incorporating knowledge, risk perception and practices to eliminate pesticide residues in food: a structural equation modelling analysis based on survey data of 986 Chinese farmers. Food Control 80:143–150

Wang R, Yuan Y, Yen H, Grieneisen M, Arnold J, Wang D, Wang C, Zhang M (2019a) A review of pesticide fate and transport simulation at water shed level using SWAT: current status and research concerns. Sci Total Environ 669:512–516

Wang X, Zhou L, Zhang X, Luo F, Chen Z (2019b) Transfer of pesticide residue during tea brewing: understanding the effects of pesticide's physico-chemical parameters on its transfer behaviour. Food Res Int 121:776–784

Wang J, Wang S, Shan F, Sun Y, Liu S (2020a) Missing data recovery combined with Parallel factor analysis model for eliminating Rayleigh scattering in the process of detecting pesticide mixture. Spectrochim Acta A Mol Biomol Spectrosc 232:118187

Wang S, Wang J, Shan F, Wang Y, Cheng Q, Liu N (2020b) A GA-BP method of detecting carbamate pesticide mixture based on three-dimensional fluorescence spectroscopy. Spectrochim Acta A Mol Biomol Spectrosc 224:117396

WHO (2017) Guidelines for drinking water quality, 4th edition incorporating the first addendum. World Health Organization, Geneva

WHO (2019) Preventing disease through healthy environments. In: Exposure to highly hazardous pesticides: a major public health concern. World Health Organization, Geneva

WHO/FAO (2014) International code of conduct on pesticide management. Food and Agriculture Organization of the United Nations/World Health Organization, Rome/Geneva

Wildemeersch M, Franklin O, Seidl R, Rogelj J, Moorthy I, Thurner S (2019) Modelling the multi-scaled nature of pest outbreaks. Ecol Model 409:108745
Wrzesien M, Treder W, Klamkowski K, Rudnicki WR (2019) Prediction of the apple scab using machine learning and simple weather stations. Comp Electron Agric 161:252–259
Xie H, Wang X, Chen J, Li X, Jia G, Zou Y, Zhang Y, Cui Y (2019) Occurrence, distribution and ecological risks of antibiotics and pesticides in coastal waters around Liaodong Peninsula, China. Sci Total Environ 656:946–951
Xu D, Li X, Jin Y, Zhuo Z, Yang H, Hu J, Wang R (2020a) Influence of climatic factors on the potential distribution of pest *Heortia vitessoides* Moore in China. Global Ecol Conserv 23: e01107
Xu Y, Liu S, Lu B, Wang Z (2020b) Acute toxicity dataset for QSAR modeling and predicting missing data of six pesticides. Data Brief 29:105150
Yadav IC, Devi NL, Syed JH, Cheng Z, Li J, Zhang G, Jones KC (2015) Current status of persistent organic pesticides residues in air, water and soil, and their possible effect on neighboring countries: a comprehensive review of India. Sci Total Environ 511:123–137
Yang LN, Peng L, Zhang LM, Zhang L, Yang S (2009) A prediction model for population occurrence of paddy stem borer (Scirpophaga incertulas), based on back propagation artificial neural network and principal components analysis. Comput Electron Agric 68:200–206
Yang L, Wang Y, Chang J, Pan Y, Wei R, Li J, Wang H (2020a) QSAR modeling the toxicity of pesticides against *Americamysis bahia*. Chemosphere 258:127217
Yang L, Wang Y, Hao W, Chang J, Pan Y, Li J, Wang H (2020b) Modeling pesticides toxicity to Sheepshead minnow using QSAR. Ecotoxicol Environ Saf 193:110352
Yuan YY, Wang ST, Liu SY, Cheng Q, Wang ZF, Kong DM (2020) Green approach for simultaneous determination of multi-pesticide residue in environmental water samples using excitation-emission matrix fluorescence and multivariate calibration. Spectrochim Acta A Mol Biomol Spectrosc 228:117801
Zdravkovic M, Antovic A, Veselinovic JB, Sokolovic D, Veselinovic AM (2018) QSPR in forensic analysis – the prediction of retention time of pesticide residues based on the Monte Carlo method. Talanta 178:656–662
Zhang X, Goh KS (2015) Evaluation of three models for simulating pesticide runoff from irrigated agricultural fields. J Environ Qual 44(6):1809–1820
Zhang B, Zhang QQ, Zhang SX, Xing C, Ying GG (2020) Emission estimation and fate modelling of three typical pesticides in Dongjiang River basin, China. Environ Pollut 258:113660
Zhan-Qi R, Zhen-Hong R, Hai-Yan J (2018) Identification of different concentrations pesticide residues of dimethoate on spinach leaves by hyperspectral image technology. IFAC-PapersOnLine 51(17):758–763

Effects of Dissolved Organic Matter on the Bioavailability of Heavy Metals During Microbial Dissimilatory Iron Reduction: A Review

Yuanhang Li and Xiaofeng Gong

Contents

1 Introduction 72
2 Overview of Dissolved Organic Matter 72
2.1 Source and Extraction of DOM 73
2.2 Structure Composition and Characterization of DOM 74
3 Interaction Mechanism Between DOM and Microbial DIR 76
3.1 Mechanism of the DIR Process 76
3.2 Effect of DIR on DOM Release 77
3.3 Effect of DOM on DIR 77
4 Effect of DIR on Bioavailability of Heavy Metals 79
5 Effects of DOM on the Bioavailability of Heavy Metals and Its Mechanism 81
5.1 Effect of DOM on Heavy Metal (Cr) 81
5.2 Effect of DOM on Heavy Metal (As) 82
5.3 Effect of DOM on Heavy Metal (Hg) 82
5.4 Effect of DOM on Heavy Metal (Cu) 83
5.5 Mechanism of DOM on Bioavailability of Heavy Metals 84
6 Conclusion 85
References 86

Abstract Dissolved organic matter (DOM), a type of mixture containing complex structures and interactions, has important effects on environmental processes such as the complexation and interface reactions of soil heavy metals. Furthermore, microbial dissimilatory iron reduction (DIR), a key process of soil biogeochemical cycle, is closely related to the migration and transformation of heavy metals and causes the release of DOM by carbon-ferrihydrite associations. This chapter considers the structural properties and characterization techniques of DOM and its interaction with microbial dissimilated iron. The effect of DOM on microbial DIR is specifically

Y. Li · X. Gong (✉)
School of Resources, Environmental and Chemical Engineering, Nanchang University, Nanchang, China

Key Laboratory of Poyang Lake Environment and Resource Utilization, Ministry of Education, Nanchang University, Nanchang, China
e-mail: xfgong@ncu.edu.cn

P. de Voogt (ed.), *Reviews of Environmental Contamination and Toxicology Volume 257*, Reviews of Environmental Contamination and Toxicology 257,
https://doi.org/10.1007/398_2020_63

manifested as driving force properties, coprecipitation, complexation, and electronic shuttle properties. The study, in addition, further explored the influence of pH, microorganisms, salinity, and light conditions, mechanism of DOM and microbial DIR on the toxicity and bioavailability of different heavy metals. The action mechanism of these factors on heavy metals can be summarized as adsorption coprecipitation, methylation, and redox. Based on the findings of the review, future research is expected to focus on: (1) The combination of DOM functional group structure analysis with high-resolution mass spectrometry technology and electrochemical methods to determine the electron supply in the mechanism of DOM action on DIR; (2) Impact of DOM on differences in structure and functions of plant rhizosphere in heavy metal contaminated soil; and (3) Bioavailability of DOM-dissociative iron-reducing bacteria-heavy metal ternary binding on rhizosphere heavy metals under dynamic changes of water level from the perspective of the differences in DOM properties, such as polarity, molecular weight, and functional group.

Keywords Bioavailability · Dissolved organic matter · Heavy metal · Microbial dissimilatory iron reduction

Abbreviations

3D-EEM Three-dimensional fluorescence excitation emission matrix
AQDS Anthraquinone-2,6-disulphonate
AQS Anthraquinone-2-sulfonate
As (III) Trivalent arsenic
As (V) Pentavalent arsenic
As Arsenic
C Carbon
C=H Carbon-hydrogen bond
C=O Carbon-oxygen bond
Cd Cadmium
CH_3Hg Monomethylmercury
Cr (III) Trivalent chromium
Cr (VI) Hexavalent chromium
Cr Chromium
Cu Copper
DIR Dissimilatory iron reduction
DOC Dissolved organic carbon
DOM Dissolved organic matter
DOM* DOM free radicals
EA Elemental analysis
EDTA Ethylene diamine tetraacetic acid
EPS Extracellular polymers substances

ESR	Electron spin resonance
FA	Fulvic acid
Fe (II)	Iron(II)
Fe (III)	Iron(III)
FeS	Iron sulfide
FFFF	Flow field flow fractionation
FRI	Fluorescence regional integration
FT-ICR-MS	Fourier transform ion cyclotron resonance mass spectrometry
FTIR	Fourier transform infrared
H	Hydrogen
H_2S	Hydrogen sulfide
HA	Humic acid
Hg	Mercury
$HgCl_2$	Mercury chloride
HgS	Mercury sulfite
$HgSO_4$	Mercury sulfate
HPIA	Hydrophilic acidic organic matter
HPIB	Hydrophilic base organic matter
HPIN	Hydrophilic neutral organic matter
HPOA	Hydrophobic acidic organic matter
HPOB	Hydrophobic basic organic matter
HPON	Hydrophobic neutral organic matter
IHSS	International Humus Association
MeHg	Methylmercury
N	Nitrogen
NaAc	Sodium acetate
N-H	Nitrogen-hydrogen bond
Ni	Nickel
NMR	Nuclear magnetic resonance spectroscopy
NO_3^-	Nitrate ion
O	Oxygen
O_2•	Molecular oxygen free radicals
P	Phosphorus
PARAFAC	Parallel factor
S	Sulfur
S^{2-}	Sulfite
SO_4^{2-}	Sulfate ion
SOM	Self-organizing map
TOC	Total organic carbon
UC	Ultracentrifugation
UV-Vis	Ultraviolet and visible spectrometry
Zn	Zinc

1 Introduction

Dissolved organic matter (DOM) has widely existed in ecosystems, and it is the most active organic component in aquatic and soil environments (Skoog and Arias-Esquivel 2009). The distribution of characteristics and environmental effects of DOM in various natural water bodies, such as lakes, rivers, oceans, and groundwater, has been fruitfully researched since the twenty-first century (Hill et al. 2009; Zhang and Weber 2009; Cinzia et al. 2008). In the soil environment, however, DOM usually acts as a blinder of organic or inorganic pollutants, especially soil heavy metals, and its molecular structure and characteristics can directly affect the migration and transformation of these pollutants. There are various active functional groups such as hydroxyl, carboxyl, and carbonyl groups in DOM, it can not only complex with heavy metals copper (Cu), nickel (Ni), and zinc (Zn) to form stable complexes, but also can change the speciation of heavy metals such as chromium (Cr) (Impellitteri et al. 2002; Huang et al. 2016), which in turn affect the morphology and bioavailability of heavy metal. Thus, the impact of DOM on the environmental behavior of heavy metals has become a hot topic in environmental science research.

Recent studies have shown that the quinone group commonly present in DOM has electron transport properties and plays an important role in microbial dissimilatory iron reduction (DIR) (Gu et al. 2016; Poggenburg et al. 2018; He et al. 2019). Microbial DIR is a process in which organic matter is oxidized, taking iron(III) (Fe (III)) as the terminal electron acceptor, to provide energy for living organisms under anaerobic conditions. It plays an important role in the biogeochemical cycle. This process not only affects the distribution and morphology of iron, but is also closely related to the toxicity, morphological transformation, and bioavailability of heavy metals (such as arsenic (As), Cr, cadmium (Cd), and mercury (Hg) in the soil (Hellal et al. 2015; Li et al. 2016; Yuan et al. 2018a).

The research on the influence of DOM by mediating DIR on the bioavailability of heavy metals has aroused great interest of scholars. Systematic exploration of the DOM-microorganism-heavy metal ternary combination on heavy metal bioavailability and on their interaction mechanism can provide important theory guidance and practical value for deep understanding of the toxicity of heavy metals in soil, and on remediation and risk assessment of heavy metal contaminated soil.

2 Overview of Dissolved Organic Matter

DOM has different structures and molecular weights (e.g., low molecular weight: free amino acids, sugars, and organic acids; large molecular weight: humic substance, proteins, lignin, and polyphenols). So it is difficult to give an accurate definition of DOM, what makes more common an operational definition: DOM refers to the homogeneous mixture obtained after water extraction through a 0.45 μm filter membrane (Kalbitz et al. 2000; Benner and Amon 2015). The content of DOM is usually expressed as dissolved organic carbon (DOC).

2.1 Source and Extraction of DOM

The sources of DOM are complex, not only affected by biological factors in the soil but also controlled by many abiotic factors (e.g., adsorption, desorption, diffusion, leaching, soil pH, temperature, humidity, ionic strength, wet and dry alternation) (Vazquez et al. 2011; Yu et al. 2012). DOM, generally, in the soil is mainly produced by endogenous and exogenous sources. Endogenous sources include residues of animal, plant and microbial, humic organic matter, root exudates, and soil animal excreta, while externally ingested organic fertilizer (e.g., urban sludge, returning straw, poultry manure) is another important source of DOM (Chantigny 2003). Judging the source of organic matter from a single indicator may produce different results. Many environmental factors will also change the content and characteristics of DOM in most cases. For example, flooding conditions are conducive to the increase of DOM concentration in soil solution, and the photo-degradation of DOM in the soil solution was significantly influenced by pH (Timko et al. 2015). Therefore, the comprehensive application of multiple new technologies and methods is the development trend to analyze and explore DOM sources.

The concentration of DOM, generally, in the aquatic and soil environment is low, which are basically 3–145 mmol·L^{-1} in soil and even lower in aquatic ecosystem (Borggaard et al. 2019). Sometimes it is below the amount needed for the analysis of elements and functional groups, which increases the requirements for the accurate and efficient methods of the DOM's separation and extraction. The current extraction methods are not very mature, and different researchers have different methods for extracting DOM from different sources. Traditional separation and extraction methods include extraction, shaking, centrifugation, filtration, solid-phase extraction, and so on (Schwede-Thomas et al. 2005). The establishment of the extracted material or soil–water ratio is a difficult point for it can greatly affect the extracted DOM content. Relevant research has shown that composition characteristics of DOM in soil, water, and plant extracts are quite different (Xie et al. 2013). Extracted samples are also affected by factors such as temperature, pH, extraction time, and concentration of extracted ions (McDowell 2003). With the development of polymer materials, filter membranes based on polymer materials have made microfiltration, nanofiltration, and reverse osmosis filtration technologies more popular. In the early 1990s, Serkiz and Perdue (1990) used reverse osmosis filtration technology to obtain DOM in water bodies. This method has also become one of the recommended methods of the International Humus Association (IHSS)· Membrane filtration (ultrafiltration, nanofiltration, reverse osmosis, and electroosmosis) (Simjouw et al. 2005; Navalon et al. 2010; Ouellet et al. 2008; Mao et al. 2012), non-ionic macroporous resins (Schwede-Thomas et al. 2005; Santos et al. 2009), ion exchange resins (Leenheer et al. 2004), freeze-drying (McCurry et al. 2012; Ma et al. 2001), vacuum rotary evaporation (Spencer et al. 2010), adsorbent enrichment, and other technologies (Schwede-Thomas et al. 2005) have also been widely utilized. However, it is difficult to efficiently analyze the components of the DOM using a single technology, a combination of these methods is generally required.

2.2 Structure Composition and Characterization of DOM

The properties and functions of different components of DOM are very diverse. Each component of DOM cannot be accurately separated using existing technology, and at least one-third of them were still difficult to confirm before the twenty-first century (Hedges et al. 2000). Furthermore, DOM from different sources (soil, river, ocean, and plant) shows significant differences in composition and structure, which justifies the research according to their characteristics. However, with the emergence of more accurate characterization and separation techniques, the majority component of DOM can be analyzed now. According to different needs, workers in this field have the following relatively recognized classifications of DOM (Table 1): (a). Elements; (b). Chemical properties; (c). Molecular weight; and (d). Properties of hydrophilic- hydrophobic and acid-base (Spencer et al. 2010; Hedges et al. 2000; Ma et al. 2001; Soong et al. 2014; Bai et al. 2019).

With the in-depth study of dissolved organic matter from the external structure to the molecular level, the characterization technology of DOM has developed rapidly, from the initial characterization of spectroscopy and chromatographic techniques to the recent specific surface area, thermal analysis, electron, and atomic force microscopy, X-ray energy spectrum, and high-resolution mass spectrometry. Specifically, Zhang et al. (2009) used elemental analysis techniques to show that DOM contains

Table 1 Classification of DOM

Classification	Material composition	Methods of classification
Element	Organic nitrogen, organic phosphorus, and organic sulfides	Elemental analysis, EA
Chemical properties	1. Chemical bond (e.g., Carbon-hydrogen bond (C-H); Carbon-oxygen bond (C=O)) 2. Functional group (e.g., hydroxyl, carboxyl, and quinine) 3. Aromatic structure 4. Humification degree: humic substance (about 80% of DOM), protein-like substances, monosaccharides, and small molecule carboxylic acids	1. Fourier transform infrared (FTIR) spectrometry 2. FTIR; Three-dimensional fluorescence excitation emission matrix (3D-EEM) 3. Ultraviolet and visible spectrometry (UV-Vis) 4. 3D-EEM
Molecular weight	Large molecular weight DOM (colloid state: 5 kDa – 0.45 μm, accounting for 55–80%) and small molecular weight DOM (soluble state: <5 kDa)	Ultracentrifugation (UC) and Flow field flow fractionation (FFFF)
Properties of hydrophilic- hydrophobic and acid-base	Hydrophobic acidic organic matter (HPOA), hydrophobic neutral organic matter (HPON), hydrophobic basic organic matter (HPOB), hydrophilic acidic organic matter (HPIA), hydrophilic neutral organic matter (HPIN), and hydrophilic base organic matter (HPIB)	Porous resin column method: XAD-8 and XAD-4

carbon(C), hydrogen (H), oxygen (O), nitrogen (N), phosphorus (P), sulfur (S), and ash matter, and its element content is C > O > H > N > P > S. Elemental analysis involves the element category, while molecular spectra such as ultraviolet-visible spectrum, infrared spectrum, and fluorescence spectrum are mainly used to study the chemical bonds, functional groups, and internal structure of DOM. UV-Vis is used to characterize the content of aromatic compounds in conjugated systems. The aromaticity can be characterized by the parameter $SUVA_{280}$, whose value is high when there is a high degree of aromatization (Helms et al. 2008). FTIR is used to analyze chemical bonds (e.g., C-H and nitrogen-hydrogen bond (N-H)) and functional groups (e.g., hydroxyl and carboxyl groups) (Ouellet et al. 2008). 3D-EEMs, involving more dimensional information and more sensitive performance in detecting low-concentration organic matter, has been widely used to quantitatively analyze the composition of DOM in water bodies, revealing the dynamic characteristics of DOM in aquatic environments such as rivers, lakes, groundwater, oceans, and sewage (Huguet et al. 2010; Chen et al. 2016; Li et al. 2019). Parallel factor (PARAFAC) analysis is a very popular statistical method based on 3D-EEMs, which is used to analyze the fluorescence components and to characterize the structure, source, and destination of DOM through ecosystems and biogeochemical processes (Stedmon and Bro 2008; Kowalczuk et al. 2009; Zhou et al. 2016). Also, the PARAFAC analysis of fluorescence has been used to link DOM to its effectiveness as complexation agent with heavy metals and subsequently ascertain the migration and transformation mechanism of heavy metals affected by DOM in aquatic and soil ecosystem (Cuss et al. 2016). Another two quantitative techniques are fluorescence regional integration (FRI) and self-organizing map (SOM) analysis. FRI can be utilized to specify the area of fluorescence intensity for volume integration which can provide information on relative compositions of various types of DOM (Song et al. 2018, 2019). Compared to PARAFAC, SOM can be capable of analyzing DOM fluorescence comprehensively and to reveal the interaction between DOM factions and fluorescence groups, which have no distribution requirement for variables (Cuss and Guéguen 2016; Yang et al. 2019). However, due to the difficulty in implementation, SOM needs to be further optimized (Cuss and Guéguen 2016). Nuclear magnetic resonance spectroscopy (NMR) is mainly focused on the study of two types of nuclei, hydrogen and carbon. It can be used to quantitatively or qualitatively determine different types of hydrogen and carbon in DOM. It has similarities to infrared spectrometry, both of them can analyze the number and types of chemistry functional groups (Nebbioso and Piccolo 2013). Electron spin resonance (ESR) spectroscopy and electrochemical analysis methods can be used to characterize the electron transfer ability of DOM. Previous works have shown that the electron transfer ability of DOM is mainly affected by the quinone group in the DOM structure. However, there may exist other redox-active ingredients, such as redox-active atoms in heterocyclic structures (e.g., nitrogen and sulfur), and the specific reasons need to be further studied (Nurmi and Tratnyek 2002; Tang et al. 2018). The comprehensive use of the abovementioned characterization methods provides a technical guarantee for studying the environmental behaviors of DOM in adsorption, binding, complexation, and photochemical reaction.

3 Interaction Mechanism Between DOM and Microbial DIR

Iron-reducing microorganisms (including bacteria, fungi, and archaea) are commonly found in the anaerobic environment of nature, most of them are acidophilic and thermophilic, and even a small part can survive in extreme conditions (Lovley et al. 2004). Iron-reducing microorganisms can mediate the reduction of Fe(III) to Fe (II) as a driving force during the metabolic process. This process is considered as one of the earliest breathing methods on the earth and plays an important role in soil biogeochemical cycles (Tor and Lovley 2001). DOM can be used by microorganisms as an electron donor or an electronic shuttle in the process of microbial iron reduction, to attend the material cycle and energy metabolism. Its specific mechanism needs further research.

3.1 Mechanism of the DIR Process

The process of DIR is closely related to the fermentation reaction, respiration and productivity processes, and its reduction mechanism has been a hot research topic. There are differences in the reduction mechanism between soluble iron and insoluble iron oxides. Insoluble iron cannot enter the cells of microorganisms and can only transfer electrons to Fe(III) through protein substances. At present, there are four main explanations for the action mechanism of DIR: (1) Direct contact mechanism, that is, extracellular polymers substances (EPS) are formed by direct contact between iron-reducing bacteria and iron oxides, which can transfer electrons and reduce iron oxide (Richter et al. 2012). Bonneville et al. (2006) studied the kinetic model of the adsorption from *S. putrefaciens* to hematite colloidal particles, and the results showed that the outer membrane reductase of bacteria can transfer electrons to Fe(III) colloid. (2) Electron shuttle substance, that is, some small molecules oxidize reductive substances secreted by dissolved organic matter in soil, sediment or some small molecule redox substances during microbial metabolism, can act as an electron shuttle substance between the dissimilatory iron-reducing bacteria and Fe (III) compounds. Previous studies have shown that quinones secreted by *Shewanella oneidensis* function as electronic shuttles during the anaerobic respiration of iron minerals (Newman and Kolter 2000; Trump et al. 2006). (3) Cell appendages, that is, certain bacteria such as *Geobacter metallireducens* will produce appendages such as flagella, which have a tropism towards Fe(III) compounds and can transfer electrons to achieve DIR under certain environmental conditions. Nanowires are considered to be a special flagellar mechanism that uses insoluble iron oxides as electron acceptors to directly transfer electrons (Childers et al. 2002). (4) Chelation-promoting mechanism, soluble iron chelates are formed by small molecule chelates (e.g., Ethylene Diamine Tetraacetic Acid (EDTA) and polyphosphates) combined with insoluble iron oxides, which thereby promoting microbial iron reduction. It was reported that *Geothrix fermentans* and *Shewanella alga BrY* can secrete Fe(III) chelate (Nevin and Lovley 2002).

3.2 Effect of DIR on DOM Release

The transition between Fe(II) and Fe(III) plays an important role in dynamic environments during alternating reducing and oxidizing conditions. The oxidation of Fe(II) and resultant precipitation of Fe(III) oxyhydroxides with varying crystallinity occur in redox dynamic environments and are key processes determining the fate of contaminants and nutrients (Chen and Thompson 2018). Research by Sodano et al. (2017) showed that Fe(III) oxidation can form coprecipitation after combining with a large amount of dissolved organic carbon, thereby preventing DOC from being mineralized by microorganisms. The study of DOM release by DIR has attracted great interest from scholars for both of them can affect soil biogeochemical cycle. Pan et al. (2016) used dissolved organic carbon from different components of forest fresh leaf layer and forest humified layer to prepare carbon-ferrihydrite association. Anaerobic bacterial culture experiments suggested that the concentrations of forest fresh leaf layer DOC and forest humified layer DOC increased by 11.2 ($\pm$0.4)% and 13.3 ($\pm$0.3)%, respectively, during the iron reduction process mediated by *Shewanella oneidensis MR-1*. The 3D-EEMs characteristics showed that the fluorescence properties and intensity of DOM had changed. The study by Vermeire et al. (2019) showed that *Shewanella putrefaciens* mediated iron reduction significantly increased the release of dissolved organic matter in ash soil (soil age 270 years), which is consistent with the study of Pan et al. (2016).

3.3 Effect of DOM on DIR

To date, the effect of DOM on microbial DIR is mainly reflected in the following aspects (Table 2): (a) The combination of DOM and Fe(II) formed DOM-Fe(II) complexes which were potentially considered as crucial natural reductants, which was of great significantly to promote Fe(III) obtaining electron and then reduction

Table 2 Influential mechanism of DOM on the DIR

Mechanism	Process	References
Driving force	DOM + Fe(II) $\rightarrow$ DOM-Fe(II) The combination of DOM and Fe (II) enhances the driving force for Fe (III) reduction.	Royer et al. (2002), Lee et al. (2016), and Zhu et al. (2013)
Coprecipitation	DOM + Fe(III) $\rightarrow$ coprecipitation	Eusterhues et al. (2008), Satoh et al. (2009), Yang et al. (2018)
Complexation	DOM + Fe(III) $\rightarrow$ DOM-Fe(III) complexes	Jones et al. (2009), Morita et al. (2017), and Chen and Thompson (2018)
Electron shuttles	Fe(III) $\leftrightarrow$ DOM $\leftrightarrow$ Fe(II) DOM, served as electron shuttles, was used by microorganism to affect DIR	Jiang and Kappler (2008), Trump et al. (2006), Huang et al. (2010), and Meng et al. (2018)

(Royer et al. 2002; Zhu et al. 2013). (b) Coprecipitation occurs by DOM binding with iron compounds widely exist in soil or sediments, controlling the structure and reactivity of natural iron oxides (Eusterhues et al. 2008). Precipitation formed when DOM was added to soil or sediments, which convert active iron ions in the soil into stable precipitation. However, another form different from coprecipitation was that (c) DOM can also be used as a complexing agent in natural environment due to the various types of organic acids, which can not only complex with heavy metals but also occur coordination reaction with active metal elements such as iron and manganese. The coordination reaction between DOM and Fe(III) can transform insoluble Fe(III) minerals to Fe(III) with low crystallinity (e.g., pyrite) (Jones et al. 2009; Chen and Thompson 2018). (d) The DOM structure contains quinones, which can act as electron shuttles (i.e., redox properties) to accelerate the process of DIR, which have dual attributes including electron acceptor and electron donor (Jiang and Kappler 2008; Trump et al. 2006).

In recent years, the electron transfer ability (redox properties) of DOM, which may determine the fate of some contaminants, especially involving the DIR, has gradually become a research hotspot (Meng et al. 2018; Wu et al. 2020). Bauer et al. (2007) used chemical reducing agents (Zn and hydrogen sulfide (H_2S)) and oxidants (trivalent iron complexes) to study the electron transfer capabilities of humic substance (DOM model) under different pH and complexed Fe(III) conditions. The results showed that the redox buffer range of DOM is $-0.9 \sim 1.0$ V, and thus the concept of DOM as a redox buffer was proposed. DOM affecting iron reduction under aerobic environment is generally considered as a chemical reaction. However, in anaerobic habitat, a large amount of DOM existing on the surface and naturally soluble humic substances can be used by iron-reducing bacteria, fermenting iron-reducing bacteria, methanogens, and other types of strains regarded as electron acceptors, to produce the enzymatic reactions driven by specific microorganisms, that accelerate the process of iron DIR (Lovley et al. 1999; Benz et al. 1998; Klüpfel et al. 2014). For example, Wolf et al. (2009) studied the role of DOM models (humic acid (HA), fulvic acid (FA)) and quinones in the iron reduction process mediated by *Geobacter metallireducens*. The results showed that HA has a stronger stimulating effect on iron reduction than FA. The kinetics of microbial DIR is mainly controlled by the redox potential of the electron shuttle, rather than dissolution and adsorption. Also, the concurrent function of DOM (electron shuttling) was proposed by the study of Valenzuela et al. (2019), in which anoxic sediment incubations performed with both iron ore and humic substances were seven times faster of anaerobic oxidation of methane rate than that only supplied with iron ore. Hence, DIR affected by DOM cannot only be a simple chemical process involved but also containing the interpenetrating effect of biology and chemistry. Exploring the electron transfer process and interaction mechanism between DOM-microbial DIR provide a theoretical basis for a deeper understanding of soil heavy metal pollution fate and elementary geochemical cycle.

4 Effect of DIR on Bioavailability of Heavy Metals

Microorganisms can transport electrons to toxic heavy metals in the process of reducing iron oxides, which significantly affect the bioavailability of heavy metals. This fact can be used to repair heavy metal pollution in soil consequently. Cr usually exists in the environment as trivalent chromium (Cr(III)) and hexavalent chromium (Cr(VI)). The toxicity of chromium is closely related to its valence. Studies have shown that both Fe and Cr can serve as electron acceptors under soil anaerobic conditions, and the reduction of microbial dissimilatory iron significantly affects the rate and extent of Cr(VI) to Cr(III) conversion (Huang et al. 2016). In addition, Wrighton et al. (2011) reported that under the mixed culture of chlorite, montmorillonite, and Cr (VI) rich in Fe(III), Cr_2O_3 produced by reduction of Cr(VI) can be combined with clay minerals. Once the agglomerate structure formed, Cr(III) was difficult to oxidize. Therefore, microbial iron reduction is considered as one of the detoxification methods of Cr(VI). On the contrary, microbial DIR may release As (V) in the soil and promote the conversion of pentavalent arsenic (As(V)) to highly toxic trivalent arsenic (As(III)).

Yamaguchi et al. (2011) studied the release of As in the soil under flooding conditions. The results showed that the redox potential, the change in pH, and the amount of iron compounds controlled the release of As existing in the form of As (III). Moreover, compared to oxic environments, As can be released from As-Fe coprecipitate mineral to a higher level in a strong iron-reducing environment with the induction of hydroquinone at alkalescent pH, as shown in the study of Yuan et al. (2018c). Although there is evidence demonstrating that the conversion of As (V) to As(III) was thermodynamically difficult, the dissolution of As (V) undoubtedly increases the ecological risk of arsenic contamination in water or soil. Kulkarni et al. (2018) found that FA (a fraction of DOM) derived from Bangladesh groundwater was highly aromatic, and therefore showed a highly electron-accepting capacity, which stimulates DIR, accelerating the release of dissolved arsenic in reducing aquifers. The more compelling evidence was proposed by Chen et al. (2017), who found that anthraquinone-2,6-disulphonate (AQDS) (DOM analog) substantially enhanced the concentrations of As(III) up to an order of magnitude higher than in the absence of AQDS. Hence, DOM is an indispensable indicator of affecting DIR and As mobilization, which provides crucial implications for our understanding of DIR mediating the release of As. In another contribution, Wang et al. (2017) show that addition of biochar caused 19.0-fold increase of slurries of respiratory As reducing genes, which indicating that microbial reduction of As(V) to As(III) during DIR process occurred, further contributing to the increase in the release of As. Previous studies have reported that the microorganisms of DIR play important role in As mobilization during the DIR process (Guo et al. 2013; Dai et al. 2016; Xue et al. 2020). The species of DIR bacteria or EPS can be the determinants, for example, EPS (mannan) excreted by *Shewanella putrefaciens* IAR-S1 and *Shewanella xiamenensis* IR-S2 suppressed As mobilization while EPS (succinic acid) excreted by *Klebsiella oxytoca* IR-ZA enhanced (Liu et al. 2020). It can be seen that the mobilization of arsenic is not just a single consequence of iron reduction, the influence of pH, DOM, and microorganisms cannot be ignored.

Hg in the soil, generally, can be divided into inorganic mercury such as mercury sulfate ($HgSO_4$), mercury chloride ($HgCl_2$), and more toxic organic mercury (such as Monomethylmercury (CH_3Hg)). Most of the inorganic Hg will be converted into methylmercury (MeHg) after entering the soil solution by the influence of soil organic matter and microorganisms, which will increase its toxicity in soil and water (Li et al. 2019; Han et al. 2014). Hg-methylating is an anaerobic microbial process that is usually driven by dissimilatory sulfate-reducing bacteria. However, Kerin et al. (2006) found that many dissimilatory iron-reducing bacteria such as *Geobacter metallireducens* can utilize Fe(III), nitrate, and fumarate significantly affecting the methylation of Hg. The methylation efficiency is the highest, reaching about 14.5%, when Fe(III) is used as the electron acceptor (Kerin et al. 2006). Moreover, iron sulfide (FeS) also has a good ability to remove inorganic Hg(II), which can completely remove Hg after sufficient contact time in the range of Hg (II) concentration of 500–1,250 $mmol \cdot L^{-1}$. The reaction rate ranges from fast to slow, from initial rapid adsorption to subsequent precipitation or polymerization (Han et al. 2014).

The migration and fixation of Cd in paddy soil have been a hot topic of research. Fe(III), Mn (IV), nitrate ion (NO_3^-), and sulfate ion (SO_4^{2-}) in the soil can be used as electron acceptors. They participate in the redox process, which significantly influences the mobility of cadmium in soil (Yuan et al. 2019). Dissimilated iron reduction has a dual character in cadmium migration (Yu et al. 2016). It has been reported that iron reduction may lead to the rerelease of cadmium adsorbed on the mineral surface, and recent studies have shown that after 20 days of contact, the secondary iron minerals formed will fix cadmium through adsorption or coprecipitation (Li et al. 2016). The key role in this process is played by dissimilatory reducing bacteria and soil organic matter that can provide carbon source. Research by Muehe et al. (2013) showed that adding lactic acid and acetate to the iron-rich soil under anaerobic conditions can promote the speed of microbial reducing iron and reduce the dissolved cadmium content. While in the control group, the reduction of dissolved cadmium is much smaller.

The microbial DIR not only affects the bioavailability of common heavy metals (e.g., Cr, As, Cd, Cu, and Hg) in the soil, but also affects some trace elements and radionuclides. As shown in previous research, the microbial reducing bacteria (*Methanosarcina mazei*) were inoculated into media with V(V) concentrations of 2, 5, and 10 $mmol \cdot L^{-1}$ (Zhang et al. 2014). After a period of chlorite decreasing, V(V) was almost completely reduced to V(IV). It has also been reported that under water environmental conditions, dissimilatory iron-reducing bacteria can effectively control the migration and diffusion of radionuclide uranium (Holmes et al. 2002). The mechanism of DIR on the bioavailability of soil heavy metals can be roughly summarized as three aspects (Fig. 1): (1) Redox; (2) methylation; (3) adsorption and coprecipitation, the mechanism of which is affected by soil DOM, pH, and Eh. There are differences in the toxicity of different heavy metals influenced by microbial DIR. The effects of iron reduction on the bioavailability of soil heavy metals should consider the types of heavy metals, and the soil physical and chemical indicators.

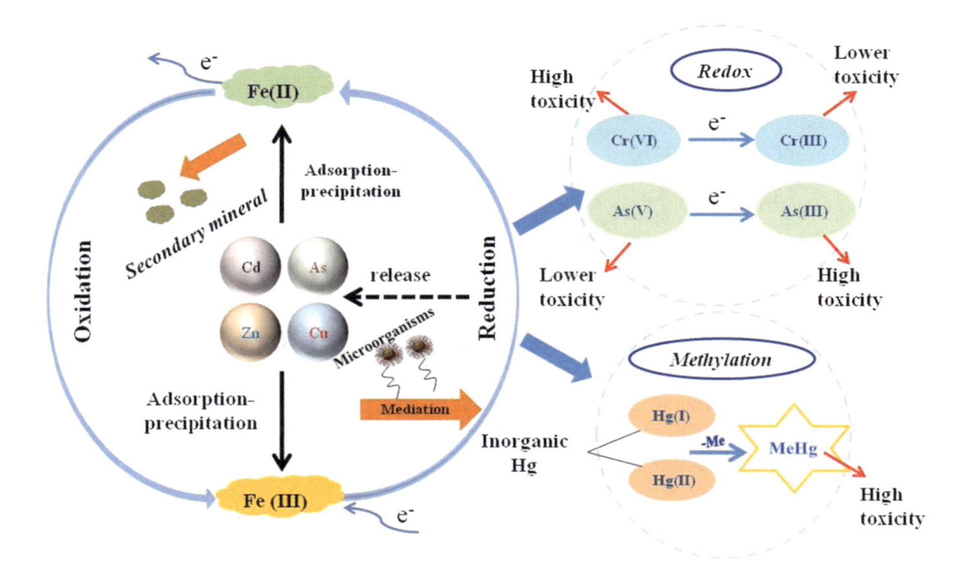

Fig. 1 Influential mechanism of DIR on heavy metals

5 Effects of DOM on the Bioavailability of Heavy Metals and Its Mechanism

5.1 *Effect of DOM on Heavy Metal (Cr)*

The environmental behavior of DOM on heavy metal chromium in soil mainly includes adsorption behavior and morphological transformation. Gong et al. (2015) investigated the kinetics of toxic Cr(VI) removal by DOM in activated sludge and results showed that the total organic carbon (TOC) concentration increased rapidly from 50.93 $mg{\cdot}L^{-1}$ to 127.40 $mg{\cdot}L^{-1}$ during the Cr(VI) removal process in the pH range of 2–9. With the increase of the initial pH, however, the removal efficiency is greatly decreased. The removal of Cr(VI) is a combination of biological adsorption and biological reduction. In addition, ultraviolet light may induce DOM to produce reaction intermediates such as molecular oxygen free radicals ($O_2\bullet$) and DOM free radicals (DOM*) as DOM excited by the absorption of light to enhance reduction of Cr(VI). Dong et al. (2014) used soil DOM as a control to evaluate the impact of DOM from two types of biochars (sugar beet tailing and Brazilian pepper) on Cr(VI) reduction in ice and water phases. The study found that Cr (VI) reduction decreased from 80 to 86% to negligible with increasing pH from 2 to 10. The results were similar to those obtained by Gong et al. (2015), who found that the increase in pH is an inhibitory indicator on the removal of Cr(VI). In complex ecosystems of soils, the reduction of Cr often does not consist of a single redox system. It was found that the concentration of sodium formate and model DOM (anthraquinone-2-sulfonate (AQS)) had apparently effect on Cr(VI) reduction in the anaerobic system of quinone-reducing bacteria/dissolved organic matter (DOM)/Fe(III). At the same time, the coexisting cycle of Fe(II)/Fe(III) and DOM (ox)/DOM (red) showed a higher redox function than a single cycle, and their abiotic coupling can significantly enhance the reduction of Cr(VI) by quinone-reducing bacteria (Huang et al. 2016).

5.2 *Effect of DOM on Heavy Metal (As)*

The toxicity of heavy metal As in soil is different from Cr. The toxicity of As(III) is much higher than that of As(V). Oxidation and methylation of arsenide can reduce the toxicity of arsenic, and its transformation is affected by soil DOM. Research has indicated that the oxidation efficiency of arsenic by DOM raises with increasing pH. Electron spin resonance and Fourier transform infrared analysis showed that the functional groups affecting As(III) oxidation are the semiquinone radicals and hydroxyl radicals in DOM, respectively (Dong et al. 2014). As(V) can be reduced by microbial action with the DOM acting as electron donors, when As(V) exists in the soil in large amounts. Chen et al. (2017) found that a more positive effect resulting from the mixed system with 0.05 mmol·L^{-1} AQDS and sodium acetate (NaAc) compared to other gradient levels (0.10 and 1.00 mmol·L^{-1}) of AQDS, whereas an inhibitory effect was observed with 1.00 mmol·L^{-1}. Furthermore, Multiple-dynamic effects of "bacteria-AQDS-DOM," which result from AQDS, shifted the microbial community. High-throughput sequencing results indicated that the addition of 0.05 mmol·L^{-1} and 0.1 mmol·L^{-1} AQDS formed microbial community with the majority of *Bacillus* and *Clostridium*, while *Desulfitobacterium* and *Alicyclobacillus* is the majority with the addition of 1 mmol·L^{-1} AQDS. In addition, DOM plays a vital role in the methylation of arsenic in paddy soils. Paddy soils are widely studied by scientists as a model system for heavy metal methylation. Studies have demonstrated that different high concentrations of DOC can increase the arsenic availability of methylated bacteria in paddy soils. On the other hand, DOM can be an effective carbon and nitrogen source for microorganisms, affecting the soil microbial community structure (Goldberg et al. 2017). DOM from different sources can be used as a nutrient for microbial growth (including arsenic methylase), both factors contribute to arsenic volatilization (Huang et al. 2012).

5.3 *Effect of DOM on Heavy Metal (Hg)*

In natural environments such as sediments, wetlands, and paddy soils, mercury can easily form Hg-DOM complexes with DOM. Hg-DOM complexes may affect Hg morphology, bioavailability, transport, and methylation (Mazrui et al. 2016). The Hg-DOM complexes caused decrease production of MeHg by 80% by *Geobacter sulphreducens* (methylated bacteria), which was attributed to the formation of Mercury sulfide (HgS) by Hg (II) and sulfite (S^{2-}). Although there are a large number of oxygen-containing functional groups such as carboxylic acids in the DOM, mercury will preferentially bind to trace amounts of thiols and other sulfur-containing groups present in organics (Ravichandran 2004). DOM can form strong complexation with Hg (II) driven by reducing sulfur or sulfhydryl functional groups, mediating the transfer of electrons and the redox effect with mercury (Lee et al. 2018). The ratio of Hg/DOM is also an important indicator to consider the activity of mercury in water. In fact, the Hg-DOM photochemical reaction may be more complete at a lower Hg–DOM ratio in a natural aquatic environment (Luo et al. 2017). Microorganisms are inclined to utilize freshly deposited Hg (II) to form MeHg in a short period of time, but the longer Hg remains in the aquatic environment, the less likely it is to be involved in methylation and bioaccumulation (Luo et al. 2017; Jonsson et al. 2014). Li et al. (2019) studied the alleviating effects of five DOM model samples on the toxicity of MeHg in embryonic zebrafish, the results indicated the function and composition of DOM significantly affected the alleviating effect of MeHg on teratogenicity of zebrafish embryos. Parallel factor analysis showed that DOM model of humus-like, large molecular weight, and hydrophobic (thiosalicylic acid, glutathione, humic acid) had the best mitigative effects. Therefore, the effects of DOM on mercury can be summarized as follows: mercury combines with sulfur-containing groups in DOM; mercury forms Hg-DOM complexes with DOM; DOM and mercury undergo a photochemical reaction to reduce Hg^{2+} and alleviate the toxicity to MeHg.

5.4 *Effect of DOM on Heavy Metal (Cu)*

Copper is different from the above heavy metals (Cr, As, Hg) in that it is a trace element of plants. Excessive copper concentration will negatively impact the metabolism of soil microorganisms and seed germination, which is toxic to plants (Araújo et al. 2019; Zitoun et al. 2019). Compost is a cost-effective soil improver and has been widely used in the rehabilitation of heavy metal contaminated soil. Studies have shown that the application of cattle manure can lead to an increase of Cu and dissolved organic matter in the soil. It was found, after 3 years of fertilizing soil observations, that DOC is a key factor affecting copper migration in alkaline soil modified by cattle manure. Once a highly stable Cu-DOC complex is formed, Cu will have greater potential to migrate to deeper soil profiles. In addition, the UV-visible spectral parameters ($SUVA_{254}$ and E_2/E_3) of DOC both showed significant correlation with the copper ion concentration thus illustrating the importance of dissolved organic matter in the copper migration process. Therefore, the ecological risks caused by the migration of Cu should be considered when applying amendments to the soil (Araújo et al. 2019). In contrast, other studies have reported a high binding affinity between copper and certain components of DOC. Absorption of copper by aquatic organisms and the toxicity of copper were weakened in this process (Cooper et al. 2014). Zitoun et al. (2019) evaluated the effects of salinity and dissolved organic matter as the two main driving factors impacting copper morphology and copper toxicity on the bioavailability of blue mussel embryos

Mytilus galloprovincialis in a 48-h bioassay. The results illustrated that DOC is positively correlated with the half-maximum effect concentration of copper for 48 h ($[Cu_{48\text{-h-EC50}}]$), and its equation can be expressed as $[Cu_{48\text{-h-EC50}}] = 121.72 \times DOC^{0.6557}$. The complexation of copper and DOM has shown the contradiction between enhanced copper migration and suppressed biological toxicity in different studies.

5.5 Mechanism of DOM on Bioavailability of Heavy Metals

The fundamental reason why DOM affects the bioavailability of heavy metals is that it contains active functional groups such as carboxyl, hydroxyl, carbonyl, quinone, and methoxy groups, which can engage in a series of reactions such as ion exchange adsorption, complexation, hydrogenation, redox and methylation with heavy metal ions in water, soil, and sediment (Fig. 2), fixing or releasing trace metal elements (Gong et al. 2015; Araújo et al. 2019; Yuan et al. 2018b), or affecting the migration and transformation of heavy metals through adsorption and analysis with common metal oxides (e.g., iron oxide, aluminum oxide) (Yuan et al. 2019; Gaberell et al. 2003; Yan et al. 2016).

The bioavailability mechanism of heavy metals is multifaceted, which is affected by the nature of the soil DOM (e.g., functional groups, molecular weights, and chemical bonds), for example, Wang et al. (2010) utilized ultrafiltration technology to separate DOM according to molecular weight. The study showed that DOM with molecular weight less than 1,000 Da can increase the bioavailability of Cu to lettuce seedlings, while DOM with molecular weight greater than 1,000 Da has the opposite

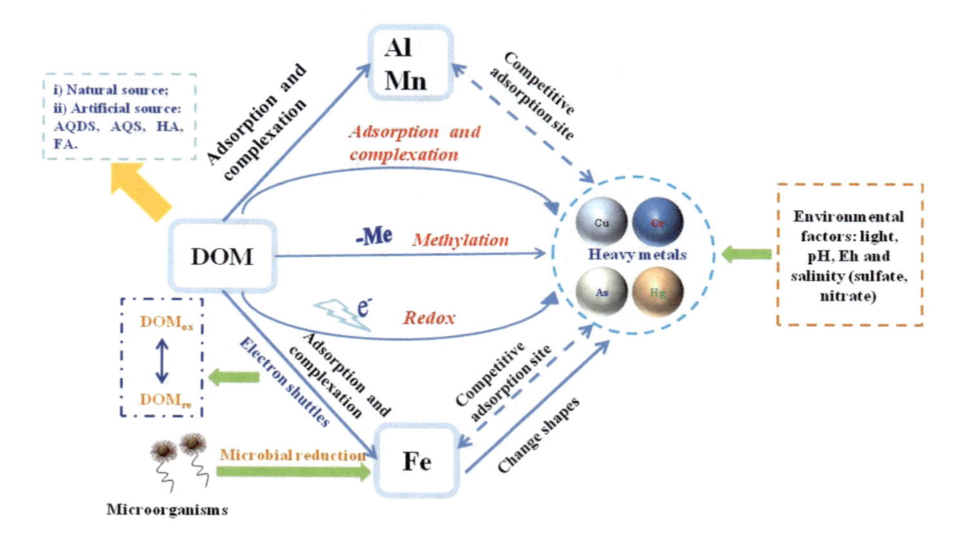

Fig. 2 Influential mechanism of DOM on heavy metals

effect (Yan et al. 2016; Wang et al. 2010). Valence and morphology of heavy metals (water-soluble state> exchangeable state> carbonate bound state> Fe-Mn oxide bound form> organic bound state> residual state) also play an important role (Gaberell et al. 2003; Fuentes et al. 2008). Furthermore, the impact of DOM on the fixation or release of heavy metals is also affected by soil physical and chemical properties (pH, Eh, salinity, and temperature) and light conditions (Fulda et al. 2013; Xu et al. 2015; Kim et al. 2019). In addition to the role of endogenous environmental factors in the bioavailability of soil heavy metals, organic fertilizer or biochar ingested externally is also an important way to change the migration and bioavailability of heavy metals. Complexation of DOM released by biochar with soil heavy metals may influence the effectiveness of soil heavy metal repair (Xu et al. 2015). Liu et al. (2019) studied the complexation of heavy metals Hg with DOM released by 36 kinds of biochar. The model was established with main functional groups in DOM (thiol, carboxyl, and phenol), and results showed that more than 99% of mercury complex with thiol in DOM, which undoubtedly changed the form and bioavailability of mercury. The study also showed that the concentration of Hg-DOM complexes is lower in wood substrate than in biochar in agricultural residues and feces, which provides a new theoretical basis for the use of biochar for soil remediation (Liu et al. 2019). It is worth noting that the mechanism of DOM on heavy metals of the same type is also inconsistent. Studies focusing on the differences in heavy metal types deem that the difference in the effect of DOM on the bioavailability of heavy metals depends more on whether ternary complexes are formed on the surface of biological ligands, but not all metal ions can form these ternary complexes. Study of Lamelas et al. (2005) substantiated that DOM promotes the absorption of lead by *Chlorella kesslerii* cells because the formed Pb-DOM complex increases the hydrophobicity of lead and continues to form a ternary complex of Pb-DOM-algae on the surface of algae, thereby enhancing the penetrability of lead in biofilms. A similar result was also found by Shi et al. (2017). However, under the same conditions, copper and cadmium only formed the Cu-DOM and Cd-DOM binary complexes with DOM (Li et al. 2013; Yuan et al. 2015; Shi et al. 2017). At present, there are few explorations on this complicated complexation mechanism, which needs further verification. Studying the complexation relationship of DOM to heavy metals is the key to reveal how heavy metals reach the plant rhizosphere environment and change their bioavailability.

6 Conclusion

DOM is a particularly active organic component in the soil environment. It is a mixture of complex structures and interactions and plays an important role in the soil ecosystem. This chapter reviewed the sources, classification, structural composition, and characterization methods of DOM, which may help reveal the implication of DOM in soil carbon cycling and the migration and transformation behavior of soil heavy metals after DOM-heavy metal compounding. Another significant process (i.e., microbial dissimilatory iron reduction) is simultaneously considered. The interaction between DOM and DIR is closely related to soil biogeochemical cycle. A large amount of carbon-ferrihydrite associations present in the soil will cause the release of DOC during DIR, which impact on the stabilization and accumulation of soil DOM. However, the endogenous DOM and externally added DOM (organic fertilizer or straw) also affect the reduction of microbial dissimilation iron, which is specifically manifested in the driving force properties, coprecipitation, complexation, and electronic shuttle properties of DOM. Future research is expected to fill the knowledge gap on the following aspects:

Although abundant functional groups (e.g., quinone, semiquinone) in DOM provide electrons and driving force for the reduction of microbial DIR, the electron transfer capability of DOM is rarely quantified accurately. Integrating high-resolution mass spectrometry techniques such as Fourier transform ion cyclotron resonance mass spectrometry (FT-ICR-MS) with electrochemical method to analyze the functional group structure and electron supply is a development trend to study the mechanism of DOM on DIR.

Both DOM and DIR will affect the environmental behavior of soil heavy metals. Dynamically reaction can be produced by DOM and different heavy metals through adsorption fractionation, coprecipitation, methylation, and redox. Microorganisms are the key media for soil DIR, and the activity and metabolism of microorganisms are usually restricted by the toxicity of heavy metals. Therefore, it is worth paying attention to the differences in the abundance, structure, and function of microbial communities in the rhizosphere of plants with DOM intake in heavy metal contaminated soil. It is necessary to further study whether heavy metals directly affect iron reduction or indirectly affect iron reduction by inhibiting microbial activity.

It is increasingly recognized that the iron cycle in wetland soil and sediment, especially in plant rhizosphere, and its coupling process with heavy metals are affected by changes in hydrological conditions. Therefore, there is a need to explore the mechanism of DOM-bacteria of DIR-heavy metal ternary combination on the bioavailability of heavy metals in rhizosphere under the dynamic change of water level from the perspective of the differences in DOM properties such as polarity, molecular weight, and functional group, as a new type of soil heavy metal pollution control and ecosystem restoration.

Acknowledgments The financial support of this study from the National Natural Science Foundation of China (No. 41761095) is gratefully acknowledged.

Conflict of Interest The authors declare that they have no conflict of interest.

References

Araújo E, Strawn DG, Morra M, Moore A, Ferracciú ALR (2019) Association between extracted copper and dissolved organic matter in dairy-manure amended soils. Environ Pollut 246:1020–1026

Bai HC, Wei SQ, Jiang ZM, He MJ, Ye BY, Liu GY (2019) Pb (II) bioavailability to algae (*Chlorella pyrenoidosa*) in relation to its complexation with humic acids of different molecular weight. Ecotoxicol Environ Saf 167:1–9

Bauer M, Heitmann T, Macalady DL, Blodau C (2007) Electron transfer capacities and reaction kinetics of peat dissolved organic matter. Environ Sci Technol 41(1):139–145

Benner R, Amon RMW (2015) The size-reactivity continuum of major bioelements in the ocean. Annu Rev Mar Sci 7:185–205

Benz M, Schink B, Brune A (1998) Humic acid reduction by *Propionibacterium freudenreichii* and other fermenting bacteria. Appl Environ Microbiol 64(11):4507–4512

Bonneville S, Behrends T, Van Cappellen P, Hyacinthe C, Röling WFM (2006) Reduction of Fe (III) colloids by *Shewanella putrefaciens*: a kinetic mode. Geochim Cosmochim Acta 70:5842–5854

Borggaard OK, Holm PE, Strobel BW (2019) Potential of dissolved organic matter (DOM) to extract As, Cd, Co, Cr, Cu, Ni, Pb and Zn from polluted soils: a review. Geoderma 343:235–246

Chantigny MH (2003) Dissolved and water-extractable organic matter in soils: a review on the influence of land use and management practices. Geoderma 113(3–4):357–380

Chen C, Thompson A (2018) Ferrous iron oxidation under varying pO_2 levels: the effect of Fe(III)/Al(III) oxide minerals and organic matter. Environ Sci Technol 52(2):597–606

Chen XS, Jiang T, Lu S, Wei SQ, Wang DY, Yan JL (2016) Three-dimensional fluorescence spectral characteristics of different molecular weight fractionations of dissolved organic matter in the water-level fluctuation zones of three gorges reservoir areas. Environ Sci 37(3):884–892. (in Chinese)

Chen Z, Wang Y, Jiang X, Fu D, Xia D, Wang HT, Dong GW, Li QB (2017) Dual roles of AQDS as electron shuttles for microbes and dissolved organic matter involved in arsenic and iron mobilization in the arsenic-rich sediment. Sci Total Environ 574(574):1684–1694

Childers SE, Ciufo S, Lovley DR (2002) *Geobacter metallireducens* accesses insoluble Fe(III) oxide by chemotaxis. Nature 416:767–769

Cinzia DV, Alessandro P, Serena FU (2008) Dissolved organic carbon variability in a shallow system (Gulf of Trieste, northern Adriatic Sea). Estuar Coast Shelf Sci 78(2):280–290

Cooper CA, Tait T, Gray H, Cimprich G, Santore RC, McGeer JC, Wood CM, Smith DC (2014) Influence of salinity and dissolved organic carbon on acute Cu toxicity to the rotifer *Brachionus plicatilis*. Environ Sci Technol 48(2):1213–1221

Cuss CW, Guéguen C (2016) Analysis of dissolved organic matter fluorescence using self-organizing maps: mini-review and tutorial. Anal Methods 8(4):716–725

Cuss CW, McConnell SM, Guéguen C (2016) Combining parallel factor analysis and machine learning for the classification of dissolved organic matter according to source using fluorescence signatures. Chemosphere 155:283–291

Dai X, Li P, Tu J, Zhang R, Wei D, Li B, Wang Y, Jiang Z (2016) Evidence of arsenic mobilization mediated by an indigenous iron reducing bacterium from high arsenic groundwater aquifer in Hetao Basin of Inner Mongolia, China. Int Biodeterior Biodegradation 128:22–27

Dong X, Ma LQ, Gress J, Harrisa W, Li YC (2014) Enhanced Cr(VI) reduction and As(III) oxidation in ice phase: important role of dissolved organic matter from biochar. J Hazard Mater 267:62–70

Eusterhues K, Wagner FE, Hausler W, Hanzlik M, Knicker H, Totsche KU, Gel-Knabner IK, Schwertmann U (2008) Characterization of ferrihydrite-soil organic matter coprecipitates by X-ray diffraction and Mossbauer spectroscopy. Environ Sci Technol 42:7891–7897

Fuentes A, Lloréns M, Sáez J, Isabel AM, Ortuño JF, Meseguer VF (2008) Comparative study of six different sludges by sequential speciation of heavy metals. Bioresour Technol 99:517–525

Fulda B, Voegelin A, Kretzschmar R (2013) Redox-controlled changes in cadmium solubility and solid-phase speciation in a paddy soil as affected by reducible sulfate and copper. Environ Sci Technol 47:12775–12783

Gaberell M, Chin YP, Hug SJ, Sulzberger B (2003) Role of dissolved organic matter composition on the Photoreduction of Cr(VI) to Cr(III) in the presence of iron. Environ Sci Technol 37 (19):4403–4409

Goldberg SJ, Nelson CE, Viviani DA, Shulse CN, Church MJ (2017) Cascading influence of inorganic nitrogen sources on DOM production, composition, lability and microbial community structure in the open ocean. Environ Microbiol 19:3450–3464

Gong YF, Song J, Ren HT, Han X (2015) Comparison of Cr(VI) removal by activated sludge and dissolved organic matter (DOM): importance of UV light. Environ Sci Pollut Res 22 (23):18487–18494

Gu LP, Huang B, Xu ZX, Ma XD, Pan XJ (2016) Dissolved organic matter as a terminal electron acceptor in the microbial oxidation of steroid estrogen. Environ Pollut 218:26–33

Guo H, Liu C, Lu H, Wanty RB, Wang J, Zhou Y (2013) Pathways of coupled arsenic and iron cycling in high arsenic groundwater of the Hetao basin, Inner Mongolia, China: an iron isotope approach. Geochim Cosmochim Acta 112:130–145

Han DS, Orillano M, Khodary A, Duan YH, Batchelor B, Abdel-Wahab A (2014) Reactive iron sulfide (FeS)-supported ultrafiltration for removal of mercury (Hg (II)) from water. Water Res 53:310–321

He QX, Yu LP, Li JB, He D, Cai XX, Zhou SG (2019) Electron shuttles enhance anaerobic oxidation of methane coupled to iron(III) reduction. Sci Total Environ 688:664–672

Hedges JI, Eglinton G, Hatcher PG, Kirchman DL, Arnosti C, Derenne S, Evershed RP, Kögel-Knabner I, de Leeuw JW, Littke R, Michaelis W, Rullkötter J (2000) The molecularly-uncharacterized component of nonliving organic matter in natural environments. Org Geochem 31:945–958

Hellal J, Guédron S, Huguet L, Schäfer J, Laperche V, Joulian C, Lanceleur L, Burnol A, Ghestem JP, Garrido F, Battaglia-Brunet F (2015) Mercury mobilization and speciation linked to bacterial iron oxide and sulfate reduction: a column study to mimic reactive transfer in an anoxic aquifer. J Contam Hydrol 180:56–68

Helms JR, Stubbins A, Ritchie JD (2008) Absorption spectral slopes and slope ratios as indicators of molecular weight, source, and photobleaching of chromophoric dissolved organic matter. Limnol Oceanogr 53(3):955–969

Hill JR, O'Driscoll NJ, Lean DRS (2009) Size distribution of methylmercury associated with particulate and dissolved organic matter in freshwaters. Sci Total Environ 408(2):408–414

Holmes DE, Finneran KT, O'Neil RA, Lovley DR (2002) Enrichment of members of the family *Geobacteraceae* associated with stimulation of dissimilatory metal reduction in uranium-contaminated aquifer sediments. Appl Environ Microbiol 68(5):2300–2306

Huang DY, Zhuang L, Cao WD, Xu W, Zhou SG, Li FB (2010) Comparison of dissolved organic matter from sewage sludge and sludge compost as electron shuttles for enhancing Fe(III) bioreduction. J Soils Sediments 10:722–729

Huang H, Jia Y, Sun GX, Zhu YG (2012) Arsenic speciation and volatilization from flooded paddy soils amended with different organic matters. Environ Sci Technol 46:2163–2168

Huang B, Gu LP, He H, Xu ZX, Pan XJ (2016) Enhanced biotic and abiotic transformation of Cr (VI) by quinone-reducing bacteria/dissolved organic matters/Fe(III) in anaerobic environment. Environ Sci Process Impacts 18:1185–1192

Huguet A, Vacher L, Saubusse S (2010) New insights into the size distribution of fluorescent dissolved organic matter in estuarine waters. Org Geochem 41(6):595–610

Impellitteri CA, Lu YF, Saxe JK, Allen HE, Peijnenburg WJGM (2002) Correlation of the partitioning of dissolved organic matter fractions with the desorption of Cd, Cu, Ni, Pb and Zn from 18 Dutch soils. Environ Int 28:401–410

Jiang J, Kappler A (2008) Kinetics of microbial and chemical reduction of humic substances: implications for electron shuttling. Environ Sci Technol 42:3563–3569

Jones AM, Collins RN, Rose J, Waite TD (2009) The effect of silica and natural organic matter on the Fe(II)-catalysed transformation and reactivity of Fe(III) minerals. Geochim Cosmochim Acta 73:4409–4422

Jonsson S, Skyllberg U, Nilsson MB, Lundberg E, Andersson A, Björn E (2014) Differentiated availability of geochemical mercury pools controls methylmercury levels in estuarine sediment and biota. Nat Commun 5:4624

Kalbitz K, Solinger S, Park JH, Michalzik B, Matzner E (2000) Controls on the dynamics of dissolved organic matter in soils: a review. Soil Sci 165(4):277–304

Kerin EJ, Gilmour CC, Roden E, Suzuki MT, Coates JD, Mason RP (2006) Mercury methylation by dissimilatory iron-reducing bacteria. Appl Environ Microbiol 72(12):7919–7921

Kim HB, Kima JG, Choi JH, Kwon E, Baek K (2019) Photo-induced redox coupling of dissolved organic matter and iron in biochars and soil system: enhanced mobility of arsenic. Sci Total Environ 689:1037–1043

Klüpfel L, Keiluweit M, Kleber M, Sander M (2014) Redox properties of plant biomass-derived black carbon (biochar). Environ Sci Technol 48(10):5601–5611

Kowalczuk P, Durako MJ, Young H (2009) Characterization of dissolved organic matter fluorescence in the South Atlantic bight with use of PARAFAC model: interannual variability. Mar Chem 113(3/4):182–196

Kulkarni HV, Mladenov N, McKnight DM, Zheng Y, Kirk MF, Nemergut DR (2018) Dissolved fulvic acids from a high arsenic aquifer shuttle electrons to enhance microbial iron reduction. Sci Total Environ 615:1390–1395

Lamelas C, Wilkinson KJ, Slaveykova VI (2005) Influence of the composition of natural organic matter on Pb bioavailability to microalgae. Environ Sci Technol 39(16):6109–6116

Lee YP, Fujii M, Terao K, Kikuchi T, Yoshimura C (2016) Effect of dissolved organic matter on Fe (II) oxidation in natural and engineered waters. Water Res 103:160–169

Lee S, Kim DH, Kim KW (2018) The enhancement and inhibition of mercury reduction by natural organic matter in the presence of *Shewanella oneidensis* MR-1. Chemosphere 194:515–522

Leenheer JA, Noyes TI, Rostad CE, Davisson ML (2004) Characterization and origin of polar dissolved organic matter from the Great Salt Lake. Biogeochemistry 69(1):125–141

Li T, Tao Q, Liang C, Shohag MJI, Yang XE, Sparks DL (2013) Complexation with dissolved organic matter and mobility control of heavy metals in the rhizosphere of hyperaccumulator *Sedum alfredii*. Environ Pollut 182:248–255

Li CC, Yi XY, Dang Z, Yu H, Zeng T, Wei CH, Feng CH (2016) Fate of Fe and Cd upon microbial reduction of Cd-loaded polyferric flocs by *Shewanella oneidensis MR-1*. Chemosphere 144:2065–2072

Li D, Xie LT, Carvan-III MJ, Guo LD (2019) Mitigative effects of natural and model dissolved organic matter with different functionalities on the toxicity of methylmercury in embryonic zebrafish. Environ Pollut 252:616–626

Liu P, Ptacek CJ, Blowes DW (2019) Mercury complexation with dissolved organic matter released from thirty-six types of biochar. Bull Environ Contam Toxicol 103:175–180

Liu H, Li P, Wang HL, Qin C, Tan T, Shi B, Zhang GL, Jiang Z, Wang YH, Hasan SZ (2020) Arsenic mobilization affected by extracellular polymeric substances (EPS) of the dissimilatory iron reducing bacteria isolated from high arsenic groundwater. Sci Total Environ 735:139501. https://doi.org/10.1016/j.scitotenv.2020.139501

Lovley DR, Fraga JL, Coates JD, Blunt-harris EL (1999) Humics as an electron donor for anaerobic respiration. Environ Microbiol 1(1):89–98

Lovley DR, Holmes DE, Nevin KP (2004) Dissimilatory Fe(III) and Mn (IV) reduction. Adv Microb Physiol 49:219–286

Luo HW, Yin X, Jubb AM, Chen HM, Lu X, Zhang WH, Lin H, Yu HQ, Liang LY, Sheng GP, Gu BH (2017) Photochemical reactions between mercury (Hg) and dissolved organic matter decrease Hg bioavailability and methylation. Environ Pollut 220:1359–1365

Ma HZ, Allen HE, Yin YJ (2001) Characterization of isolated fractions of dissolved organic matter from natural waters and a wastewater effluent. Water Res 35(4):985–996

Mao JD, Kong XQ, Schmidt-Rohr K, Pignatello JJ, Perdue EM (2012) Advanced solid-state NMR characterization of marine dissolved organic matter isolated using the coupled reverse osmosis/ electrodialysis method. Environ Sci Technol 46(11):5806–5814

Mazrui NM, Jonsson S, Thota S (2016) Enhanced availability of mercury bound to dissolved organic matter for methylation in marine sediments. Geochim Cosmochim Acta 194:153–162

McCurry DL, Speth TF, Pressman JG (2012) Lyophilization and reconstitution of reverse-osmosis concentrated natural organic matter from a drinking water source. J Environ Eng 138 (4):402–410

McDowell WH (2003) Dissolved organic matter in soils future directions and unanswered questions. Geoderma 113:179–186

Meng Y, Zhao ZW, Burgos WD, Li Y, Zhang B, Wang YH, Liu WB, Sun LJ, Lin LM, Luan FB (2018) Iron(III) minerals and anthraquinone-2,6-disulfonate (AQDS) synergistically enhance bioreduction of hexavalent chromium by *Shewanella oneidensis MR-1*. Sci Total Environ 640–641:591–598

Morita Y, Yamagata K, Oota A, Ooki A, Isoda Y, Kuma K (2017) Subarctic wintertime dissolved iron speciation driven by thermal constraints on Fe(II) oxidation, dissolved organic matter and stream reach. Geochim Cosmochim Acta 215:33–50

Muehe EM, Adaktylou IJ, Obst M, Zeitvogel F, Behrens S, Planer-Friedrich B, Kraemer U, Kappler A (2013) Organic carbon and reducing conditions lead to cadmium immobilization by secondary Fe mineral formation in a pH-neutral soil. Environ Sci Technol 47:13430–13439

Navalon S, Alvaro M, Alcaina I, Garcia H (2010) Multi-method characterization of DOM from the Turia river (Spain). Appl Geochem 25(11):1632–1643

Nebbioso A, Piccolo A (2013) Molecular characterization of dissolved organic matter (DOM): a critical review. Anal Bioanal Chem 405(1):109–124

Nevin KP, Lovley DR (2002) Mechanisms for accessing insoluble Fe(III) oxide during dissimilatory Fe(III) reduction by *Geothrix fermentans*. Appl Environ Microbiol 68(5):2294–2299

Newman DK, Kolter R (2000) A role for excreted quinones in extracellular electron transfer. Nature 405:94–97

Nurmi JT, Tratnyek PG (2002) Electrochemical properties of natural organic matter (NOM), fractions of NOM, and model biogeochemical electron shuttles. Environ Sci Technol 36:617–624

Ouellet A, Catana D, Plouhinec JB, Lucotte M, Gélinas Y (2008) Elemental, isotopic, and spectroscopic assessment of chemical fractionation of dissolved organic matter sampled with a portable reverse osmosis system. Environ Sci Technol 42(7):2490–2495

Pan WN, Kan JJ, Inamdar S, Chen CM, Sparks D (2016) Dissimilatory microbial iron reduction release DOC (dissolved organic carbon) from carbon-ferrihydrite association. Soil Biol Biochem 103:232–240

Poggenburg C, Mikutta R, Schippers A, Dohrmann R, Guggenberger G (2018) Impact of natural organic matter coatings on the microbial reduction of iron oxides. Geochim Cosmochim Acta 224:223–248

Ravichandran M (2004) Interactions between mercury and dissolved organic matter – a review. Chemosphere 55:319–331

Richter K, Schicklberger M, Gescher J (2012) Dissimilatory reduction of extracellular electron acceptors in anaerobic respiration. Appl Environ Microbiol 78(4):913–921

Royer RA, Burgos WD, Fisher AS, Unz RF, Dempsey BA (2002) Enhancement of biological reduction of hematite by electron shuttling and Fe(II) complexation. Environ Sci Technol 36:1939–1946

Santos PS, Otero M, Duarte RM, Duarte AC (2009) Spectroscopic characterization of dissolved organic matter isolated from rainwater. Chemosphere 74(8):1053–1061

Satoh Y, Kikuchi K, Satoh Y, Fujimoto H (2009) Reverse process of coprecipitation of dissolved organic carbon with Fe(III) precipitates in a lake. Limnology 10:131–134

Schwede-Thomas SB, Chin YP, Dria KJ, Hatcher P, Kaiser E, Sulzberger B (2005) Characterizing the properties of dissolved organic matter isolated by XAD and C18 solid phase extraction and ultrafiltration. Aquat Sci 67(1):61–71

Serkiz SM, Perdue EM (1990) Isolation of dissolved organic matter from the suwannee river using reverse-osmosis. Water Res 24(7):911–916

Shi W, Jin ZF, Hu SY, Fang XM, Li FL (2017) Dissolved organic matter affects the bioaccumulation of copper and lead in *Chlorella pyrenoidosa*: a case of long-term exposure. Chemosphere 174:447–455

Simjouw JP, Minor EC, Mopper K (2005) Isolation and characterization of estuarine dissolved organic matter: comparison of ultrafiltration and C_{18} solid-phase extraction techniques. Mar Chem 96(3/4):219–235

Skoog AC, Arias-Esquivel VA (2009) The effect of induced anoxia and reoxygenation on benthic fluxes of organic carbon, phosphate, iron, and manganese. Sci Total Environ 407 (23):6085–6092

Sodano M, Lerda C, Nistico R, Martin M, Magnacca G, Celi L, Said-Pullicino D (2017) Dissolved organic carbon retention by coprecipitation during the oxidation of ferrous iron. Geoderma 307:19–29

Song FH, Wu FC, Feng WY, Tang Z, Giesy JP, Guo F, Shi D, Liu XF, Qin N, Xing BH, Bai YC (2018) Fluorescence regional integration and differential fluorescence spectroscopy for analysis of structural characteristics and proton binding properties of fulvic acid sub-fractions. J Environ Sci (China) 75:116–125

Song FH, Wu FC, Feng WY, Liu SS, He J, Li TT, Zhang J, Wu AM, Amarasiriwardena D, Xing BH, Bai YC (2019) Depth-dependent variations of dissolved organic matter composition and humification in a plateau lake using fluorescence spectroscopy. Chemosphere 225:507–516

Soong JL, Calderón FJ, Betzen J, Francesca Cotrufo M (2014) Quantification and FTIR characterization of dissolved organic carbon and total dissolved nitrogen leached from litter: a comparison of methods across litter types. Plant Soil 385:125–137

Spencer RGM, Aiken GR, Dyda RY, Butler KD, Bergamaschi BA, Hernes PJ (2010) Comparison of XAD with other dissolved lignin isolation techniques and a compilation of analytical improvements for the analysis of lignin in aquatic settings. Org Geochem 41(5):445–453

Stedmon CA, Bro R (2008) Characterizing dissolved organic matter fluorescence with parallel factor analysis : a tutorial. Limnol Oceanogr Methods 6:572–579

Tang ZR, Huang CH, Tan WB, He XS, Zhang H, Li D, Xi BD (2018) Electron transfer capacities of dissolved organic matter derived from swine manure based on electrochemical method. Chin J Anal Chem 46(3):422–431

Timko SA, Gonsior M, Cooper WJ (2015) Influence of pH on fluorescent dissolved organic matter photo-degradation. Water Res 85:266–274

Tor JM, Lovley DR (2001) Anaerobic degradation of aromatic compounds coupled to Fe(III) reduction by Ferroglobus placidus. Environ Microbiol 3(4):281–287

Trump VJI, Sun Y, Coates JD (2006) Microbial interactions with humic substances. Adv Appl Microbiol 60:55–96

Valenzuela EI, Avendaño KA, Balagurusamy N, Arriaga S, Nieto-Delgado C, Thalasso F, Cervantes FJ (2019) Electron shuttling mediated by humic substances fuels anaerobic methane oxidation and carbon burial in wetland sediments. Sci Total Environ 650:2674–2684

Vazquez E, Amalfitano S, Fazi S, Butturini A (2011) Dissolved organic matter composition in a fragmented Mediterranean fluvial system under severe drought conditions. Biogeochemistry 102(1–3):59–72

Vermeire ML, Bonneville S, Stenuit B, Delvaux B, Cornélis JT (2019) Is microbial reduction of Fe (III) in podzolic soils influencing C release? Geoderma 340:1–10

Wang XD, Chen XN, Liu S, Ge XZ (2010) Effect of molecular weight of dissolved organic matter on toxicity and bioavailability of copper to lettuce. J Environ Sci (China) 22(12):1960–1965

Wang N, Xue XM, Juhasz AL, Chang ZZ, Li HB (2017) Biochar increases arsenic release from an anaerobic paddy soil due to enhanced microbial reduction of iron and arsenic. Environ Pollut 220:514–522

Wolf M, Kappler A, Jiang J, Meckenstock RU (2009) Effects of humic substances and quinones at low concentrations on ferrihydrite reduction by *Geobacter metallireducens*. Environ Sci Technol 43(15):5679–5685

Wrighton KC, Thrash JC, Melnyk RA, Bigi JP, Byrne-Bailey KG, Remis JP, Schichnes D, Auer M, Chang CJ, Coates JD (2011) Evidence for direct electron transfer by a gram-positive bacterium isolated from a microbial fuel cell. Appl Environ Microbiol 77(21):7633–7639

Wu C, An WH, Liu ZY, Lin J, Qian ZY, Xue SG (2020) The effects of biochar as the electron shuttle on the ferrihydrite reduction and related arsenic (As) fate. J Hazard Mater 390:121391. https://doi.org/10.1016/j.jhazmat.2019.121391

Xie L, Yang H, Qu XX, Zhu YR, Zhang ML, Wu FC (2013) Spectral analysis of dissolved organic matter of typical land and aquatic plants in dianchi lake. Environ Sci Res 26(1):72–79. (in Chinese)

Xu S, Gong XF, Zou HL, Liu CY, Chen CL, Zeng XX (2015) Mechanism of the adsorption process: Cu^{2+} adsorpted by ramie stalk. J Chin Chem Soc 62:1072–1078

Xue SG, Jiang XX, Wu C, Hartley W, Qian ZY, Luo XH, Li WC (2020) Microbial driven iron reduction affects arsenic transformation and transportation in soil-rice system. Environ Pollut 260:114010. https://doi.org/10.1016/j.envpol.2020.114010

Yamaguchi N, Nakamura T, Dong D, Takahashi Y, Amachi S, Makino T (2011) Arsenic release from flooded paddy soils is influenced by speciation, Eh, pH, and iron dissolution. Chemosphere 83(7):925–932

Yan M, Ma J, Ji G (2016) Examination of effects of Cu(II) and Cr(III) on Al(III) binding by dissolved organic matter using absorbance spectroscopy. Water Res 93:84–90

Yang R, Li ZW, Huang B, Luo NL, Huang M, Wen JJ, Zhang Q, Zhai XQ, Zeng GM (2018) Effects of Fe(III)-fulvic acid on Cu removal via adsorption versus coprecipitation. Chemosphere 197:291–298

Yang XF, Meng L, Meng FG (2019) Combination of self-organizing map and parallel factor analysis to characterize the evolution of fluorescent dissolved organic matter in a full-scale landfill leachate treatment plant. Sci Total Environ 654:1187–1195

Yu HB, Song YH, Xi BD, Xia XH, He XS, Tu X (2012) Application of chemometrics to spectroscopic data for indicating humification degree and assessing salinization processes of soils. J Soils Sediments 12(3):341–353

Yu HY, Liu C, Zhu JS, Li FB, Deng DM, Wang Q, Liu CS (2016) Cadmium availability in rice paddy fields from a mining area: the effects of soil properties highlighting iron fractions and pH value. Environ Pollut 209:38–45

Yuan DH, Guo XJ, Wen L, He LS, Wang JG, Li JQ (2015) Detection of Copper (II) and Cadmium (II) binding to dissolved organic matter from macrophyte decomposition by fluorescence excitation-emission matrix spectra combined with parallel factor analysis. Environ Pollut 204:152–160

Yuan CL, Liu TX, Li FB, Liu CS, Yu HY, Sun WM, Huang WL (2018a) Microbial iron reduction as a method for immobilization of a low concentration of dissolved cadmium. J Environ Manag 2018:747–753

Yuan DH, An YC, He XC, Yan CL, Jia YP, Wang HT, He LS (2018b) Fluorescent characteristic and compositional change of dissolved organic matter and its effect on heavy metal distribution in composting leachates. Environ Sci Pollut Res 25:18866–18878

Yuan ZD, Ma X, Wu X, Wang X, Wang XF, Jia YF (2018c) Effect of hydroquinone-induced iron reduction on the stability of Fe(III)-As(V) Co-precipitate and arsenic mobilization. Appl Geochem 97:1–10

Yuan CL, Li FB, Cao WH, Yang Z, Hu M, Sun WM (2019) Cadmium solubility in paddy soil amended with organic matter, sulfate, and iron oxide in alternative watering conditions. J Hazard Mater 378:120672. https://doi.org/10.1016/j.jhazmat.2019.05.065

Zhang HC, Weber EJ (2009) Elucidating the role of electron shuttles in reductive transformations in anaerobic sediments. Environ Sci Technol 43:1042–1048

Zhang J, Dai JL, Wang RQ, Li FS, Wang WX (2009) Adsorption and desorption of divalent mercury (Hg^{2+}) on humic acids and fulvic acids extracted from typical soils in China. Colloids Surf A Physicochem Eng Aspects 335:194–201

Zhang J, Dong HL, Zhao LD, McCarrick R, Agrawal A (2014) Microbial reduction and precipitation of vanadium by mesophilic and thermophilic *methanogens*. Chem Geol 370:29–39

Zhou Z, Guo L, Minor EC (2016) Characterization of bulk and chromophoric dissolved organic matter in the Laurentian Great Lakes during summer 2013. J Great Lakes Res 42:789–801

Zhu ZK, Tao L, Li FB (2013) Effects of dissolved organic matter on adsorbed Fe(II) reactivity for the reduction of 2-nitrophenol in TiO_2 suspensions. Chemosphere 93:29–34

Zitoun R, Clearwater SJ, Hassler C, Thompson KJ, Albert A, Sander SC (2019) Copper toxicity to blue mussel embryos (*Mytilus galloprovincialis*) the effect of natural dissolved organic matter on copper toxicity in estuarine waters. Sci Total Environ 653:300–314

The Toxic Effect of Silver Nanoparticles on Nerve Cells: A Systematic Review and Meta-Analysis

Atousa Janzadeh, Michael R. Hamblin, Narges Janzadeh, Hossein Arzani, MahsaTashakori-Miyanroudi, Mahmoud Yousefifard, and Fatemeh Ramezani

Contents

1 Introduction 94
2 Material and Methods 95
2.1 Search Strategy 95
2.2 Eligibility Criteria 97
2.3 Data Extraction 97
2.4 Quality Control 97
2.5 Data Analysis 98

A. Janzadeh
Radiation Biology Research Center, Iran University of Medical Sciences, Tehran, Iran
e-mail: atousajanzadeh@gmail.com

M. R. Hamblin
Wellman Center for Photomedicine, Massachusetts General Hospital, Harvard Medical School, Boston, USA

Laser Research Centre, Faculty of Health Science, University of Johannesburg, Doornfontein, South Africa
e-mail: Hamblin@helix.mgh.harvard.edu

N. Janzadeh
Occupational Medicine Research Center (OMRC), Iran University of Medical Sciences (IUMS), Tehran, Iran
e-mail: Na.janzadeh@gmail.com

H. Arzani
Department of Medical Physics and Biomedical Engineering, Shahid Beheshti University of Medical Sciences, Tehran, Iran
e-mail: hossin.arzani@yahoo.com

MahsaTashakori-Miyanroudi
Department of Medical Physiology, Faculty of Medicine, Iran University of Medical Sciences, Tehran, Iran
e-mail: mhs_tashakori@yahoo.com

M. Yousefifard (✉) · F. Ramezani (✉)
Physiology Research Center, Iran University of Medical Sciences, Tehran, Iran
e-mail: yousefifard.M@iums.ac.ir; ramezani.f@iums.ac.ir

P. de Voogt (ed.), *Reviews of Environmental Contamination and Toxicology Volume 257*,
Reviews of Environmental Contamination and Toxicology 257,
https://doi.org/10.1007/398_2021_67

3 Results 99
3.1 Article Characteristics 99
3.2 Meta-Analysis 100
4 Discussion 111
5 Conclusion 114
References 115

Abstract Despite the increasing use of silver nanoparticles in medical sciences, published studies on their interaction with nerve cells and evaluation of risks are dispersed. This systematic review and meta-analysis could be used to devise safety guidelines for the use of silver nanoparticles in industry and medicine to reduce adverse effects on the CNS.

After extensive searches, the full text of 30 related studies was reviewed and data mining completed. Data were analyzed by calculating the mean of different ratios between treated and untreated groups. Linear regression between variables was evaluated by meta-regression. Subgroup analysis was also performed due to heterogeneity.

Treatment with silver nanoparticles significantly reduced cell viability (SMD $= -1.79\%$; 95% CI: -2.17 to -1.40; $p < 0.0001$). Concentration > 0.1 μg/mL could kill neurons, while lower concentration would not (SMD -0.258; 95% CI: -0.821 to 0.305; $p = 369$). In addition to the concentration, the coating, size of the nanoparticles, and cell type are also factors that influence SNP nerve cell toxicity. Measurement of apoptosis (SMD $= 2.21$; 95% CI: 1.62 to 2.80; $p=0.001$) and lactate dehydrogenase release rate (SMD $= 0.9$; 95% CI: 0.33 to 1.47; $p < 0.0001$) also confirmed the destructive effect of silver nanoparticles on nerve cells.

Keywords Apoptosis · MTT assay · Nerve cell · Neuron · Silver nanoparticle · Toxic

1 Introduction

The scientific interest and available funding for nanotechnology is progressively expanding in many fields (Jeevanandam et al. 2018; Thiruvengadam et al. 2018; Wennersten et al. 2008). Nanoparticles with defined variables, including size, shape, charge, surface chemistry, coating, and solvents are now used in many applications (Gatoo et al. 2014; Heinz et al. 2017; Jazayeri et al. 2016; Khan et al. 2019b). The widespread use of various nanoparticles has also increased concerns about their possible toxic effects to both humans and the environment (Ajdary et al. 2018; Tuncsoy 2018). The release of nanoparticles into the environment and the consequent exposure of humans or other organisms could have adverse effects on the cell membrane and function and damage tissues or organs (Brandelli 2019; Wang et al. 2018). Because various types of nanoparticles induce different and unpredictable effects on human health, while their wider application has been increasing rapidly, nanotoxicology studies continue to attract enormous interest.

Due to their antimicrobial properties, silver nanoparticles (SNPs) have become one of the most widely used types of metal nanoparticles (Sharma et al. 2018; Slavin et al. 2017). SNPs are increasingly being used for the treatment of cancer (Aziz et al. 2019; Bin-Meferij and Hamida 2019; Yesilot and Aydin 2019), wounds and burns (Kalantari et al. 2020; Rani et al. 2018; Tejada et al. 2016), coatings for implants and surgical supplies, and also in many consumer products such as cosmetics (Gajbhiye and Sakharwade 2016), deodorants (Gajbhiye and Sakharwade 2016; Salvioni et al. 2017), textiles (Giannossa et al. 2013), bandages (Boroumand et al. 2018; Kalantari et al. 2020), catheters (Mala et al. 2017; Rupp et al. 2004), and topically applied antimicrobial agents (Boroumand et al. 2018; Lingabathula and Yellu 2017; Rai et al. 2009). SNPs now are the most widely commercialized among all types of nanoparticles, and the market value is considerable (Neelakandan and Thomas 2018; Soleimani and Habibi-Pirkoohi 2017). The properties of SNPs vary according to the source and particle type, and variations in particle size, shape and surface chemistry can have significant effects on their toxicity (Stensberg et al. 2012).

Several studies have reported the failure of the blood brain barrier (BBB) to protect against the entry of SNPs into the brain, and that SNPs could induce astrocyte swelling and neuronal degeneration. It is known that the biological half-life of silver within the central nervous system is longer than that in other organs, therefore there might be some physiological consequences and risks to the brain due to extended exposure (Z. Yang et al. 2010).

Compared to other types of cells, neurons are generally more sensitive to toxins due to their limited ability to regenerate. Therefore, the analysis of the possible toxic effects of SNPs on the nervous system is important. This knowledge could be used to improve safety guidelines for the potential use of SNPs in industry and medicine, to reduce any adverse effects on the CNS (Z. Yang et al. 2010).

Numerous articles have investigated the toxicity of SNP on different cells, and some have had contradictory results, but there are no systematic review and meta-analysis of these studies. In this article we attempt to answer the question of whether SNPs have any effect on the survival of nerve cells, and besides to answer other questions such as "At what concentration are the nanoparticles more toxic to nerve cells? Can the type of coating affect the nanoparticle toxicity to nerve cells?"

2 Material and Methods

2.1 Search Strategy

We searched articles without limits on language concerning the effect of SNPs on the viability of neural cells which were published up to December 01, 2019 in MEDLINE, SCOPUS, EMBASE, and Web of Science. The keywords used for the literature search are shown in Table 1.

Table 1 Strategy of search for Medline database

((("Neurons" [mh] or "Neuron" [tiab] or "Neurons" [tiab] or "Nerve Cells" [tiab] or "Cell, Nerve" [tiab] or "Cells, Nerve" [tiab] or "Nerve Cell" [tiab] or "Dendrites" [mh] or "Dendrites" [tiab] or "Neurites" [mh] or "Neurites" [tiab] or "Nerve Fibers" [mh] or "Nerve Fibers" [tiab] or "Cerebellar Mossy Fibers" [tiab] or "Cerebellar Mossy Fiber" [tiab] or "Mossy Fiber, Cerebellar" [tiab] or "Mossy Fibers, Cerebellar" [tiab] or "Axons" [mh] or "Axon" [tiab] or "Neurofibrils" [mh] or "Neurofibrils" [tiab] or "neural" [tiab] "Nissl Bodies" [mh] or "Nissl Bodies" [tiab] or "Bodies, Nissl" [tiab] or "Nissl Granules" [tiab] or "Granule, Nissl" [tiab] or "Granules, Nissl" [tiab] or "Nissl Granule" [tiab] or "Nitrergic Neurons" [mh] or "Nitrergic Neurons" [tiab] or " Lewy Bodies" [mh] or "Bodies, Lewy" [tiab] or " Lewy Body" [tiab] or " Body, Lewy" [tiab] or "Cholinergic Fibers" [mh] or " Cholinergic Fibers" [tiab] or "Cholinergic Fiber" [tiab] or "Fiber, Cholinergic" [tiab] or "Fibers, Cholinergic" [tiab] or "Nerve Fibers, Myelinated" [mh] or "Fiber, Myelinated Nerve" [tiab] or "Fibers, Myelinated Nerve" [tiab] or "Myelinated Nerve Fiber" [tiab] or "Myelinated Nerve Fibers" [tiab] or "Nerve Fiber, Myelinated" [tiab] or "A Fibers" [tiab] or "A Fiber" [tiab] or "Fiber, A" [tiab] or "Fibers, A" [tiab] or "B Fibers" [tiab] or "B Fiber" [tiab] or "Fiber, B" [tiab] or "Fibers, B" [tiab] or "Nerve Fibers, Unmyelinated" [mh] or "Nerve Fiber, Unmyelinated" [tiab] or "Unmyelinated Nerve Fiber" [tiab] or "Unmyelinated Nerve Fibers" [tiab] or "C Fibers" [tiab] or "C Fiber" [tiab] or "Myelin Sheath" [mh] or "Myelin Sheaths" [tiab] or "Sheath, Myelin" [tiab] or "Sheaths, Myelin" [tiab] or "Myelin" [tiab] or "Neurofibrils" [mh] or "Neurofibrillary Tangles" [mh] or "Neurofibrillary Tangle" [tiab] or "Tangle, Neurofibrillary" [tiab] or "Tangles, Neurofibrillary" [tiab] or "Nitrergic Neurons" [mh] or "Neuron, Nitrergic" [tiab] or "Nitrergic Neuron" [tiab] or "Neurons, Nitroxidergic" [tiab] or "Neuron, Nitroxidergic" [tiab] or "Nitroxidergic Neuron" [tiab] or "Nitroxidergic Neurons" [tiab] or "Neurons, Nitrergic" [tiab] or "Nitrergic Nerves" [tiab] or "Nitroxidergic Nerves" [tiab] or "Pyramidal Cells" [mh] or "Cell, Pyramidal" [tiab] or "Cells, Pyramidal" [tiab] or "Pyramidal Cell" [tiab] or "Pyramidal Neurons" [tiab] or "Neuron, Pyramidal" [tiab] or "Neurons, Pyramidal" [tiab] or "Pyramidal Neuron" [tiab] or "Serotonergic Neurons" [mh] or "Neuron, Serotonergic" [tiab] or "Neurons, Serotonergic" [tiab] or "Serotonergic Neuron" [tiab] or "Serotoninergic Neurons" [tiab] or "Neuron, Serotoninergic" [tiab] or "Neurons, Serotoninergic" [tiab] or "Serotoninergic Neuron" [tiab] or "Growth Cones" [mh] or "Growth Cone" [tiab] or "neural cells" [tiab] or "Neural Stem Cells" [mh] or "Cell, Neural Stem" [tiab] or "Cells, Neural Stem" [tiab] or "Neural Stem Cell" [tiab] or "Stem Cell, Neural" [tiab] or "Stem Cells, Neural" [tiab] or "Oligodendrocyte Precursor Cells" [mh] or "Cell, Oligodendrocyte Precursor" [tiab] or "Cells, Oligodendrocyte Precursor" [tiab] or "Oligodendrocyte Precursor Cell" [tiab] or "Precursor Cell, Oligodendrocyte" [tiab] or "Precursor Cells, Oligodendrocyte" [tiab] or "Pre-Oligodendrocytes" [tiab] or "Pre Oligodendrocytes" [tiab] or "Pre-Oligodendrocyte" [tiab] or "Oligodendrocyte Progenitors" [tiab] or "Oligodendrocyte Progenitor" [tiab] or "Progenitor, Oligodendrocyte" [tiab] or "Progenitors, Oligodendrocyte" [tiab] or "Preoligodendrocytes" [tiab] or "Preoligodendrocyte" [tiab] or "Oligodendrocyte Precursors" [tiab] or "Oligodendrocyte Precursor" [tiab] or "Precursor, Oligodendrocyte" [tiab] or "Precursors, Oligodendrocyte" [tiab] or "Oligodendrocyte Progenitor Cells" [tiab] or "Cell, Oligodendrocyte Progenitor" [tiab] or "Cells, Oligodendrocyte Progenitor" [tiab] or "Oligodendrocyte Progenitor Cell" [tiab] or "Progenitor Cell, Oligodendrocyte" [tiab] or "Progenitor Cells, Oligodendrocyte" [tiab] or "Astrocytes" [mh] or "Astrocyte" [tiab] or "Astrocytes" [tiab] or " Astroglia" [tiab] or "Astroglias" [tiab] or "Neuroglia" [mh] or " Glial Cells" [tiab] or " Cell, Glial" [tiab] or "Cells, Glial" [tiab] or "Glial Cell" [tiab] or " Neuroglial Cells" [tiab] or " Cell, Neuroglial" [tiab] or "Cells, Neuroglial" [tiab] or "Neuroglial Cell" [tiab] or "Glia" [tiab] or "Schwann Cells" [mh] or "Cells, Schwann" [tiab] or " Cell, Neuroglial" [tiab] or " Cell, Neuroglial" [tiab] or "Satellite Cells, Perineuronal" [mh] or "Cell, Perineuronal Satellite" [tiab] or "Cells, Perineuronal Satellite" [tiab] or "Perineuronal Satellite Cell" [tiab] or "Satellite Cell, Perineuronal" [tiab] or "Perineuronal Satellite Cells" [tiab] or "Ganglia" [mh] or "Microglia" [mh] or "Microglia" [tiab]))) AND (("Silver"[mh] or "Silver Compounds"[mh] or "Silver Compound*"[tiab] or "Silver nanoparticle" [tiab] or "Silver nanoparticles" [tiab] or "Silver nano particle" [tiab] or "Silver nano

(continued)

Table 1 (continued)

particles" [tiab] or "Ag nanoparticle" [tiab] or "Ag-NPs" [tiab] or "Ag nano particle" [tiab] or "Ag nano particles" [tiab] or "Ag nanoparticles" [tiab] or "Silver colloid*" [tiab] or "Silver nanomaterial" [tiab] or "Silver nano material" [tiab] or "Silver nano composit" [tiab] or "Silver nanorod" [tiab] or "Silver nano rod" [tiab] or "Silver nanopowder" [tiab] or "Silver nano powder" [tiab] or "Silver nanocompound" [tiab] or "Silver nano compound" [tiab] or "colloidal silver" [tiab]))

2.2 *Eligibility Criteria*

After potentially relevant articles were obtained from all four databases, review articles, commentaries, and non-relevant articles based on title and abstract were excluded.

Only studies reporting the results of experimental studies measuring the toxic effects of SNPs on any type of nerve cells, using in-vitro experiments that compared the cell viability with untreated control cells, were included for analysis. Cellular viability was measured by a range of different tests, including Alamar blue, fluorescent calcein release, Trypan blue, MTT, WST-1, WST-8, MTS, crystal violet, and Annexin V, and the results were extracted for analysis. For measurement of apoptosis the extent of cleaved caspase 3, the TUNEL assay, and measurement of the Bax/Bcl2 ratio were reported. Lactate dehydrogenase (LDH) release was considered an appropriate test for necrosis. Studies lacking a control that did not receive any treatment with SNPs, studies without toxicity data, and those without the number of repeats for each group were not considered.

2.3 *Data Extraction*

Two investigators independently reviewed the full text of all articles. The following information was extracted from the complete version of each study: article metadata (author name and year of publication), cell name, cell source, nanoparticle zeta potential, SNP size, drug attached to SNPs, SNP concentration, SNP coating, exposure time, outcome, number of control wells, number of treatment wells, and the results of the tests. If necessary, data from figures were extracted by digitizing software (Rohatgi 2010).

2.4 *Quality Control*

The quality assessment of the included studies was conducted according to methods described in reference (Y. Liu et al. 2018). There were eight groups of criteria employed (exclusions, randomization, blinding, sample size, figures and statistical representation of data, definition of statistical methods and measures, implementation of statistical methods and measures, reagents and cells). These criteria were expanded to 20 separate items listed in Table 3.

1. Samples that were excluded from the analysis.
2. Which method of randomization was used to determine how samples were allocated to experimental groups.
3. Whether the investigator was blinded to the group allocation during the experiment and/or when assessing the outcome.
4. How the sample size was chosen to ensure adequate power to detect a pre-specified effect size.
5. Exact sample size (n) for each experimental group/condition was given as a number, not a range.
6. Whether the samples represented technical or biological replicates.
7. A statement of how many times the experiment was replicated.
8. Results were defined as a median or average.
9. Error bars were defined as s.d., s.e.m., or c.i.
10. Common statistical tests (such as t-test, simple $\chi 2$ tests, Wilcoxon and Mann-Whitney tests, or any form of ANOVA testing). If not a common test, the test is described in the methods section.
11. If the statistical test used was a t or z test, it was reported as one sided or two sided.
12. Adjustments for multiple comparisons were applied where appropriate.
13. The statistical test results (e.g., p-values, F statistic, etc.) were presented.
14. The authors show that their data met the assumptions of the tests.
15. An estimate of variation is reported for each group of data.
16. The variance between the groups that were statistically compared was comparable (difference less than twofold).
17. Every antibody used in the manuscript has been characterized by either citation, catalog number, clone number or validation profile.
18. The source of all cell lines was provided.
19. The authors reported whether the cell lines used have been recently authenticated.
20. The authors reported whether the cell lines have recently been tested for contamination (within 6 months of use).

2.5 Data Analysis

Cell viability data were obtained from 20 articles and 145 experiments. The analysis was performed with STATA 14. Data for apoptosis were obtained from 9 articles and 23 experiments and LDH release data were obtained from 11 articles and 65 experiments. Data were expressed as mean ratios between the experimental and control groups. If there was a significant heterogeneity ($p < 0.05$) by the fixed model, a random-effect model was applied. Linear regression between variables such as concentration of SNPs, size of SNPs, coating of SNPs, and duration of exposure with the cellular viability was evaluated by "metareg command."

Subgroup analysis was performed for SNP size (size $\leq$20 nm and size >20 nm), SNPs coating (PVP, citrate, uncoated, and others), SNP concentration, exposure time ($T \leq 24$ h, $T > 24$ h), cell source (rat, mouse, human), cell type (normal, cancerous). p-values were reported by statistical hypothesis testing at the two-sided 0.05 level. Publication bias was explored by a funnel plot.

3 Results

3.1 Article Characteristics

The search strategy identified 1,073 articles (Fig. 1). 445 articles from PUBMED, 342 from Scopus, 208 articles from Embase, and 78 from Web of Science were

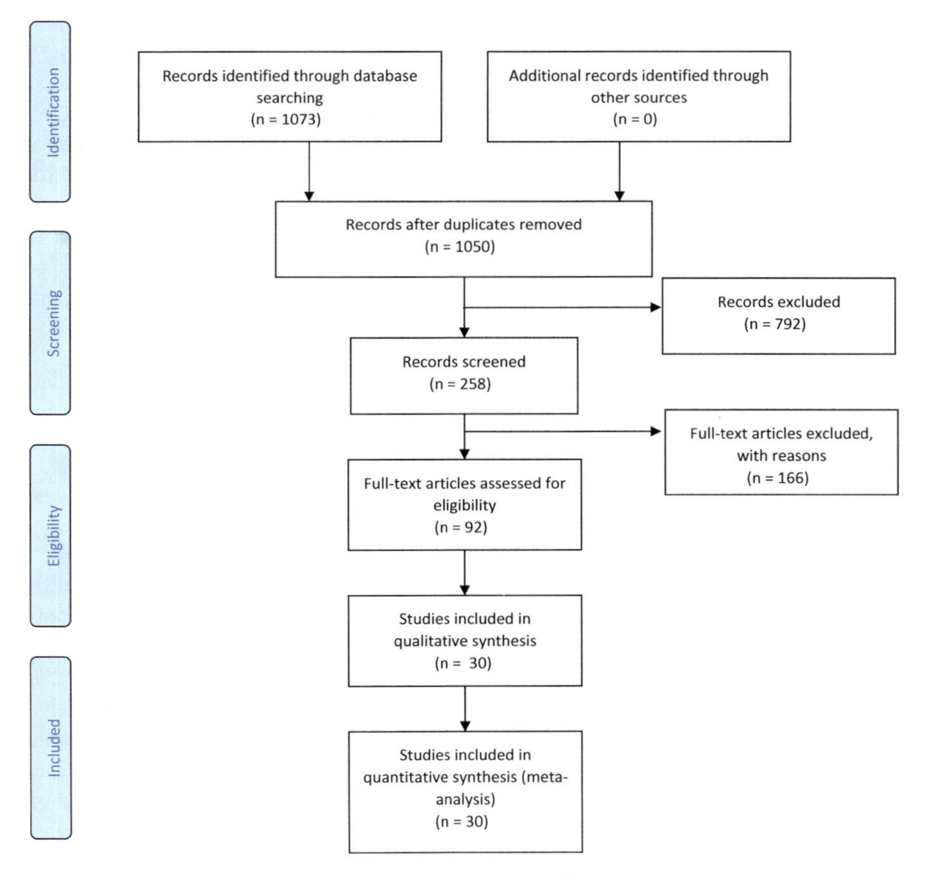

Fig. 1 PRISMA flow diagram for a systematic review detailing the database searches, the number of abstracts screened, and the full texts retrieved

obtained. After applying the eligibility criteria, 92 articles were identified to retrieve the full text. After assessment of the full texts, 30 articles contained enough of the data required for further analysis. The source species of the cells included mouse, rat, and human. The studies that investigated the effect of SNPs on nerve cell lines employed sizes ranging from 1 to 100 nm. The SNPs were categorized into non-coated, coated with PVP, citrate, or other coatings. The data from the 30 articles are summarized in Table 2.

The relative quality of the 30 articles was assessed (Table 3) using 20 individual criteria. Study design features which helped to reduce bias, such as randomization, blinding, cause of exclusion of samples, how the sample size was chosen, adjustments for multiple comparisons, similarity of variance between groups, antibody characterization, cell authentication, and cell contamination, were often poorly reported.

Three publications among the total of 30 publications reported cell line authentication (10%). None of the publications reported the cause of exclusion of samples, blinding, and similarity of variance between groups, how the sample size was chosen, and lack of cell contamination in their full text.

All the articles in the meta-analysis described how many times the experiment was replicated, whether the samples represented technical or biological replicates, error bars were defined, common statistical tests were used, data met the assumptions of the tests, an estimate of variation was reported within each group of data, and the source of cell lines was provided.

In the present study no publication bias was observed concerning the effect of silver nanoparticles on the number of viable neural cells ($p = 0.182$) (Fig. 2a), on apoptosis in neural cells (Fig. 2b) ($p = 0.994$), and on the release of lactate dehydrogenase (Fig. 2c) ($p = 0.182$).

3.2 *Meta-Analysis*

3.2.1 Effect of SNP Exposure on Neural Cell Viability

Treatment of neural cells with silver nanoparticles in cell culture significantly reduced the number of living cells (SMD $= -1.79$; 95% CI: -2.17 to -1.40; $p < 0.0001$) depending on the concentration (Fig. 3). The results of the meta-analysis showed significant heterogeneity between studies ($I^2 = 76.2\%$, $p < 0.0001$). Subgroup analysis was performed to find the cause of the heterogeneity.

After subgroup analysis, a comparison of SNP-treated neural cells with non-treated neural cells revealed that the concentration of SNPs can be the source of heterogeneity (Table 4). In the analysis of concentration of nanoparticles, a concentration equal to or less than 0.01 μg/mL ($I^2 = 0.0\%$; $p = 0.404$) and in the group receiving 0.1–0.01 μg/mL the heterogeneity was lower ($I^2 = 40.9\%$; $p = 0.037$).

Table 2 Characteristics of the studies included in the meta-analysis

Number	Root/cell name	SNPs Size (nm)	Preparation method	SNPs concentration (μg/mL)	Coating	Exposure time/h	Criteria checked	Reference
1	Rat/cerebellum granule cells	26	HuZheng Co.	0.01–2.5	PVP	24 h	Cell viability (AB staining assay), Caspase-3, LDH	Yin et al. (2013)
2	Mouse/embryonic stem cell	34	HuZheng Co.	0.01–1	Uncoated	24 h	Cell viability (Alamar blue assay)	Yin et al. (2018)
3	Rat/cerebral astrocytes	24	HuZheng Co.	0.01–80	PVP	24 h	Cell viability (Alamar blue), LDH, apoptosis (caspase activity)	Sun et al. (2016)
4	Human/embryonic neural cells	20, 80	VWR International, Radnor	0.00022–0.22	Non-coated	14 days	Apoptosis (TUNEL)	Söderstjerna et al. (2013)
5	Mouse/retina neuron	20, 80	VWR International, Radnor	0.0035–0.22	Non-coated	72 h	Apoptosis (TUNEL)	Söderstjerna et al. (2014)
6	Human/ glutamatergic neurons	20	Hand made	0.1–5.0	Citrate, PVP	72 h	Cell viability, LDH	Begum et al. (2016)
7	Mouse/endothelial (bEnd.3) and astrocyte-like (ALT)	8	Sigma-Aldrich Co.	4	PVP	48 h	Cell viability (Alamar blue)	Chen et al. (2016)
8	Human/cancerous U251-MG and T98G	20	Sigma-Aldrich Co.	0–1	Citrate	48 h	Cell viability (WST-1 reagent)	Choi et al. (2018)
9	Human/non-cancerous BEAS-2B cells	<20	Sigma-Aldrich Co.	0.13, 1.33	Non-coated	CV; 144 h, CFE; 192 h, Caspase3; 4 months	Cell viability (crystal violet assay and colony forming efficiency assay), Caspase-3	Choo et al. (2016)

(continued)

Table 2 (continued)

Number	Root/cell name	SNPs Size (nm)	Preparation method	SNPs concentration (μg/mL)	Coating	Exposure time/h	Criteria checked	Reference
10	Human/neuro-blastoma SH-SY5Y	30	Synthesized by *E. coli*	0.1–0.4 μM	Non-coated	24 h	Cell viability	Barhoum et al. (2019)
11	Mouse/N9 microglial cells	49.7 ± 10.5	Handmade: Reduction method	50	Citrate	24 h	Cell viability (MTS assay, LDH)	Gonzalez-Carter et al. (2017)
12	Human/neurons and astrocytes	20	NIEHS Co.	0.1–50	Citrate	72 h	Caspase-3	Repar et al. (2018)
13	Rat/cortical neurons and astrocyte	20 and 40	Bachem	5–100	CKK peptide	24 h	Cytotoxicity	Haase et al. (2012)
14	Rat/cortical neurons	<10	Nanopoly Co.	0.4–10	Non-coated	24 h	Cell viability (LDH assay), Caspase3, tunnel	Kim et al. (2014)
15	Murine/n2a and bv2	10	Gold Nanotech, Inc.	0.25–3	Non-coated	24 h	Cell viability (Alamar blue)	Hsiao et al. (2017)
16	Human/embry-onic stem cell (hESC)	7.9	ABC Nanotech	25 and 50	Citrate	24 h	Cell viability assay by CCK-8, LDH	Oh et al. (2016)
17	Rat/PC12	14	AgNP preparation by reduction method	0.06–5	PVP	48 h	TUNEL, cell viability (trypan blue)	Hadrup et al. (2012)
18	Human/glioblastoma U87 cells	Not reported	ATCC	25–250	Non-coated	14 days	MTT	Gao et al. (2014)
19	Human dental-pulp-stem-cells-derived neurons	85	Handmade	10–100	Citrate	72 h	Cell viability	Bonaventura et al. (2018)

20	Rat/cerebellar granule cells	<100	Sigma-Aldrich Co.	2.5–75	PVP	24 h	Cell viability	Ziemińska et al. (2014)
21	Human/embry-onic neural stem cells (NSCs)	23	Handmade	1–20	Non-coated	24 h	Cell viability by MTT, LDH, TUNEL	Liu et al. (2015)
22	Murine/hippocampal neuronal HT22 cells	30	Sun Nano, California	5	Non-coated	24 h	Caspase-3, cell viability (Alamar blue assay)	Ma et al. (2015)
23	Murin/HT22 hippocampal neuronal cells	3–15	NR	5	Non-coated	48 h	MTT	Mytych et al. (2017)
24	Murine/neural stem cells	<20	Handmade	0.1–5	AOT, CTAB, PVP, PLL, BSA	24 h	Cell viability (WST-8) TEST (CCK8)	Pongrac et al. (2018)
25	Rat/primary astrocytes	7.8	Handmade	0.1–100	Gallic acid	24 h	MTT, LDH, apoptosis	Salazar-García et al. (2015)
26	Mouse/neural stem cell	10	Nanocomposix	1–50	Non-coated	24 h	LDH	Braun et al. (2013)
27	Human/brain endothelial cells and astrocytes	87	Nanocomposix	5	PVP	4 h	LDH	Khan et al. (2019a)
28	Rat-Astroglia	75 ± 20	Handmade	10–100	PVP	24 h	LDH release	Luther et al. (2011)
29	Rat-Astroglia	75 ± 20	Handmade	10–300		4 h	LDH release	Luther et al. (2012)
30	Murin/neural stem cell	10	Reduction of silver nitrate	0.1–5	PVP, CTAB, AOT, PLL	24 h	Viability by CCK-8, apoptosis	Pavi et al. (2019)

Table 3 Articles score based on agency for healthcare research and quality's methods guide for effectiveness of reviews

Author/year	1	2	3	4	5	6	7	8	9	10	11	12	13	14	15	16	17	18	19	20
Yin/2013	N	N	N	N	Y	Y	Y	Y	Y	Y	N	N	Y	Y	Y	N	Y	Y	N	N
Yin/2018	N	N	N	N	Y	Y	Y	Y	Y	N	Y	N	N	Y	Y	N	Y	Y	N	N
Sun/2016	N	N	N	N	Y	Y	Y	Y	Y	Y	Y	N	Y	Y	Y	N	Y	Y	Y	N
Derstjerna/2013	N	N	N	N	Y	Y	Y	Y	Y	Y	Y	N	Y	Y	Y	N	Y	Y	N	N
Derstjerna/2015	N	N	N	N	Y	Y	Y	Y	Y	Y	N	N	N	Y	Y	N	Y	Y	N	N
Begum/2016	N	N	N	N	Y	Y	Y	Y	Y	Y	Y	N	Y	Y	Y	N	Y	Y	Y	N
Chen/2016	N	N	N	N	Y	Y	Y	Y	Y	Y	Y	Y	Y	Y	Y	N	N	Y	N	N
Choi/2018	N	N	N	N	Y	Y	Y	Y	Y	N	N	Y	N	Y	Y	N	Y	Y	N	N
Choo/2016	N	N	N	N	Y	Y	Y	Y	Y	Y	Y	N	Y	Y	Y	N	Y	Y	N	N
Dayem/2014	N	N	N	N	Y	Y	Y	Y	Y	Y	Y	N	Y	Y	Y	N	NA	Y	N	N
Gonzalez/2017	N	N	N	N	Y	Y	Y	Y	Y	Y	Y	N	Y	Y	Y	N	N	Y	N	N
Repar/2018	N	N	N	N	N	Y	Y	Y	Y	Y	Y	N	Y	Y	Y	N	Y	Y	N	N
Haase/2012	N	N	N	N	Y	Y	Y	Y	Y	Y	Y	N	Y	Y	Y	N	N	Y	N	N
Kim/2014	N	N	N	N	Y	Y	Y	Y	Y	Y	Y	Y	Y	Y	Y	N	N	Y	N	N
Hsiao/2016	N	N	N	N	Y	Y	Y	Y	Y	Y	Y	Y	Y	Y	Y	Y	Y	Y	N	N
Oh/2016	N	N	N	N	Y	Y	Y	Y	Y	Y	N	N	Y	Y	Y	N	N	Y	N	N
Hadrup/2012	N	N	N	N	Y	Y	Y	Y	Y	Y	N	Y	Y	Y	Y	N	Y	Y	N	N
Gao/2014	N	N	N	N	Y	Y	Y	Y	Y	N	N	N	N	Y	Y	Y	NA	Y	N	N
Bonaventuraa/2018	N	N	N	N	Y	Y	Y	Y	Y	Y	Y	Y	Y	Y	Y	Y	Y	Y	N	N
Ziemi/2014	N	N	N	N	Y	Y	Y	Y	Y	Y	Y	N	Y	Y	Y	N	NA	Y	N	N
Liu/2015	N	N	N	N	Y	Y	Y	Y	Y	Y	Y	N	Y	Y	Y	N	N	Y	N	N
MA/2015	N	N	N	N	Y	Y	Y	Y	Y	Y	Y	Y	Y	Y	Y	N	Y	Y	N	N
Mytych/2016	N	N	N	N	Y	Y	Y	Y	Y	Y	Y	Y	Y	Y	Y	N	N	Y	N	N
Pongrac/2017	N	N	N	N	Y	Y	Y	Y	Y	Y	Y	Y	Y	Y	Y	N	NA	Y	N	N
García/2015	N	N	N	N	Y	Y	Y	Y	Y	Y	Y	Y	Y	Y	Y	Y	Y	Y	N	N
Braun/2013	N	N	N	N	Y	Y	Y	Y	Y	Y	Y	Y	Y	Y	Y	Y	N	Y	N	N

Khan/2019	N	N	N	N	Y	Y	Y	Y	Y	Y	Y	Y	Y	Y	Y	Y	Y	Y	N	N
Luther/2011	N	N	N	N	Y	Y	Y	Y	Y	Y	Y	Y	Y	Y	Y	Y	N	Y	N	N
Pavičić/2020	N	N	N	N	Y	N	Y	Y	Y	Y	Y	Y	Y	Y	Y	Y	Y	Y	Y	N
Percent	0	0	0	0	96	96	100	100	100	90	80	46	90	100	100	30	53	100	10	0

(1) Samples were excluded from the analysis, (2) Which method of randomization was used to determine how samples were allocated to experimental groups, (3) Whether the investigator was blinded, (4) How the sample size was chosen (5)The exact sample size (6) Whether the samples represent technical or biological replicates, (7) How many times the experiment shown was replicated, (8) The summary estimates are defined as a median or average, (9) The error bars are defined as s.d., s.e.m., or c.i., (10) Common statistical test, or the test is described, (11) *t* or *z* test reported as one sided or two sided, (12) Adjustments for multiple comparisons are applied, (13) The statistical test results are presented, (14) The authors show that their data meet the assumptions of the tests, (15) An estimate of variation is reported within each group of data, (16) The variance is similar between the groups that are being statistically compared, (17) Antibody citation, catalog number, (18) The source of cell lines, (19) Whether the cell lines used have been authenticated recently, (20) Whether the lines used have been tested for contamination recently

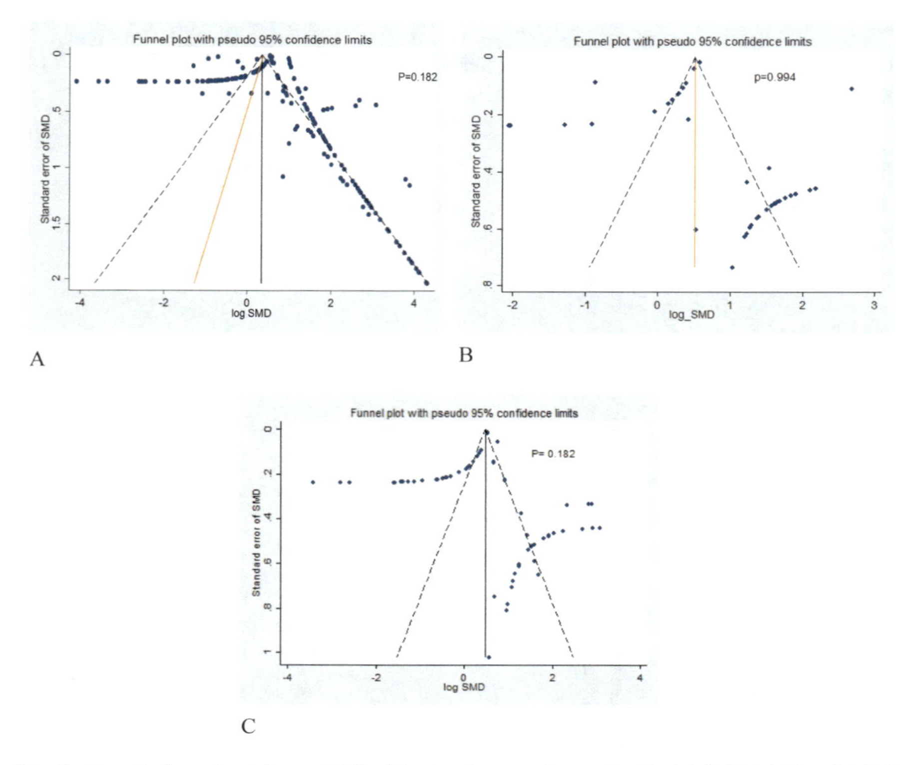

Fig. 2 Funnel plots of articles on SNPs effect on the treated neural cells. (**a**) Cell viability, (**b**) Cell apoptosis, (**c**) LDH release. SMD: Standardized mean difference

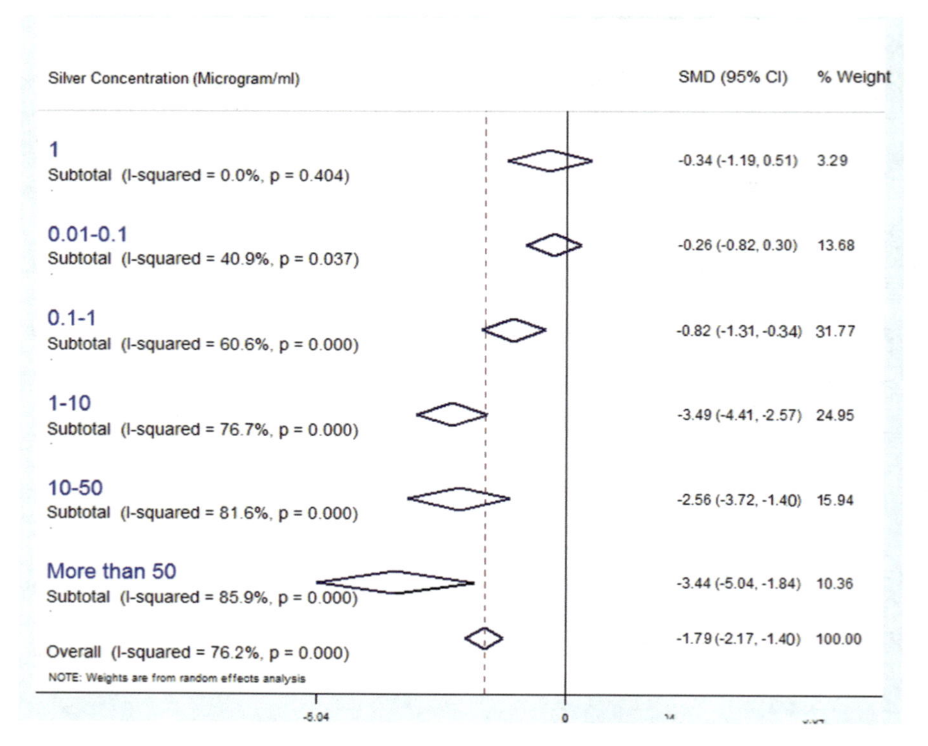

Fig. 3 Forest plot of SNPs treatment on neural cell viability compared to non-treated cells categorized based on the SNP concentration

The heterogeneity decreased in studies with PVP-coated nanoparticles when the analysis was performed by coating type ($I^2 = 67.3\%$; $p < 0.0001$) (Table 4). SNP treatment in all subgroups significantly reduced the number of living cells compared to non-treated neural cells. The SNP size, SNP concentration, cell species, SNP coating, and the cell type were factors that influenced cell survival (Table 4). The number of surviving cells treated with SNPs ≤20 nm (SMD = −2.330; 95% CI: −2.948 to −1.713; $p < 0.0001$) was less than the number of surviving cells treated with SNPs >20 nm (SMD = −1.034; 95% CI: −1.506 to −0.562; $p < 0.0001$).

Concentrations less than 0.01 μg/mL (SMD = −0.338; 95% CI: −1.186 to 0.510; $p = 435$) and also 0.01–0.1 μg/mL) SMD = −0.258; 95% CI: −0.821 to 0.305; $p = 0.369$ (showed no significant toxic effects compared to the control group (Table 4). When the nanoparticle concentration increased up to 0.1–1 μg/mL (SMD = −1.064; 95% CI: −1.716 to −0.412; $p = 0.001$), 1–10 μg/mL (SMD = −2.715; 95% CI: −3.465 to −1.965; $p < 0.0001$), and between 10–50 μg/mL (SMD = −2.159; 95% CI: −3.209 to −1.109; $p < 0.0001$) and above 50 μg/mL (SMD = −3.438; 95% CI: −5.036 to −1.839; $p = 0.025$) the number of living cells steadily decreased.

The cell species of origin also affected its resistance to SNPs. Cells of human origin were more sensitive (SMD = −2.310; 95% CI: −2.981 to −1.639; $p < 0.0001$) compared to rat cells (SMD = −0.900; 95% CI: −1.572 to −0.228; $p = 0.009$) and mouse cells (SMD = −1.948; 95% CI: −2.552to −1.345; $p < 0.0001$). After exposure to SNPS, the viability of normal cells (SMD = −2.058; 95% CI: −2.497 to −1.618; $p < 0.0001$) was less than that of cancer cells (SMD = −0.861; 95% CI: −1.679 to −0.043; $p = 0.039$).

The intra-group regression between subgroups showed SNPs concentration (p=0.043) can be an effective factor on neural cell toxicity (Table 4).

A linear regression between the concentration of nanoparticles as a quantitative variable and the effect on cell viability is shown in Fig. 4. As the concentration of nanoparticles was increased, the number of living cells decreased.

3.2.2 SNPs Effect on Apoptosis in Neural Cells

Ten studies and 38 experiments examined the effect of SNPs on apoptosis by measuring caspase-3 activity, TUNEL assay, and measurement of the Bax/Bcl2 ratio. The results in Fig. 5 showed that SNPs increased apoptosis in neural cells (SMD = 2.21; 95% CI: 1.62 to 2.80; p 0.001). There is no heterogeneity between experiments ($I^2 = 58.6\%$, $p = 0.000$).

3.2.3 SNP Effect on LDH Release in Neural Cells

In 11 studies and 65 separate experiments, the effect of SNPs with different coatings on LDH release was investigated using different concentrations and sizes. The results showed that SNPs had a moderate effect on the release of lactate

Table 4 Overall and subgroup analysis of SNP treatment effect on neural cell viability

Subgroup	Number of experiments	Heterogeneity (*p*-value)	SMD (95% CI)	*p*-value
Cell species				
Rat	33	65.4% (<0.0001)	−0.900 (−1.572 to −0.228)	0.009
Mouse	93	77.2% (<0.0001)	−1.948 (−2.552 to −1.345)	<0.0001
Human	47	77.4% (<0.0001)	−2.310 (−2.981 to −1.639)	<0.0001
	Overall significant *p*-value among subgroups: $p = 0.07$			
Cell type				
Normal neural cell	141	74.7% (<0.0001)	−2.058 (−2.497 to −1.618)	<0.0001
Cancerous cell	32	81.2% (<0.0001)	−0.861 (−1.679 to −0.043)	0.039
	Overall significant *p*-value among subgroups: $p = 0.069$			
SNPs size				
≤20 nm	104	78.1% (<0.0001)	−2.330 (−2.948 to −1.713)	<0.0001
>20 nm	57	65.3% (<0.0001)	−1.034 (−1.506 to −0.562)	<0.0001
	Overall significant *p*-value among subgroups: $p = 0.19$			
SNPs coat				
Uncoated	63	77.6% (<0.0001)	−2.009 (−2.594 to −1.424)	<0.0001
PVP	34	67.3% (<0.0001)	−1.488 (−2.161 to −0.815)	<0.0001
Citrate	10	80.9% (<0.0001)	−0.814 (−2.075 to 0.447)	0.206
Others	54	78.0% (<0.0001)	−2.055 (−2.953 to −1.157)	<0.0001
	Overall significant *p*-value among subgroups: $p = 0.25$			
Exposure time				
≤24 h	133	73.8% (<0.0001)	−1.714 (−2.176 to −1.252)	<0.0001
>24 h	40	81.9% <0.0001	−2.000 (−2.714 to −1.285)	<0.0001
	Overall significant *p*-value among subgroups: $p = 0.41$			

SNP concentration				
≤ 0.01 μg/mL	4	0.0% (0.404)	−0.338 (−1.186 to 0.510)	0.435
0.01–0.1 μg/mL	18	40.9% (0.037)	−0.258 (−0.821 to 0.305)	0.369
0.1–1 μg/mL	34	68.1% (<0.0001)	−1.064 (−1.716 to −0.412)	0.001
1–10 μg/mL	69	76.4% (<0.0001)	−2.715 (−3.465 to −1.965)	<0.0001
10–50 μg/mL	28	80.1% (<0.0001)	−2.159 (−3.209 to −1.109)	<0.0001
>50 μg/mL	20	85.9% (<0.0001)	−3.438 (−5.036 to −1.839)	0.025
	Overall significant *p*-value among subgroups: $p = 0.004$			
Overall	173	76.2% (<0.0001)	−1.787 (−2.173 to −1.401	<0.0001

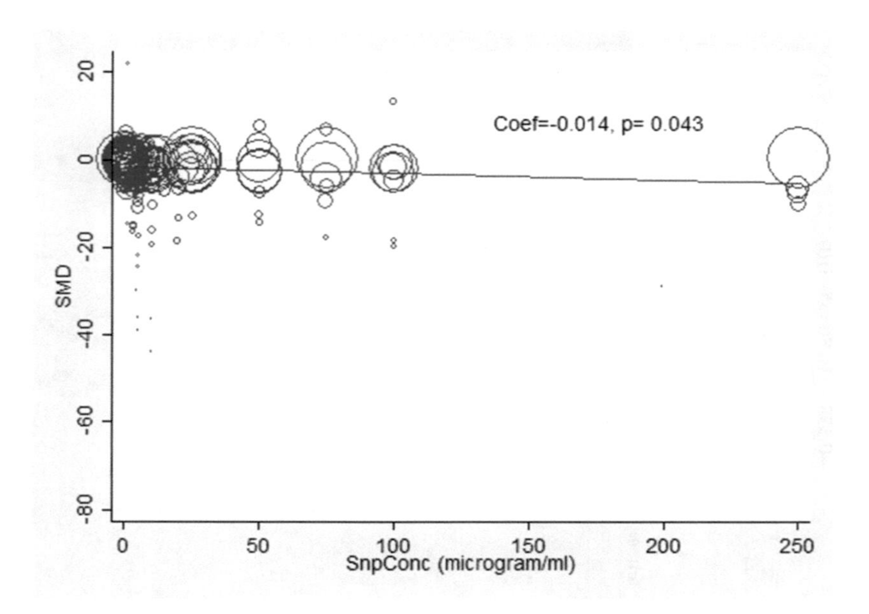

Fig. 4 Linear regression between SNP concentration and its effect on cell viability

dehydrogenase in neurons (SMD =0.9; 95% CI: 0.33 to 1.47; $p < 0.0001$) (Fig. 6). Due to the high overall heterogeneity ($I^2 = 74.7\%$, $p < 0.0001$) in the studies, subgroup analysis was performed to find the source of heterogeneity.

According to Table 5, the concentration of SNPs, cell species, SNP size, SNP coating were heterogeneous factors. At 0–0.01 μg/mL concentration, ($I^2 = 0.0\%$, $p < 0.948$), in human cells ($I^2 = 0.0\%$, $p = 0.986$), SNP size ≤20 nm ($I^2 = 46.0\%$, $p = 0.003$), SNP coating = PVP ($I^2 = 51.5\%$, $p = 0.009$) the heterogeneity was decreased.

The amount of LDH release was increased in mouse cells exposed to SNPs (SMD = 6.222; 95% CI: 3.745 to 8.698; $p < 0.0001$), but in human cells (SMD = −0.121; 95% CI: −0.924 to 0.682; $p = 0.768$) and rat cells (SMD = 0.380; 95%). CI: −0.179 to 0.940; $p = 0.183$) there was no increase in LDH release.

SNPs nanoparticles equal to and smaller than 20 nm caused LDH release in neurons (SMD = 1.569; 95% CI: 0.387 to 2.751; $p = 0.009$), whereas with SNPs larger than 20 nm there was a weak effect (SMD = 0.597; 95% CI: 0.131 to 1.062; $p = 0.012$) on LDH release.

SNPs nanoparticles without any coating (SMD = 2.182; 95% CI: 0.365 to 3.999; $p = 0.019$) and with a citrate coating (SMD = 2.351; 95% CI: 0.041 to 4.661; $p = 0.046$) had a large effect on LDH release. However, PVP-coated nanoparticles had a small effect (SMD = 0.264; 95% CI: −0.377 to 0.906; $p = 0.419$) on LDH release.

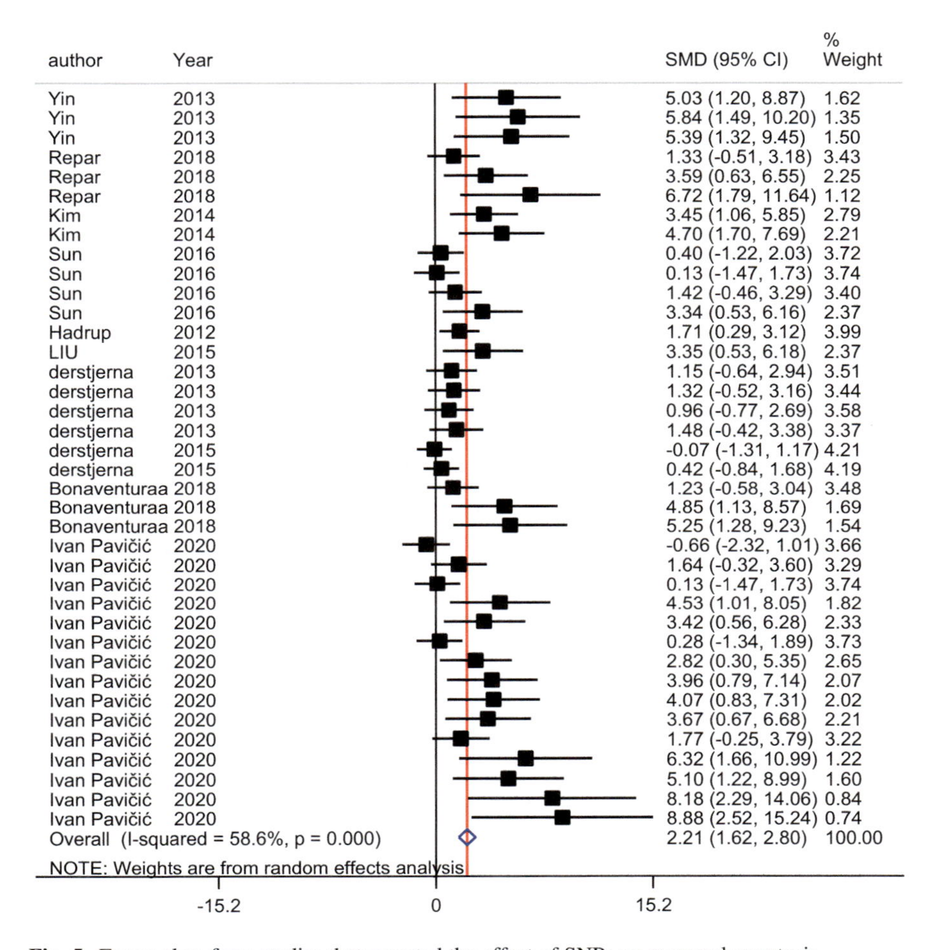

Fig. 5 Forest plots from studies that reported the effect of SNPs on neuronal apoptosis

4 Discussion

According to the results, treatment of nerve cells with silver nanoparticles significantly reduced the number of living cells. This finding is consistent with that of review articles by Sawicki et al. (2019) and Ferdous and Nemmar (2020).

The results showed that a concentration >0.1 μg/mL killed the nerve cells while a concentration <0.1 μg/mL did not kill the cells. The results of linear regression showed that there was a linear relationship between increasing SNP concentration from 0 to 250 μg/mL that decreased the number of living cells with a 1.4% increase in cytotoxicity per unit of SNPs concentration.

In addition to the concentration of SNPs, the type of coating, and the size of the nanoparticles, cell type and cell species were also factors that influenced the extent of

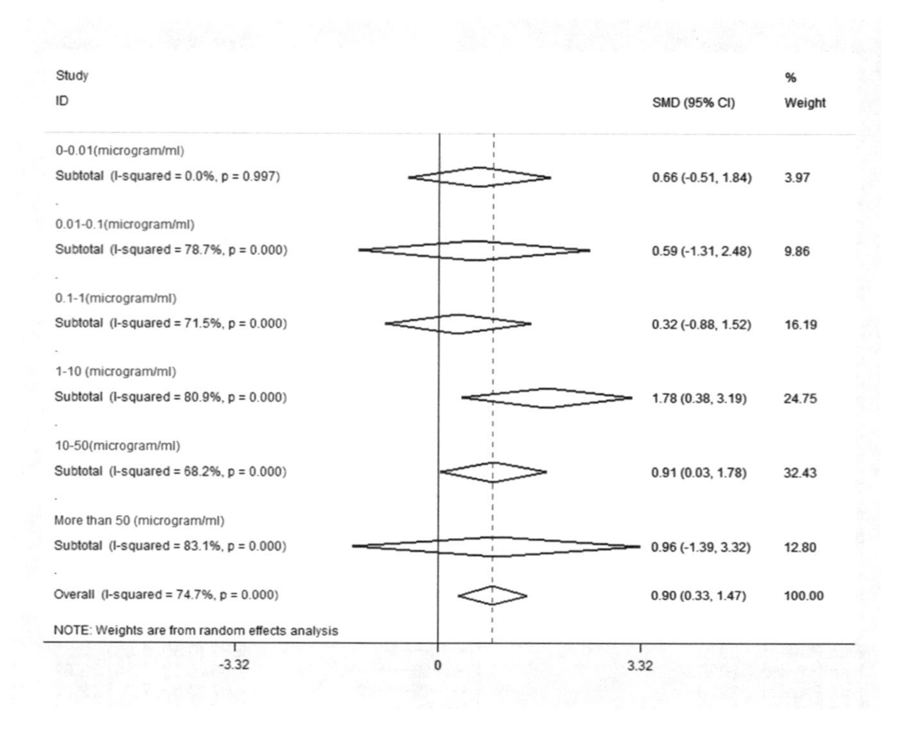

Fig. 6 Forest plots from studies that reported the effect of SNPs on LDH release in neural cells

SNPs toxicity. This finding is consistent with the finding of Strużyńska and Skalska (2018) and Ferdous and Nemmar (2020) that showed the size and surface coating are the key factors determining the toxicity of AgNPs.

Nanoparticles smaller than 20 nm had a more lethal effect. Obviously, the surface area, volume ratio, and surface reaction all vary with the particle size. The rate of deposition onto and diffusion into biological surfaces and the attachment efficiency are also affected by the size of the nanoparticles (Akter et al. 2018; Nel et al. 2006). Smaller SNPs exert stronger cytotoxicity due to their larger specific surface area that regulates the level of oxidative stress, and also the faster dissolution rate of SNPs to produce silver ions which depends on interfacial interactions (Zhang et al. 2014). Increasing the size of the SNPs will change the way they interact with cells and reduce the effects on cell viability.

Uncoated and PVP-coated nanoparticles had a higher lethal effect than citrate coated SNPs. This agrees with the findings of studies performed on macrophages, epithelial cells (Nguyen et al. 2013), and *Caenorhabditis elegans* (X. Yang et al. 2012). These studies also showed that uncoated nanoparticles had higher toxicity than coated nanoparticles, and between PVP and citrate coatings, PVP was more toxic. The protective effect of a citrate coating could be due to the formation of citrate chelation complexes with free Ag^+ ions and consequently reduced Ag^+

Table 5 Overall and subgroup analysis of SNPs treatment effect on LDH release

Subgroup	Number of experiments	Heterogeneity (*p*-value)	SMD (95% CI)	*p*-value
Cell species				
Rat	46	69.2% (<0.0001)	0.380 (−0.179 to 0.940)	0.183
Mouse	15	82.6% (<0.0001)	6.222 (3.745 to 8.698)	<0.0001
Human	4	0.00% (0.986)	−0.121 (−0.924 to 0.682)	0.768
SNPs size				
≤20 nm	31	46.0% (0.003)	1.569 (0.387 to 2.751)	0.009
>20 nm	34	83.2% (<0.0001)	0.597 (0.131 to 1.062)	0.012
SNPs coating				
Uncoated	16	83.1% (<0.0001)	2.182 (0.365 to 3.999)	0.019
PVP	16	51.5% (0.009)	0.264 (−0.377 to 0.906)	0.419
Citrate	7	82.7% (<0.0001)	2.351 (0.041 to 4.661)	0.046
Others	26	74.5% (<0.0001)	0.830 (−0.091 to 1.752)	0.077
Exposure time				
≤24 h	59	71.2% (<0.0001)	0.588 (0.048 to 1.128)	0.033
>24 h	6	84.8%(<0.0001)	8.280 (4.070 to 12.490)	<0.0001
SNPs concentration				
≤ 0.01 μg/mL	2	0.0% (0.997)	0.664 (−0.513 to 1.841)	0.269
0.01–0.1 μg/mL	6	78.7% (<0.0001)	0.587 (−1.309 to 2.483)	0.544
0.1–1 μg/mL	9	71.5% (<0.0001)	0.320 (−0.880 to 1.519)	0.601
1–10 μg/mL	17	80.9% (<0.0001)	1.784 (0.383 to 3.185)	0.013
10–50 μg/mL	21	68.2% (<0.0001)	0.907 (0.033 to 1.780)	0.042
>50 μg/mL	10	83.1% (<0.0001)	0.964 (−1.390 to 3.318)	0.422
Overall	65	74.7% (<0.0001)	0.900 (0.331 to 1.469)	0.002

availability (X. Yang et al. 2012). Neural cells from mouse and rat origin were more resistant to SNP-mediated toxicity compared to human cells. Therefore, caution should be used in extrapolating the results of studies carried out in rat and mouse cells to those expected with human cells.

The results of this study suggest that the toxicity of SNPs against normal nerve cells is greater than that against cancer cells, which should be considered in the use of these nanoparticles for cancer treatment.

Cell death due to SNPs exposure can be largely attributed to apoptosis. However, in subgroup analysis we observed that necrosis increased at concentrations greater than 1 μg/mL. This result is in agreement with published studies on non-neural cancer cells such as HePG-2 (Zhu et al. 2016), HeLa (Yuan et al. 2018), and MCF-7 cells (George et al. 2018) that all suggested that the SNP mechanism was via the induction of apoptosis. In comparison, the proportion of necrosis or apoptosis in breast cancer cells exposed to SNPs showed that there was a mixture of apoptosis and necrosis at lower concentrations, but only necrosis was induced at higher concentrations (ÇIFTÇI et al. 2013).

In most articles, SNPs were purchased and in the articles not mentioned how nanoparticles were synthesized (Table 2). Therefore, in subgroup analysis, the effect of nanoparticle synthesis method was not investigated and further studies are needed to explore the effect of synthesis method on neurons viability. In conclusion, as the use of SNPs in medical and industrial applications continues to increase, guidelines that recommend exposure limits depending on size, coating and other parameters will need to be laid down. These guidelines should focus on possible brain exposure to SNPs to reassure stakeholders against any risk of brain damage, especially for developing children who may be the most vulnerable.

5 Conclusion

Treatment of neurons with SNPs significantly reduces the number of living cells. There is a linear relationship between increasing the concentration of SNPs from 0 to 250 μg/mL and the survival of neurons. As each unit of SNPs concentration increases, the number of living cells decreases by 1.4%. SNPs smaller than 20 nm compared to the larger nanoparticles, as well as uncoated nanoparticles and PVP-coated SNPs compared to the citrate coated SNPs had a more lethal effect on neural cells.

The effect of SNPs on human cells is more toxic than its effect on mouse and rat cells. SNPs are also more toxic to normal neurons than cancer cells. These findings, based on studies to date, are crucial for future use of SNPs in biology and medicine.

Declarations

Funding FR was supported by IRAN University of Medical Sciences, Grants 97-3-75-13323. MRH was supported by US NIH Grants R01AI050875 and R21AI121700.

Conflicts of Interest/Competing Interests MRH declares the following potential conflicts of interest. Scientific Advisory Boards: Transdermal Cap Inc., Cleveland, OH; BeWell Global Inc., Wan Chai, Hong Kong; Hologenix Inc. Santa Monica, CA; LumiThera Inc., Poulsbo, WA; Vielight, Toronto, Canada; Bright Photomedicine, Sao Paulo, Brazil; Quantum Dynamics LLC, Cambridge, MA; Global Photon Inc., Bee Cave, TX; Medical Coherence, Boston MA; NeuroThera, Newark DE; JOOVV Inc., Minneapolis-St. Paul MN; AIRx Medical, Pleasanton CA; FIR Industries, Inc. Ramsey, NJ; UVLRx Therapeutics, Oldsmar, FL; Ultralux UV Inc., Lansing MI; Illumiheal & Petthera, Shoreline, WA; MB Lasertherapy, Houston, TX; ARRC LED, San Clemente, CA; Varuna Biomedical Corp. Incline Village, NV; Niraxx Light Therapeutics, Inc., Boston, MA. Consulting; Lexington Int, Boca Raton, FL; USHIO Corp, Japan; Merck KGaA, Darmstadt, Germany; Philips Electronics Nederland B.V. Eindhoven, Netherlands; Johnson & Johnson Inc., Philadelphia, PA; Sanofi-Aventis Deutschland GmbH, Frankfurt am Main, Germany. Stockholdings: Global Photon Inc., Bee Cave, TX; Mitonix, Newark, DE.

Availability of Data and Material The data that support the findings of this study are available from the corresponding author (FR) on request.

References

Ajdary M, Moosavi MA, Rahmati M, Falahati M, Mahboubi M, Mandegary A, Varma RS (2018) Health concerns of various nanoparticles: a review of their in vitro and in vivo toxicity. Nanomaterials 8. https://doi.org/10.3390/nano8090634

Akter M, Sikder MT, Rahman MM, Ullah AKMA, Hossain KFB, Banik S, Kurasaki M (2018) A systematic review on silver nanoparticles-induced cytotoxicity: physicochemical properties and perspectives. J Adv Res 9:1–16. https://doi.org/10.1016/j.jare.2017.10.008

Aziz N, Faraz M, Sherwani MA, Fatma T, Prasad R (2019) Illuminating the Anticancerous efficacy of a new fungal chassis for silver nanoparticle synthesis. Front Chem 7. https://doi.org/10.3389/fchem.2019.00065

Barhoum A, El-Hout SI, Ali GAM, Abu Serea ES, Ibrahim AH, Pal K, Abdelbasir SM (2019) A broad family of carbon nanomaterials: classification, properties, synthesis, and emerging applications. In: Handbook of nanofibers, pp 451–490. https://doi.org/10.1007/978-3-319-53655-2_59

Begum AN, Aguilar JS, Elias L, Hong Y (2016) Silver nanoparticles exhibit coating and dose-dependent neurotoxicity in glutamatergic neurons derived from human embryonic stem cells. Neurotoxicology 57:45–53. https://doi.org/10.1016/j.neuro.2016.08.015

Bin-Meferij MM, Hamida RS (2019) Biofabrication and antitumor activity of silver nanoparticles utilizing novel Nostoc sp. Bahar M. Int J Nanomedicine 14:9019–9029. https://doi.org/10.2147/IJN.S230457

Bonaventura G, La Cognata V, Iemmolo R, Zimbone M, Contino A, Maccarrone G, Cavallaro S (2018) Ag-NPs induce apoptosis, mitochondrial damages and MT3/OSGIN2 expression changes in an in vitro model of human dental-pulp-stem-cells-derived neurons. Neurotoxicology 67:84–93. https://doi.org/10.1016/j.neuro.2018.04.014

Boroumand Z, Golmakani N, Boroumand S (2018) Clinical trials on silver nanoparticles for wound healing. Nanomed J 5(4):186–191. https://doi.org/10.22038/nmj.2018.05.00001

Brandelli A (2019) The interaction of nanostructured antimicrobials with biological systems: cellular uptake, trafficking and potential toxicity. Food Sci Human Wellness. https://doi.org/10.1016/j.fshw.2019.12.003

Braun NJ, Comfort KK, Schlager JJ, Hussain SM (2013) Partial recovery of silver nanoparticle-induced neural cytotoxicity through the application of a static magnetic field. BioNanoScience 3 (4):367–377. https://doi.org/10.1007/s12668-013-0109-2

Chen IC, Hsiao IL, Lin HC, Wu CH, Chuang CY, Huang YJ (2016) Influence of silver and titanium dioxide nanoparticles on in vitro blood-brain barrier permeability. Environ Toxicol Pharmacol 47:108–118. https://doi.org/10.1016/j.etap.2016.09.009

Choi JH, Min WK, Gopal J, Lee YM, Muthu M, Chun S, Oh JW (2018) Silver nanoparticle-induced hormesis of astroglioma cells: a Mu-2-related death-inducing protein-orchestrated modus operandi. Int J Biol Macromol 117(2017):1147–1156. https://doi.org/10.1016/j.ijbiomac.2018.05.234

Choo WH, Park CH, Jung SE, Moon B, Ahn H, Ryu JS, Oh SM (2016) Long-term exposures to low doses of silver nanoparticles enhanced in vitro malignant cell transformation in non-tumorigenic BEAS-2B cells. Toxicol In Vitro 37:41–49. https://doi.org/10.1016/j.tiv.2016.09.003

Çiftçi H, Türk M, Tamer U, Karahan S, Menemen Y (2013) Silver nanoparticles: cytotoxic, apoptotic, and necrotic effects on MCF-7 cells. Turk J Biol 37(5):573–581. https://doi.org/10.3906/biy-1302-21

Ferdous Z, Nemmar A (2020) Health impact of silver nanoparticles: a review of the biodistribution and toxicity following various routes of exposure. Int J Mol Sci 21(7):2375

Gajbhiye S, Sakharwade S (2016) Silver nanoparticles in cosmetics. J Cosmetics Dermatol Sci Appl 6(1):48–53. https://doi.org/10.4236/jcdsa.2016.61007

Gao W, Leung SW, Bhushan A, Lai JCK (2014) Cytotoxic effects of silver and gold nanoparticles in human glioblastoma U87 cells. Technical proceedings of the 2014 NSTI nanotechnology conference and expo, NSTI-Nanotech, vol 3, pp 134–137

Gatoo MA, Naseem S, Arfat MY, Mahmood Dar A, Qasim K, Zubair S (2014) Physicochemical properties of nanomaterials: implication in associated toxic manifestations. Biomed Res Int 2014:1–8. https://doi.org/10.1155/2014/498420

George PAB, Kumar N, Abrahamse H, Ray S (2018) Apoptotic efficacy of multifaceted biosynthesized silver nanoparticles on human adenocarcinoma cells. Sci Rep 8. https://doi.org/10.1038/s41598-018-32480-5

Giannossa LC, Longano D, Ditaranto N, Nitti MA, Paladini F, Pollini M, Cioffi N (2013) Metal nanoantimicrobials for textile applications. Nanotechnol Rev 2(3). https://doi.org/10.1515/ntrev-2013-0004

Gonzalez-Carter DA, Leo BF, Ruenraroengsak P, Chen S, Goode AE, Theodorou IG, Porter AE (2017) Silver nanoparticles reduce brain inflammation and related neurotoxicity through induction of H 2 S-synthesizing enzymes. Sci Rep 7(January):1–14. https://doi.org/10.1038/srep42871

Haase A, Rott S, Mantion A, Graf P, Plendl J, Thünemann AF, Reiser G (2012) Effects of silver nanoparticles on primary mixed neural cell cultures: uptake, oxidative stress and acute calcium responses. Toxicol Sci 126(2):457–468. https://doi.org/10.1093/toxsci/kfs003

Hadrup N, Loeschner K, Mortensen A, Sharma AK, Qvortrup K, Larsen EH, Lam HR (2012) The similar neurotoxic effects of nanoparticulate and ionic silver in vivo and in vitro. Neurotoxicology 33(3):416–423. https://doi.org/10.1016/j.neuro.2012.04.008

Heinz H, Pramanik C, Heinz O, Ding Y, Mishra RK, Marchon D, Ziolo RF (2017) Nanoparticle decoration with surfactants: molecular interactions, assembly, and applications. Surf Sci Rep 72 (1):1–58. https://doi.org/10.1016/j.surfrep.2017.02.001

Hsiao IL, Hsieh YK, Chuang CY, Wang CF, Huang YJ (2017) Effects of silver nanoparticles on the interactions of neuron- and glia-like cells: toxicity, uptake mechanisms, and lysosomal tracking. Environ Toxicol 32(6):1742–1753. https://doi.org/10.1002/tox.22397

Jazayeri MH, Amani H, Pourfatollah AA, Pazoki-Toroudi H, Sedighimoghaddam B (2016) Various methods of gold nanoparticles (GNPs) conjugation to antibodies. Sensing Bio-Sensing Res 9:17–22. https://doi.org/10.1016/j.sbsr.2016.04.002

Jeevanandam J, Barhoum A, Chan YS, Dufresne A, Danquah MK (2018) Review on nanoparticles and nanostructured materials: history, sources, toxicity and regulations. Beilstein J Nanotechnol 9(1):1050–1074. https://doi.org/10.3762/bjnano.9.98

Kalantari K, Mostafavi E, Afifi AM, Izadiyan Z, Jahangirian H, Rafiee-Moghaddam R, Webster TJ (2020) Wound dressings functionalized with silver nanoparticles: promises and pitfalls. Nanoscale 12(4):2268–2291. https://doi.org/10.1039/C9NR08234D

Khan AM, Korzeniowska B, Gorshkov V, Tahir M, Schrøder H, Skytte L, Kjeldsen F (2019a) Silver nanoparticle-induced expression of proteins related to oxidative stress and neurodegeneration in an in vitro human blood-brain barrier model. Nanotoxicology 13 (2):221–239. https://doi.org/10.1080/17435390.2018.1540728

Khan I, Saeed K, Khan I (2019b) Nanoparticles: properties, applications and toxicities. Arab J Chem 12(7):908–931. https://doi.org/10.1016/j.arabjc.2017.05.011

Kim SH, Ko JW, Koh SK, Lee IC, Son JM, Moon C, Kim JC (2014) Silver nanoparticles induce apoptotic cell death in cultured cerebral cortical neurons. Mol Cell Toxicol 10(2):173–179. https://doi.org/10.1007/s13273-014-0019-6

Lingabathula H, Yellu N (2017) Extra pulmonary toxicity assessment of gold and silver nanorods following intra tracheal instillation in rats. Drug Res 67(10):606–612. https://doi.org/10.1055/s-0043-113255

Liu F, Mahmood M, Xu Y, Watanabe F, Biris AS, Hansen DK, Wang C (2015) Effects of silver nanoparticles on human and rat embryonic neural stem cells. Front Neurosci 9:1–9. https://doi.org/10.3389/fnins.2015.00115

Liu Y, Eaton ED, Wills TE, McCann SK, Antonic A, Howells DW (2018) Human ischaemic cascade studies using SH-SY5Y cells: a systematic review and meta-analysis. Transl Stroke Res 9:564–574. https://doi.org/10.1007/s12975-018-0620-4

Luther EM, Koehler Y, Diendorf J, Epple M, Dringen R (2011) Accumulation of silver nanoparticles by cultured primary brain astrocytes. Nanotechnology 22(37). https://doi.org/10.1088/0957-4484/22/37/375101

Luther EM, Schmidt MM, Diendorf J, Epple M, Dringen R (2012) Upregulation of metallothioneins after exposure of cultured primary astrocytes to silver nanoparticles. Neurochem Res 37(8):1639–1648. https://doi.org/10.1007/s11064-012-0767-4

Ma W, Jing L, Valladares A, Mehta SL, Wang Z, Andy Li P, Bang JJ (2015) Silver nanoparticle exposure induced mitochondrial stress, caspase-3 activation and cell death: amelioration by sodium selenite. Int J Biol Sci 11(8):860–867. https://doi.org/10.7150/ijbs.12059

Mala R, Annie Aglin A, Ruby Celsia AS, Geerthika S, Kiruthika N, VazagaPriya C, Srinivasa Kumar K (2017) Foley catheters functionalised with a synergistic combination of antibiotics and silver nanoparticles resist biofilm formation. IET Nanobiotechnol 11(5):612–620. https://doi.org/10.1049/iet-nbt.2016.0148

Mytych J, Zebrowski J, Lewinska A, Wnuk M (2017) Prolonged effects of silver nanoparticles on p53/p21 pathway-mediated proliferation, DNA damage response, and methylation parameters in HT22 hippocampal neuronal cells. Mol Neurobiol 54(2):1285–1300. https://doi.org/10.1007/s12035-016-9688-6

Neelakandan MS, Thomas S (2018) Applications of silver nanoparticles for medicinal purpose. JSM Nanotechnol Nanomed 6(1):1063. Retrieved from https://www.jscimedcentral.com/Nanotechnology/nanotechnology-6-1063.pdf

Nel A, Xia T, Mädler L, Li N (2006) Toxic potential of materials at the nanolevel. Science 311:622–627. https://doi.org/10.1126/science.1114397

Nguyen KC, Seligy VL, Massarsky A, Moon TW, Rippstein P, Tan J, Tayabali AF (2013) Comparison of toxicity of uncoated and coated silver nanoparticles. J Phys Conf Ser 429(1). https://doi.org/10.1088/1742-6596/429/1/012025

Oh JH, Son MY, Choi MS, Kim S, Choi AY, Lee HA, Yoon S (2016) Integrative analysis of genes and miRNA alterations in human embryonic stem cells-derived neural cells after exposure to silver nanoparticles. Toxicol Appl Pharmacol 299:8–23. https://doi.org/10.1016/j.taap.2015.11.004

Pavi I, Mili M, Pongrac IM, Brki L, Matijevi T, Ili K, Vinkovi I (2019) Neurotoxicity of silver nanoparticles stabilized with different coating agents: in vitro response of neuronal precursor cells. https://doi.org/10.1016/j.fct.2019.110935

Pongrac IM, Ahmed LB, Mlinarić H, Jurašin DD, Pavičić I, Marjanović Čermak AM, Milić M, Gajović S, Vinković Vrček I (2018) Surface coating affects uptake of silver nanoparticles in neural stem cells. J Trace Elem Med Biol 50:684–692. https://doi.org/10.1016/j.jtemb.2017.12.003

Rai M, Yadav A, Gade A (2009) Silver nanoparticles as a new generation of antimicrobials. Biotechnol Adv 27(1):76–83. https://doi.org/10.1016/j.biotechadv.2008.09.002

Rani GN, Rao BN, Shamili M, Jyothi Padmaja I (2018) Combined effect of silver nanoparticles and honey in experimental wound healing process in rats. Biomed Res 29:3074–3078. https://doi.org/10.4066/biomedicalresearch.29-18-898

Repar N, Li H, Aguilar JS, Li QQ, Drobne D, Hong Y (2018) Silver nanoparticles induce neurotoxicity in a human embryonic stem cell-derived neuron and astrocyte network. Nanotoxicology 12(2):104–116. https://doi.org/10.1080/17435390.2018.1425497

Rohatgi A (2010) WebPlotDigitizer – extract data from plots, images, and maps. Arohatgi. Retrieved from http://arohatgi.info/WebPlotDigitizer/

Rupp ME, Fitzgerald T, Marion N, Helget V, Puumala S, Anderson JR, Fey PD (2004) Effect of silver-coated urinary catheters: efficacy, cost-effectiveness, and antimicrobial resistance. Am J Infect Control 32(8):445–450. https://doi.org/10.1016/j.ajic.2004.05.002

Salazar-García S, Silva-Ramírez AS, Ramirez-Lee MA, Rosas-Hernandez H, Rangel-López E, Castillo CG, Gonzalez C (2015) Comparative effects on rat primary astrocytes and C6 rat glioma cells cultures after 24-h exposure to silver nanoparticles (AgNPs). J Nanopart Res 17(11):1–13. https://doi.org/10.1007/s11051-015-3257-1

Salvioni L, Galbiati E, Collico V, Alessio G, Avvakumova S, Corsi F, Colombo M (2017) Negatively charged silver nanoparticles with potent antibacterial activity and reduced toxicity for pharmaceutical preparations. Int J Nanomedicine 12:2517–2530. https://doi.org/10.2147/IJN.S127799

Sawicki K, Czajka M, Matysiak-kucharek M, Fal B, Męczyńska-wielgosz S, Sikorska K, Kruszewski M (2019) Toxicity of metallic nanoparticles in the central nervous system. Nanotechnol Rev:175–200

Sharma G, Nam J-S, Sharma AR, Lee S-S (2018) Antimicrobial potential of silver nanoparticles synthesized using medicinal herb coptidis rhizome. Molecules 23(9):2269. https://doi.org/10.3390/molecules23092269

Slavin YN, Asnis J, Häfeli UO, Bach H (2017) Metal nanoparticles: understanding the mechanisms behind antibacterial activity. J Nanobiotechnol 15(1):65. https://doi.org/10.1186/s12951-017-0308-z

Söderstjerna E, Johansson F, Klefbohm B, Englund Johansson U (2013) Gold- and silver nanoparticles affect the growth characteristics of human embryonic neural precursor cells. PLoS One 8(3):1–13. https://doi.org/10.1371/journal.pone.0058211

Söderstjerna E, Bauer P, Cedervall T, Abdshill H, Johansson F, Johansson UE (2014) Silver and gold nanoparticles exposure to in vitro cultured retina – studies on nanoparticle internalization, apoptosis, oxidative stress, glial- and microglial activity. PLoS One 9(8). https://doi.org/10.1371/journal.pone.0105359

Soleimani M, Habibi-Pirkoohi M (2017) Biosynthesis of silver nanoparticles using Chlorella vulgaris and evaluation of the antibacterial efficacy against *Staphylococcus aureus*. Avicenna J Med Biotechnol 9:120–125

Stensberg MC, Wei Q, Mclamore ES, Marshall D, Wei A, Porterfield DM, Sepulveda MS (2012) Toxicological studies on silver nanoparticles. Nanomedicine (Lond) 6(5):879–898. https://doi.org/10.2217/nnm.11.78.Toxicological

Strużyńska L, Skalska J (2018) Mechanisms underlying neurotoxicity of silver nanoparticles. Cell Mol Toxicol Nanoparticles 1048:227–250

Sun C, Yin N, Wen R, Liu W, Jia Y, Hu L, Jiang G (2016) Silver nanoparticles induced neurotoxicity through oxidative stress in rat cerebral astrocytes is distinct from the effects of silver ions. Neurotoxicology 52:210–221. https://doi.org/10.1016/j.neuro.2015.09.007

Tejada S, Manayi A, Daglia M, Nabavi SF, Sureda A, Hajheydari Z, Nabavi SM (2016) Wound healing effects of curcumin: a short review. Curr Pharm Biotechnol 17(11):1002–1007. https://doi.org/10.2174/1389201017666160721123109

Thiruvengadam M, Rajakumar G, Chung I-M (2018) Nanotechnology: current uses and future applications in the food industry. 3 Biotech 8(1):74. https://doi.org/10.1007/s13205-018-1104-7

Tuncsoy BS (2018) Toxicity of nanoparticles on insects: a review. Adana Bilim ve Teknoloji Üniversitesi Fen Bilimleri Dergisi 1(2):49–61. Retrieved from https://dergipark.org.tr/download/article-file/614363

Wang R, Song B, Wu J, Zhang Y, Chen A, Shao L (2018) Potential adverse effects of nanoparticles on the reproductive system. Int J Nanomedicine 13:8487–8506. https://doi.org/10.2147/IJN.S170723

Wennersten R, Fidler J, Spitsyna A (2008) Nanotechnology: a new technological revolution in the 21st century. In: Handbook of performability engineering, pp 943–952. https://doi.org/10.1007/978-1-84800-131-2_57

Yang Z, Liu ZW, Allaker RP, Reip P, Oxford J, Ahmad Z, Ren G (2010) A review of nanoparticle functionality and toxicity on the central nervous system. J R Soc Interface 7:S411–S422. https://doi.org/10.1098/rsif.2010.0158.focus

Yang X, Gondikas AP, Marinakos SM, Auffan M, Liu J, Hsu-Kim H, Meyer JN (2012) Mechanism of silver nanoparticle toxicity is dependent on dissolved silver and surface coating in *caenorhabditis elegans*. Environ Sci Technol 46:1119–1127. https://doi.org/10.1021/es202417t

Yesilot S, Aydin C (2019) Silver nanoparticles; a new hope in cancer therapy? Eastern J Med 24(1):111–116. https://doi.org/10.5505/ejm.2019.66487

Yin N, Liu Q, Liu J, He B, Cui L, Li Z, Jiang G (2013) Silver nanoparticle exposure attenuates the viability of rat cerebellum granule cells through apoptosis coupled to oxidative stress. Small 9:1831–1841. https://doi.org/10.1002/smll.201202732

Yin N, Hu B, Yang R, Liang S, Liang S, Faiola F (2018) Assessment of the developmental neurotoxicity of silver nanoparticles and silver ions with mouse embryonic stem cells in vitro. J Interdiscip Nanomed 3(3):133–145. https://doi.org/10.1002/jin2.49

Yuan YG, Zhang S, Hwang JY, Kong IK (2018) Silver nanoparticles potentiates cytotoxicity and apoptotic potential of camptothecin in human cervical cancer cells. Oxid Med Cell Longev 2018. https://doi.org/10.1155/2018/6121328

Zhang T, Wang L, Chen Q, Chen C (2014) Cytotoxic potential of silver nanoparticles. Yonsei Med J 55:283–291. https://doi.org/10.3349/ymj.2014.55.2.283

Zhu B, Li Y, Lin Z, Zhao M, Xu T, Wang C, Deng N (2016) Silver nanoparticles induce HePG-2 cells apoptosis through ROS-mediated signaling pathways. Nanoscale Res Lett 11. https://doi.org/10.1186/s11671-016-1419-4

Ziemińska E, Stafiej A, Struzyńska L (2014) The role of the glutamatergic NMDA receptor in nanosilver-evoked neurotoxicity in primary cultures of cerebellar granule cells. Toxicology 315(1):38–48. https://doi.org/10.1016/j.tox.2013.11.008

A Systematic Review on Occurrence and Ecotoxicity of Organic UV Filters in Aquatic Organisms

Ved Prakash and Sadasivam Anbumani

Contents

1 Introduction 122
2 Occurrence of Organic UV Filters 125
 2.1 In Freshwater Ecosystems 125
 2.2 In Wastewater Treatment Plants (WWTPs) 126
 2.3 In Marine Ecosystems 127
3 Ecotoxicity of Organic UV Filters on Biota 129
 3.1 Effects on Freshwater Organisms 134
 3.2 Effects on Marine Organisms 137
4 Toxicity Mechanism of Organic UV Filters 140
5 Knowledge Gaps and Future Research Directions 154
6 Conclusion 155
References 155

Abstract The growing production of cosmetic products such as organic UV filters (OUVFs) in recent years has raised concern regarding their safety to human and environmental health. The inability of wastewater treatment plants in removing these chemical entities and their high octanol-water partition coefficient values tend to result in the persistence of OUVFs in several environmental matrices, leading these to be categorized as "emerging environmental contaminants" because of their unknown risk. Besides aquatic ecosystem contamination, the application of sludge disposal equally threatens terrestrial biota. Besides, the available reviews focusing on levels of OUVFs in aqueous systems (freshwater and marine), instrumental analysis from various samples, and specific toxicity effects, compiled information on the ecotoxicity of OUVFs is currently lacking. Hence, the present manuscript systematically reviews the ecotoxicity of OUVFs in freshwater and marine organisms occupying lower to higher trophic levels, including the underlying mechanisms

V. Prakash · S. Anbumani (✉)
Ecotoxicology Laboratory, Regulatory Toxicology Group, CSIR-Indian Institute of Toxicology Research, Lucknow, India

Academy of Scientific and Innovative Research (AcSIR), Ghaziabad, India
e-mail: vedglee880@gmail.com; anbumani@iitr.res.in; aquatox1982@gmail.com

P. de Voogt (ed.), *Reviews of Environmental Contamination and Toxicology Volume 257*,
Reviews of Environmental Contamination and Toxicology 257,
https://doi.org/10.1007/398_2021_68

of action and current knowledge gaps. The available scientific evidence suggests that OUVFs are a prime candidate for environmental concern due to their potential toxic effects. To the best of our knowledge, this is the first document detailing the toxicological effects of OUVFs in aquatic organisms.

Keywords Accumulation · Ecotoxicity · Occurrence · Organic UV filters (OUVFs) · Trophic level

Abbreviations

4-MBC	3-(4-methylbenzylidene) camphor
BM-DBM	butyl methoxydibenzoylmethane
BP-3	Benzophenone-3
BP-4	Benzophenone-4
BP-8	Benzophenone-8
DHHB	Diethylamino Hydroxybenzoyl Hexyl Benzoate
EHMC	2-ethyl-hexyl-4-trimethoxycinnamate
HMS	homosalate
OC	Octocrylene
OD-PABA	ethylhexyl dimethyl p-aminobenzoic acid
OMC	Octyl methoxycinnamate
OUVFs	Organic UV filters
PBSA	Phenylbenz-imidazole sulfonic acid

1 Introduction

Ultraviolet (UV) radiation plays a crucial role in maintaining the proper functioning of biotic communities in a dynamic ecosystem. The level of UV radiation also determines the extent of adverse effects on humans and other organisms, as well as beneficial effects, such as vitamin D synthesis upon short-term exposure (Cadena-Aizaga et al. 2019). Uncertainties under the current changing climatic conditions and increased UV radiation levels in certain parts of the world have forced the human population to prevent the deleterious effects of ultraviolet radiation. Since the stratospheric ozone layer that functions as a shield is unable to protect humans from the UV rays reaching the earth surface, UVA radiation induces direct tanning, photo-oxidation of melanin, premature skin aging, and skin cancer or melanoma and UVB induces sunburn and immune-suppression, etc. (Frederick et al. 1989; McKenzie et al. 2003). Besides, the skin itself is a good barrier for preventing the entry of UV radiation and subsequent deleterious effects by thickening and tanning of stratum corneum (uppermost layer of the skin) but the degree of photo-protection is not sufficient to do the same (Diffey 1998; Sheehan et al. 1998). Hence, many chemical substances are used to absorb the deleterious UV rays in sunscreen lotions and these are known as organic UV filters (OUVFs) (WHO 2002). The cosmetic industry sector is mushrooming after the enormous usage of sunscreens by humans across the world. To date, more than 27 different OUVFs are used in cosmetics either in individual or mixture combinations (Kumar and Gupta 2013) and grouped under 12 different categories based on their chemical functional groups. These include benzophenone derivatives (18 OUVFs), p-aminobenzoic acid derivatives (4 OUVFs), camphor derivatives (6 OUVFs), benzotriazole derivatives (4 OUVFs), salicylate derivatives (3 OUVFs), triazine derivatives (3 OUVFs), cinnamate derivatives (2 OUVFs), dibenzoyl methane derivative (1 OUVF), benzimidazole derivatives (2 OUVFs), benzalmalonate derivative (1 OUVF), crylene derivative (1 OUVF), benzhydrol, etc. (Ramos et al. 2015). The authorized concentrations of OUVFs in sunscreen products are shown graphically in Fig. 1. Generally, OUVFs are used in combination to enhance the sun protection factor (SPF) values that eventually lead to increased concentration in the final products posing a higher level of contamination. These chemicals are also employed in plastics, adhesives, shampoos, creams, and rubber as sun-blocking agents (Ramos et al. 2015). Evidence shows that OUVFs have the potential to enter the food web of an ecosystem (Schmid et al. 2007; Nakata et al. 2007; Kupper et al. 2006; Balmer et al. 2005). Among the OUVFs, Octocrylene (OC), 3-(4-methylbenzylidene) camphor (4-MBC), and benzophenone-3 (BP-3) (oxybenzone) are widely reported contaminants of the aquatic ecosystem (Gago-Ferrero et al. 2012; Kameda et al. 2011; Plagellat et al. 2006; Ramos et al. 2015). Because of high photostability and distribution coefficient value between octanol and water (log Kow), OUVFs bioaccumulation in aquatic fauna as well as in the human population (Hagedorn-Leweke and Lippold 1995) has been warranted. Wastewater treatment plants (WWTPs) contribute with a maximum load of OUVFs in aquatic and terrestrial ecosystems. On the other hand, direct discharge by humans in lakes, rivers, and coastal regions through recreational activities and waste from industries plays a major role in contaminating water bodies (Kim and Choi 2014). Moreover, continuous release of these chemical entities at low concentrations causes concern among ecotoxicologists about their safety and toxicity (Sharifan et al. 2016a; Ruszkiewicz et al. 2017; Ramos et al. 2016).

Briefly, once UV filters enter water bodies, they inhibit growth and photosynthesis in primary producers like green algae (Sieratowicz et al. 2011; Mao et al. 2018a; Molins-Delgado et al. 2016; Du et al. 2017; Tovar-Sánchez et al. 2013), impair growth rate and reproduction in primary consumers, such as daphnids (Sieratowicz et al. 2011; Molins-Delgado et al. 2016; Du et al. 2017), caddish fly (Campos et al. 2017a), developmental anomalies, oxidative stress (Quintaneiro et al. 2019; Zhou et al. 2019; Fong et al. 2016; Du et al. 2017), and neuronal defects (Li et al. 2016) in zebrafish, multi-xenobiotic resistance activity (MXR) in *Tetrahymena thermophila* (Gao et al. 2016) and perturbations in the expression profile of genes involved in hormone metabolism and reproduction (Zhang et al. 2016; Kim et al. 2014; Blüthgen et al. 2012, 2014; Zucchi et al. 2011b), antiandrogenicity (Liang et al. 2020), reproduction toxicity and estrogenicity (Yan et al. 2020) in secondary consumers

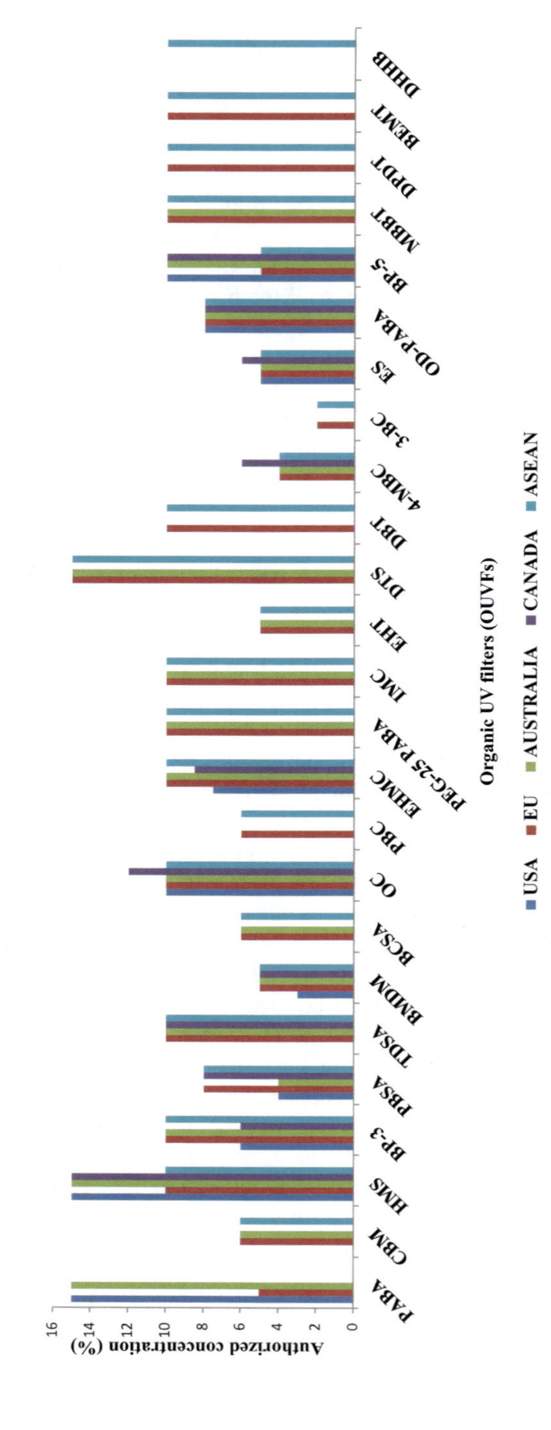

Fig. 1 Graphical presentation of authorized concentration of organic UV filters (OUVFs) in different continents

like zebrafish (*Danio rerio*) and Japanese medaka (*Oryzias latipes*) are also effects that have been observed.

The available reviews on organic UV filters mainly focused on their occurrence in freshwater (including WWTPs), marine, and terrestrial ecosystems (Brausch and Rand 2011; Kim and Choi 2014; Hopkins and Blaney 2016; Ramos et al. 2016) and validation of analytical methods for OUVFs quantification in various environmental matrices (Ramos et al. 2015; Cadena-Aizaga et al. 2019); specific toxicity effects like neuro- and reproduction toxicity in mammals, fate, and transformation in aqueous systems, among others (Ghazipura et al. 2017; Juliano and Magrini 2017; Ruszkiewicz et al. 2017; Mao et al. 2019; Schneider and Lim 2019) with less attention paid to the ecotoxicological effects on aquatic organisms. Hence, the present review is focused on covering the following aspects: (1) occurrence of OUVFs in aqueous systems; (2) potential adverse effects of OUVFs on aquatic biota inhabiting freshwater and marine ecosystems at different trophic levels; (3) toxicity mechanism; and (4) knowledge gaps and future research directions.

2 Occurrence of Organic UV Filters

2.1 *In Freshwater Ecosystems*

The structural and functional integrity of an ecosystem is maintained by the coordinating action of abiotic and biotic components. The prevalence of natural and anthropogenic stressors results in perturbed integrity, affecting the respective ecological niche. It has been noted that benzophenone toxicity profiling has been done extensively and its derivatives persist in freshwater ecosystems at variable concentrations in different parts of the globe. In Switzerland, estrogenic UV filters such as BP-3, 4-MBC, and ethylhexyl methoxycinnamate (EHMC) have been found up to 6–68 ng/L concentrations (Fent et al. 2010). Besides, BP-3, the most frequently used organic UV filter, has been detected in a concentration of 125 ng/L in Swiss rivers and lakes (Fent et al. 2010; Poiger et al. 2004) and also in Japan (Kameda et al. 2011). In Spain, the UV filter BP-3 derived metabolite Benzophenone-1 (BP-1) was detected at a concentration of 52 and 37 ng/L (Negreira et al. 2009). Additionally, 4-MBC and OC were also detected in seven small rivers in Switzerland, and contamination by 4-MBC and OC probably originating in WWTPs (Buser et al. 2006). Zwiener et al. (2007) detected the occurrence of OUVFs, including BP-3, 4-MBC, and EHMC up to 40 mg/L in a swimming pool.

Mao et al. (2018b) investigated the presence and providence of seven benzophenone type OUVFs in 5 major tributaries of a tropical urban watershed in Singapore. Concentrations in surface waters ranged from 0.019 to 0.2308 μg/L, 0.048 to 0.115 μg/L in pore water, 0.295 to 5.813 mg/Kg dw in suspended solids, and 0.006 to 0.037 mg/Kg dw in surface sediments. Samples collected from the Chesapeake Bay revealed the prevalence of BP-3, OC, and homosalate in 14 sites, in addition to EHMC with concentration ranges between 0.072 and 0.188 μg/L, 0.031

and 0.114 μg/L, and 0.008 and 0.029 μg/L, respectively. Furthermore, the relative concentration of OUVFs (hydrophobic) in the aquatic environment suggests the potential for accumulation in tissues and sediment (He et al. 2019).

2.2 *In Wastewater Treatment Plants (WWTPs)*

WWTPs play a major role in recycling wastewater through various stages of treatment before en-route to rivers and streams. OUVFs enter WWTPs through routine human activities such as bathing, after dermal application, and swimming. Samples from WWTPs analyzed for the occurrence of OUVFs in influent (untreated water) and effluent (treated water) revealed the occurrence of OUVFs in the effluent water at ng/L to μg/L concentrations and shows the inability of sewage treatment plants in removing these chemical toxicants that are contaminating the aquatic bodies. For instance, in India, OUVFs like BP-3 and BP were detected from WWTPs located at Udupi, Mangalore, Coimbatore, Manipal, Saidpur, Nagpur, and others at variable concentrations. For example, the influent concentration of oxybenzone was around 0.005–0.0856 μg/L followed by an effluent concentration of 0.0011–0.0412 μg/L, whereas benzophenone was at higher concentrations in influent and effluent water with 3.96 μg/L and 1.5 μg/L, respectively (Balakrishna et al. 2017). In other parts of the globe, water samples collected from the up- and downstream of the river and WWTPs in Catalonia (Spain) showed that the effluents were the major driver for OUVFs contamination in the aquatic environment (Gago-Ferrero et al. 2013a). The concentration of BP-4 in sewage influent was reported at 572–1,420 ng/L concentrations in Germany and Spain, respectively (Wick et al. 2010; Gago-Ferrero et al. 2013a) and in South Wales (UK) concentrations up to 13.3 and 6.3 mg/L in sewage influent and effluent water were reported (Kasprzyk-Hordern et al. 2009). In Hong Kong, BP-3 was detected at concentrations ranging between 541 and 700 ng/L, whereas in Taiwan, China, and Spain, the occurrence of BP-8 was reported at 0.01, 0.084, and 0.055 μg/L in WWTP effluent respectively (Wu et al. 2013; Tsui et al. 2014; and Pedrouzo et al. 2010).

Balmer et al. (2005) reported the occurrence of 4 commonly used OUVFs, such as EHMC, OC, 4-MBC, and BP-3 in WWTPs from Swiss lakes. A maximum of 19 μg/L concentration of EHMC was detected and the order of occurrence was OC $<$ BP-3 $\sim$ 4-MBC $<$ EHMC. Among the detected OUVFs, 4-MBC was the most frequently occurring UV filter with a maximum concentration of 2.7 μg/L. Some authors also reported the prevalence of OC in influent water from 0.018 to 0.461 μg/L concentrations in the USA, Europe, and Japan (Ramos et al. 2016) and Langford et al. (2015) reported 0.7 μg/L concentration of OC in Norway. It has also been reported that the influent concentration of 4-MBC ranges between 0.003 and 6.5 μg/L, whereas in effluent water it was reported at concentrations varying between 0.024 and 2.7 μg/L in various countries like Portugal, Switzerland, China, Spain, Germany, and Australia (Tsui et al. 2014; Liu et al. 2012; Cunha et al. 2015a; Li et al. 2007; Kupper et al. 2006; Balmer et al. 2005).

2.3 In Marine Ecosystems

Contamination of marine ecosystems by OUVFs is primarily through recreational activities by beachgoers, along with the contamination of freshwater bodies like lakes, rivers, and ponds. In the Gulf of Mexico, among the 14 geographical regions from Texas coastal zone, one of the geographical areas – Nueces – with 43 beaches – has the highest contamination of four OUVFs (BP, EHMC, BM-DBM, and OC). The input load of OUVFs into the aquatic environment is 477 kg/yr. for EHMC, followed by 318 kg/yr., 258 kg/yr., and 159 kg/yr. for OC, BM-DBM, and BP, respectively (Sharifan et al. 2016b). Different classes of OUVFs with varied concentrations are detected in marine ecosystems. For example, BP-3 was found at concentrations ranging from few parts per trillion (ng/L) to parts per billion (μg/L) (Silvia Díaz-Cruz et al. 2008; Ramos et al. 2015; Sánchez-Quiles and Tovar-Sánchez 2015). Near Majorca Island (in Spain), the BP-3 and 4-MBC concentrations ranged between 0.0536 μg/L and 0.5775 μg/L and 0.0514 and 0.1134 μg/L, respectively (Tovar-Sánchez et al. 2013) and in US Virgin Islands (in coral reefs), BP-3 was also detected at 75–1400 μg/L concentrations; in Hawaii, the measured concentration reached 19.2 μg/L values (Downs et al. 2016). It was also reported that concentrations of 227 ng/L of oxybenzone (BP-3), 100 ng/L of 2-ethyl-hexyl-4-trimethoxycinnamate (EHMC), and 153 ng/L of octocrylene (OC) in Los Angeles, New York, and Bangkok were observed (Tsui et al. 2014).

Even though the flow of OUVFs from freshwater to marine water is high, this is the indirect pathway of contamination, whereas site-specific contamination is through the recreational activities across the world on the seashores, resulting in significant contamination. OUVFs such as 4-methylbenzylidene camphor (4-MBC) oxybenzone, homosalate (HMS), BP-4, ethylhexyl methoxycinnamate (EHMC), diethylamino hydroxybenzoyl hexyl benzoate (DHHB) were identified in coastal waters (Daughton and Ternes 1999; Giokas et al. 2007, Danovaro et al. 2008; Tovar-Sánchez et al. 2013; Sánchez-Quiles and Tovar-Sánchez 2015; Downs et al. 2016). Information on how much quantity of sunscreens that enter into the marine environment is not clear, with very few data have been available. Annually, about 20,000 tons of sunscreen was estimated to release into the Northern Mediterranean (Corinaldesi 2001). Approximately 6,000–14,000 tons of sunscreen and lotions entered into coral reef areas per year (Downs et al. 2016). A variety of body lotion contains 1–10% of oxybenzone and suggest that at least 40% of coastal reefs and 10% of global reefs are at risk owing to OUVFs contamination. However, the level of OUVFs in coral reef sites quantified till date was very few (Downs et al. 2016).

It has been thought that OUVFs are generally found at ng/L detectable levels, but concentrations several orders of magnitude higher (mg/L) have been detected (around 1.395 mg/L oxybenzone) in the US Virgin Islands showing the continuous use and discharge of these cosmetic entities in aqueous systems (Downs et al. 2016). It is imperative that despite the low concentrations, the persistency of these chemicals may have additive effects if these OUVFs are sequestered by marine biota. Analysis of seawater at Trunk Bay on St. John Island in the US Virgin Islands

(Caribbean) indicated the presence of oxybenzophenone between 1 mg/L and 90 μg/L (Downs et al. 2011). However, repeated sampling showed that the concentration level of oxybenzone was at 1.39 mg/L near the rim of the Trunk Island indicating a threat to the coral communities. It has been observed that the concentration of oxybenzone shows a linear trend (75–95 μg/L) in a time span of 48 h with respect to the total number of swimmers (Downs et al. 2016).

In South Carolina's Folly Bay, OUVFs concentrations between 10 and 2013 ng/L were detected during the summer season (Bratkovics and Sapozhnikova 2011). In general, higher concentrations of OUVFs in the aquatic environment are correlated with sites having the highest usage of personal care products. The highest concentration found was 2013 ng/L for oxybenzone and 1,409 ng/L for OC. Concentrations of padimate-O, octinoxate, and avobenzone were observed between 62 and 321 ng/L, 30 and 264 ng/L concentrations, respectively and Carolina sites had the highest concentrations of OUVFs with more than 3,700 ng/L OC and ~ 2,200 ng/L oxybenzone at a local beach indicating its extensive use by beachgoers (Bratkovics et al. 2015).

Screening of OUVFs such as BP3, OC, OD-PABA, EHMC, HMS, BMDBM, 4-MBC, and DHHB in water samples from six seashores around the Gran Canaria island revealed 99% of its occurrence with an average concentration of 3,316 ng/L (Sánchez-Quiles and Tovar-Sánchez 2015). Studies have shown that five out of seven OUVFs were found in five different species of corals. This included BP-1, BP-3, BP-8, OC, and OD-PADA. Concentrations of 0.0318 and 0.0247 mg/Kg wet weight for BP-3 and BP-8 were found with 65% maximum detection frequency. During the wet season, coral tissues accumulate higher concentrations of BP-3 due to elevated recreational activities resulting in greater sunscreen discharge in coastal waters. The risk assessment suggested that more than 20% of coral samples have oxybenzone with concentrations higher than the threshold values. Coral communities are predicted to be more susceptible to the harmful effects of Benzophenone-3 during the wet season (Tsui et al. 2017).

Water samples from Maunalua bay public swimming sites in Hawaii contained oxybenzone at 100 ng/L to 19.2 μg/L as minimum and maximum concentrations. Similarly, oxybenzone at 5 μg/L concentration was detected at Maui island in Kapalua beach, in the USA. On the other hand, traceable concentrations of oxybenzone (lower than 5 μg/L) in water sample (71 swimmers within 200 m of sampling site) were detected from Hawaii Kahekili Beach Park, a major tourist spot destiny (Downs et al. 2016). In Palau (Pacific Ocean), 23 OUVFs were found in marine sediment out of which 22 were detected in aqueous samples. Both parent and metabolite chemicals were widely observed in a long stretch of the Jellyfish Lake and found at lower concentrations in the tissue of jellyfish (Cnidarians) indicating the bioaccumulation potential of OUVFs (Bell et al. 2017). From the North and Baltic Sea, the concentration of total UV filters ranges between from 0.12 to 11.2 μg/g dw. Among the four samples analyzed, all concentration was <MQL (Method Quantitation Limit) and one sample with <MDL (Method Detection Limit). Samples from the Skagerrak and Kattegat sites recorded 0.5 μg/g dw and the

German Baltic Sea with 1.7 μg/g dw. OC was found to be the dominating UV filter at 9.7 μg/g dw near to the harbor area (Apel et al. 2018b).

Apel et al. (2018a) first time analyzed the availability of OUVFs and UV stabilizers in yellow Seas and Chinese Bohai surface sediments. Among the 21 samples tested, 16 were positively detected for the occurrence of OUVFs and Stabilizers. All the UV stabilizer concentrations were below MQL (method quantitation limit) from the 16% analyzed samples, and the total UV stabilizer concentrations were ranged from 0.06–25.7 μg/g dw (mean value 3.9 μg/g dw) for Laizhou Bay, 2.1 μg/g dw for the Yellow Sea, and 0.6 μg/g dw for other parts of the Bohai Sea, respectively. There was no statistically significant contamination observed, and UV stabilizers such as UV-350, UV-PS, IAMC, 4-MBC, and BP-3 were detected below their MDLs (method detection limit) in all the samples analyzed. In sediments, OC and UV-329 are the predominant contaminants which account for 15% and 50% of the total contamination, respectively. About 69% of UV-234 stabilizer was detected, followed by other benzotriazole derivatives, and > 60% of hindered amine light stabilizers (HALS) were reported. About 46% of the samples were contaminated with OC up to 25 ng/g dw at the Lanzhou Bay, with a lower concentration of 0.3 ng/g dw in other parts of the Bohai Sea.

The concentration of EHS from Laizhou Bay reached up to 0.0012 mg/Kg dw, whereas UV-327 and UV-326 were detected up to 0.0005 and 0.0008 mg/Kg dw, respectively. Compared to Laizhou Bay, samples from the North Yellow Sea showed an elevated trend. The concentrations of EHS and OC observed were 0.0076 ± 0.0036 mg/Kg dw and 0.0014 ± 0.0071 mg/Kg dw, respectively. At the Atlantic Coast, higher concentration of EHS were detected, whereas Hong Kong and Tokyo Bay sediments were devoid of EHS. OC and EHS were detected at a concentration of 0.00425 mg/Kg dw and 0.00095 mg/Kg dw, respectively, in the South Yellow Sea, and the benzotriazole type UV stabilizers were the most prominent organic chemical entity around the Shandong Peninsula (Apel et al. 2018a). A diagrammatic representation of the occurrence of OUVFs in aquatic ecosystems and proportion of studies on OUVFs in different phyla of animal kingdom is shown in Figs. 2 and 3. Table 1 shows the levels of OUVFs in different regions of the globe.

3 Ecotoxicity of Organic UV Filters on Biota

The ubiquitous presence of OUVFs in various matrices threatens the biota across trophic levels with potential toxic effects. This section provides detailed information on the adverse effects of OUVFs noted in freshwater and marine organisms representing the trophic niche.

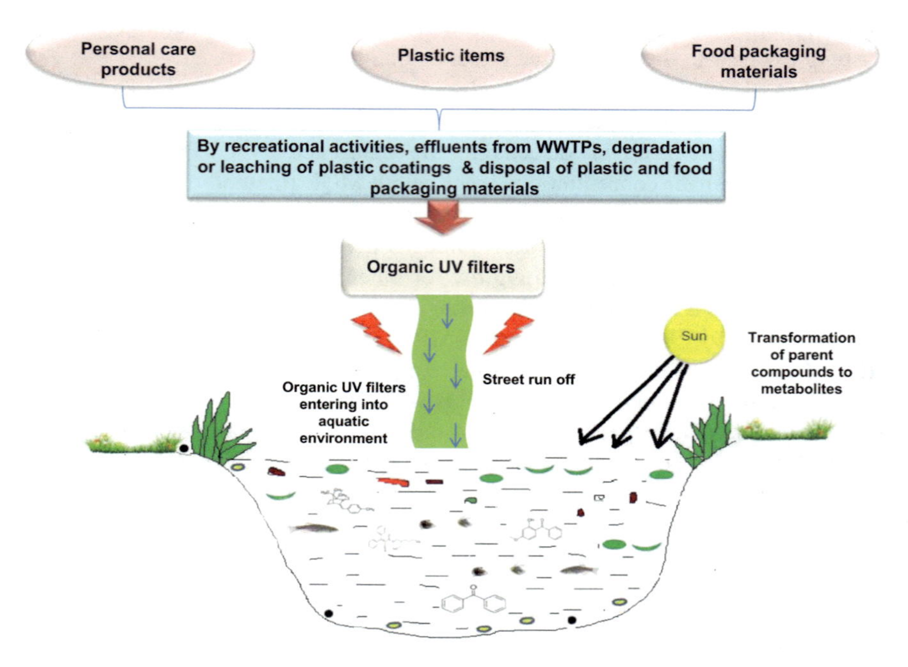

Fig. 2 Diagrammatic representation of aquatic ecosystem contamination by OUVFs

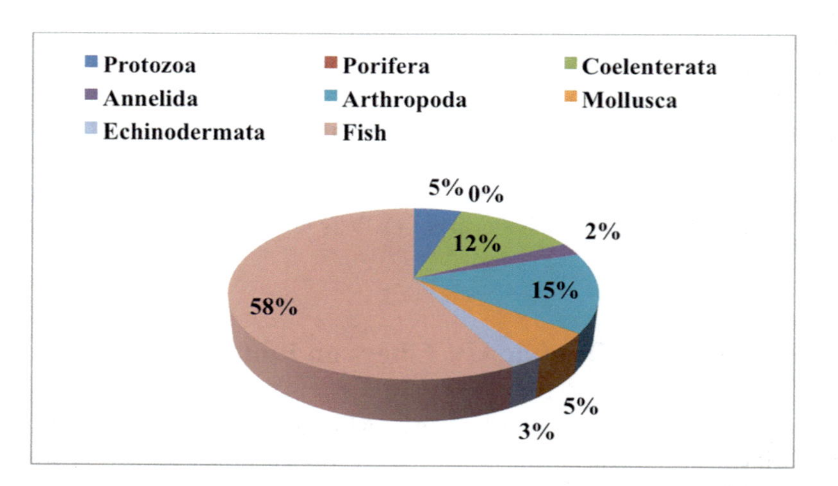

Fig. 3 Data showing the percentage of studies on organic UV filters (OUVFs) in different phyla of animal kingdom

Table 1 Concentrations of OUVFs in aquatic and sediment samples in different regions of the globe

Name of UV filter/s	Freshwater	Marine water	Sewage/ WWTPs influent	Sewage/ WWTPs effluent	Sediment	Location	Instrument used for analysis	References
BP-3	–	–	0.005–0.0856 μg/L	0.0011–0.0412 μg/L	–	India – WWTPs located in Udupi, Mangalore, Coimbatore, Manipal, Saidpur, Nagpur	HPLC-MS/MS	Balakrishna et al. (2017)
BP	–	–	3.96 μg/L	1.5 μg/L	–			
BP-4	–	–	13.3 mg/L	6.3 mg/L	–	South Wales (UK)	LC–ESI-(QqLIT) MS/MS	Kasprzyk-Hordern et al. (2009)
BP-4	–	–	1,420 ng/l	–	–	Llobregat River (NE Spain)	On line-SPE–LC–MS/MS	Gago-Ferrero et al. (2013a)
BP-4	–	–	572 ng/l	–	–	Germany and Spain	LC–MS/MS	Wick et al. (2010)
BP-3	–	–	541 ng/l	–	–	Swiss lakes, Switzerland	GC-MS	Balmer et al. (2005)
BP-3	–	–	700 ng/l	–	–	Hong Kong	HPLC-ESI-MS/MS.	Tsui et al. (2014)
BP-3	–	–	–	–	0.029 μg/g	Spain	UPLC–ESI-MS/MS	Gago-Ferrero et al. (2012)
BP-3	–	–	–	–	0.007–0.029 μg/g	Belgium, Germany	TD–GC-MS and LDLC-MS	Rodil et al. (2009a)
BP-3	–	0.0536–0.5775 μg/L	–	–	–	Nearby Majorca Island, Spain	GC-MS	Tovar-Sánchez et al. (2013)

(continued)

Table 1 (continued)

Name of UV filter/s	Freshwater	Marine water	Sewage/ WWTPs influent	Sewage/ WWTPs effluent	Sediment	Location	Instrument used for analysis	References
BP-3	–	75–1,400 μg/L	–	–	–	U.S Virgin Islands (in coral reefs)	GC–MS and LC–MS	Downs et al. (2016)
BP-3	–	0.8–19.2 μg/L	–	–	–	Hawaii		
BP-3	125 ng/L–0.3 mg/L	–	10,400 ng/L	–	–	Lake Huttnersee, Switzerland	GC-MS	M. Silvia Díaz-Cruz et al. (2008); S. Ramos et al. (2015); S. Kim et al. (2014); Sánchez-Quiles and Tovar-Sánchez (2015)
BP-8	–	–	–	10 ng/l	–	Taiwan	LC-MS	Wu et al. (2013)
BP-8	–	–	–	84 ng/l	–	China	HPLC-ESI-MS/MS	Tsui et al. (2014)
Benzimidazole, PBSA	–	–	275–3,615 ng/L	316–1,504 ng/L	–	Europe, USA, and Japan	LC–MS/MS	Wick et al. (2010)
BP-4	–	–	0.572–1.42 μg/L	–	–	Germany and Spain	LC–MS/MS	Wick et al. (2010), Gago-Ferrero et al. (2013a)
OC	–	–	Around 1,200 ng/l	–	–	Swiss lakes, Switzerland	LC-MS	Balmer et al. (2005)
OC	–	–	–	0.7 μg/L	–	Norway	LC-HRMS and GC-HRMS	Langford et al. (2015)
OC	–	–	18–461 ng/l	–	–	USA, Europe, and Japan	GC-MS, GC-MS/MS, and	Cunha et al. (2015a, b); Kupper et al. (2006); Li et al. (2007);

							LC-MS/MS	Magi et al. (2013); Moeder et al. (2010); Rodil and Moder, (2008a); Rodil et al. (2009a, b, c)
4-MBC	–	0.0514–0.1134 μg/L	–	–	–	Around Majorca Island, Spain	GC-MS	Tovar-Sánchez et al. (2013)
4-MBC	–	–	394–406 ng/l		–	South Australia	GC-MS/MS	Liu et al. (2012)
4-MBC	–	–	3 ng/l	207 ng/l	–	Hong Kong, China	HPLC-ESI-MS/MS	Tsui et al. (2014)
4-MBC	–	–	2,128 ng/l	1851 ng/l	–	China	GC–MS	Li et al. (2007)
4-MBC	–	–	45.8–154.9 ng/l	–	–	Portugal	GC-MS	Cunha et al. (2015a)
4-MBC	–	–	600–6,500 ng/l	110 ng/l 2,700 ng/l	–	Switzerland	GC-EI-SIM-MS and GC-MS	Kupper et al. (2006)
EHMC	–	–	19 μg/L	–	–			Balmer et al. (2005)
4-MBC			10–90 ng/l	24 ng/l	–	Spain	LC–ESI-(QqLIT) MS/MS	Gago-Ferrero et al. (2013a)
4-MBC	–	–	–	30–60 ng/l	–	Germany	TD–GC–MS	Rodil and Moeder, (2008b); Rodil et al. (2009a, b, c)
BP-3	6–68 ng/l	–	–	–	–	Switzerland	LC-MS/MS and GC-MS	Karl Fent et al. (2010)
4-MBC								
EHMC								
BP-3	–	227 ng/l	–	–	–	Los Angeles	HPLC-ESI-MS/MS	Tsui et al. (2014)
EHMC	–	100 ng/l	–	–	–	New York		
OC	–	153 ng/l	–	–	–	Bangkok		

BP-3, Benzophenone-3; BP-4, Benzophenone-4; BP-8, Benzophenone-8; OC, Octocrylene; 4-MBC-3-(4-methylbenzylidene) camphor); EHMC, 2-ethyl-hexyl-4-trimethoxycinnamate; OMC, Octyl methoxycinnamate; PBSA, Phenylbenzimidazole sulfonic acid

3.1 Effects on Freshwater Organisms

3.1.1 Effects on Producers

Producers are those organisms which synthesize their own food material in the presence of sunlight through the process of photosynthesis for survival and also help to sustain the next trophic level in the aquatic food chain. These organisms constitute the basal part of the ecological food chain. Exposure of the freshwater algae, *Desmodesmus subspicatus*, to different types of OUVFs such as BP-3, EHMC, 3-BC, and 4-MBC resulted in algal cell decrement representing growth inhibition phenomenon. Among the exposed OUVFs, 4-MBC was found to be more toxic during the 72-h exposure period and BP-3 least toxic (Sieratowicz et al. 2011). On the other hand, OUVFs like BP-1, EHMC, and OD-PABA induced 50% growth inhibition at 10.5 mg/L, 1.8 mg/L, and 0.03 mg/L, respectively, in *Raphidocelis subcapitata,* twisted microalgae inhabiting freshwater bodies. The order of OUVFs toxicity is OD-PABA>EHMC>BP-1 during the 3-day treatment period (Molins-Delgado et al. 2016).

Treatment with BP-3 at 4 mg/L concentrations in *Chlorella vulgaris* inhibited the growth between 48 and 96 h, whereas exposure to 220 mg/L BP-4 concentration resulted in hormesis at the initial phase followed by complete growth inhibition. Combined exposure of BP-3 and BP-4 resulted in antagonistic effects on *C. vulgaris* (Du et al. 2017). The wide occurrence of OUVFs including BP type OUVFs and their derivatives in the aquatic environment poses a serious toxic impact on the producer and consumer level. For instance, the combined toxicity of benzophenone-1 (BP-1) and benzophenone-3 (BP-3) was assessed in the algae, *Chlamydomonas reinhardtii* which is the part of producer trophic level (Mao et al. 2018a). It has been found that both BP-1 and BP-3 exert the toxicity on growth and chlorophyll content of *C. reinhardtii* exposed between 1 and 5 mg/L concentration. OUVFs inhibited the growth by interfering the photosynthesis followed by chlorophyll (*Chl-a* and *Chl-b*) and carotenoid levels reduction, suggesting the plausible mechanism of action of OUVFs at primary producer level (Mao et al. 2018a). The molecular effects of OUVFs in primary producers are ascertained from the findings of Lee et al. (2020). BP-3 exposed cultures of *Scenedesmus obliquus* induced a dose-response relationship and gene expression perturbations. *Tas* and *ATPFoC* genes involved in the oxidoreductase and ATP-synthase encoding pathways were significantly overexpressed, while *Lhcb1* and *HydA*, genes encoding light-harvesting chlorophyll B binding *protein (LHCB1)* and iron hydrogenase respectively, were downregulated. *Lhcb1* has a protective role in photooxidative damage and conversion of photons to energy and biomass, etc. (Lee et al. 2020).

3.1.2 Effects on Consumers

Consumers are organisms which depend on producers as food source for survival. The presence of OUVFs in the aquatic ecosystem inevitably results in harmful effects at the consumer trophic level and perturbing the niche of particular species, affecting the eco-dynamics of a particular ecosystem. Crustacean, *Daphnia magna* are found to be 10% and 50% sensitive to OUVFs exposure between 140–2,450 μg/L and 0507–3,610 μg/L concentrations, respectively. Daphnids were most affected by the EHMC exposure than 3-BC resulting in immobilization. On the other hand, long-term exposure to EHMC (5–500 μg/L) for 21 days eventually resulted in a significant reduction in length of parent daphnids and neonates number compared to BP-3 exposed individuals (Sieratowicz et al. 2011).

Studies on combined OUVFs exposure in aquatic consumers have shown to induce mixture toxicity. For instance, daphnids exposed to BP-4 and BP-3 resulted in the antagonistic relationship between 0.25 and 3 mg/L concentrations (Du et al. 2017), and similar observations of immobilization, reproductive impairment were confirmed in the daphnids exposed to BP-1, OD-PABA, and EHMC OUVFs (Molins-Delgado et al. 2016). The omnipresence of OUVFs affects many other organisms in the consumer trophic level and impairs their survival. For example, 4-MBC and BP-3 induced feeding inhibition in *Sericostoma vittatum* (Caddish fly) at 3.55 mg/Kg of BP-3 and 2.57 mg/Kg of 4-MBC. Biochemistry and energy metabolism of the *S. vittatum* along with the interference in the antioxidant defense system induced oxidative stress and neurotoxicity, after 6-day exposure, was noted (Campos et al. 2017a). Exposure to environmental concentrations of widely used OUVFs such as BP-3, 4-MBC, and OC resulted in decreased catalase activity and elevated glutathione-S-transferase activity in the aquatic benthic species, *Chironomus riparius*. These all three compounds negatively affect the growth of *C. riparius* larvae, ultimately resulted in impaired development (Campos et al. 2017b).

OUVFs tend to induce long-term effects like reproduction in higher trophic consumers like fishes. Exposure of Japanese medaka (*Oryzias latipes*) to BP-3 results in endocrine disruption and reproduction effects like clutch size. In this study, the adult Japanese medaka of F0 generation exposed to BP-3 for 14 days at different nominal concentrations (0.015–0.5 mg/L) has been shown to inhibit the sex steroid hormone and genes for reproduction and observed a significant down-regulation of *cyp11a, hsd17b3, cyp17, cyp19a,* and *star* genes, yielding declined egg formation rate (Kim et al. 2014).

In assessing the toxic potential of OUVFs, exposure of adult zebrafish (*Danio rerio*) and their embryos to different concentrations of BP-3 (0.0024–0.312 mg/L for adult zebrafish and 0.0082–0.00438 mg/L for embryos) resulted in a significant down-regulation of *esr1, ar, cyp19b* genes in the adult male fish brain (14-day exposure) and embryos (120-h post fertilization, hpf) along with the down-regulation of some genes in male fish testis including *cyp11b2, hsd11b2, hsd17b3,* and *hsd3b* indicating the anti-androgenic property of BP-3. However, histological analysis did not show any effect on the testis after BP-3 exposure indicating the adverse effects at gene level (Blüthgen et al. 2012). Microarray studies in adult male zebrafish exposed to 0.383 mg/L of OC showed alteration in 628 and 136 transcripts in brain and liver, respectively, representing various vital physiological processes like development, hematopoiesis, angiogenesis, organ development, cell differentiation or fat metabolism, besides developmental toxicity in embryos (Blüthgen et al. 2014).

Zucchi et al. (2011b) assessed the toxicity potential of EHMC in zebrafish during a 14-day exposure period. Exposure at 0.0022 and 0.89 mg/L concentrations showed differences in 1096 and 1,137 transcripts belonging to many pathways like lipid metabolism, estrogenic pathways (*vtg1*), vitamin A metabolism, DNA damage, apoptosis, and regulation in cell growth. Around 1.5-fold change of genes like *vtg1, esr1, esr2b, ar, cyp19b,* and *hsd17β3* upon EHMC exposure was noted (Zucchi et al. 2011b). In spite of that, complete toxicity profiling of some organic UV filters for environmental and human health concern is completely lacking. For instance, 4-MBC, a widely used OUF in many cosmetics, had limited toxicity data on vertebrate modeled. It induces axial curvature followed by neuromuscular abnormalities at 15 μM concentration of 4-MBC in zebrafish. The exposure triggers the expression of some genetic markers including *F59, znp1*, and *zn5* responsible for slow muscle differentiation and neuronal innervations of primary and secondary motor neurons during zebrafish development (Li et al. 2016). A recent study has shown that exposure of zebrafish larvae to 4-MBC sub-lethal concentrations results in developmental deformities, reduced heart and hatching rate, axial curvature, etc. The exposure eventually induced oxidative stress in a concentration-dependent manner and expression profile of genes involved in the hypothalamus-pituitary-gonadal (HPG) axis and hypothalamus-pituitary-thyroid (HPT) axis (Quintaneiro et al. 2019). Regarding the mixture toxicity, BP-3 and BP-4 were found to induce antagonistic effects in zebrafish (Du et al. 2017).

Phototransformation of OUVFs plays a major role in eliciting toxicity to non-target sentinels. OMC is photo transformed to 4-MBA and 2-EH and its toxicity potential was assessed using zebrafish embryos. Upon exposure, developmental toxicity, oxidative stress, neurotoxicity, and histological changes were noticed at 96 hpf. The findings of the study showed that the photoproducts were potentially more toxic than the parent compound. OMC and 2-EH induced the tail deformation, axial curvature, and pericardial edema, whereas 4-MBA exposed individuals exhibited pericardial yolk sac edema, spinal curvature, deformed tail, among others. The derivative of OMC, 4-MBA, and 2-EH reduced hatching rate at a significant level, while OMC did not display any effects and the heartbeat reduced with increased concentration of OMC, 4-MBA, and 2-EH at 48 hpf. These OUVFs had a very significant role in exerting oxidative stress markers including CAT, SOD, LOP, GPx, GST, GSH, and neurotoxicity marker AChE. Moreover, histological studies revealed that 4-MBA induced severe damage in the muscle fiber and yolk sac area (Nataraj et al. 2019) indicating organismal level responses toward OUVFs contamination.

Bioaccumulation of OUVFs in aquatic organisms is of major concern from an ecotoxicological point of view. The process of bioaccumulation, an event of long-term exposure results in the reduction of reproductive as well as genetic fitness of the affected organisms because of endocrine-disrupting and DNA damaging potential. To date, few experimental studies have been focused on the bioaccumulation of OUVFs in aquatic organisms. For example, exposure of octocrylene (22, 209, 383 μg/L) to adult zebrafish results in OC accumulation of 17 μg/kg and perturbed gene expression at the transcriptional level in the brain and liver resulted in abnormal developmental processes (Blüthgen et al. 2014). In another study, zebrafish exposed to OC (28.61, 505.62, and 1248.70 μg/L) for 28 days resulted in accumulation at 2.32 mg/Kg, 31.23 mg/Kg, and 70.59 mg/Kg concentrations with a significant up-regulation of *esr1* and *cyp19b* genes in gonads and *vtg1* gene in the liver of both sexes besides an apparent down-regulation of androgen receptor *(ar)* gene in gonads (Zhang et al. 2016). Under field conditions, 4-MBC and OC from seven small rivers in Switzerland have the potential to accumulate in Brown trout fish, *Salmo trutta fario* with lipid-weight based concentrations of 1800 ng/g of 4-MBC and 2,400 ng/g of OC, respectively (Buser et al. 2006). The above findings shed light on the bioaccumulation potential of OUVFs in aquatic sentinels with devastating population perturbations and more attention is required on long-term low-level exposure studies covering the entire life cycle of organisms inhabiting such kind of contaminated water bodies.

3.2 *Effects on Marine Organisms*

The marine ecosystem is much more complex with intricacies at different levels of the ecological pyramid and physico-chemical interaction. The natural and anthropogenic perturbations eventually dwindled the unidirectional energy flow channelized by different biota. This section provides details on the adverse effects of OUVFs on marine organisms occupying different trophic levels.

3.2.1 Effects on Producers

Exposure to inorganic (TiO_2 and ZnO) and organic (BP-3 and 4-MBC) UV filters negatively affects the growth of the marine phytoplankton, *Chaetoceros gracilis* after 72 h exposure. The effective growth inhibition concentrations of selected UV filters ranged between 45 and 218 mg/L. The appearance of inorganic micronutrients such as NH_4^+, NO_3, and PO_4^{-3} from the sunscreen formulation could be the plausible reason for the observed growth retardation (Tover-Sanchez et al. 2013). Toxicity assessment conducted on 4 OUVFs such as BP-3, BP-4, 4-MBC and EHMC on the marine microalgal species, *Isochrysis galbana*, determined that BP-3 was most toxic followed by EHMC, 4-MBC, and BP-4 with an EC_{50} (72 h) values of 13.87, 74.73, 171.45 and >10,000 μg/L, respectively, for growth inhibition endpoint (Paredes et al. 2014).

On the other hand, BP-3 exposure in marine algal species, *Tetraselmis suecica*, at 5, 10, 20, and 40 pg/cell for 24 h resulted in decreased cell volume, elevated growth rate by 65%, metabolic activity, and no toxic effect on cell viability. The exposure lead to increased forward scatter signal followed by cell autofluorescence (chlorophyll-a content), esterase activity, and depolarization of cells were also observed with a decreased forward scatter signal at higher concentrations. BP-3 exposure induced 65% of hormesis in terms of enhanced growth rate and 1.5 to 2.5-fold elevated esterase activity leading to metabolic perturbations resulting in cellular hyperpolarization (Seoane et al. 2017).

Recently, experiments performed on multiple OUVFs including benzophenone-3 (BP-3), homosalate (HS), ethylhexyl triazone (ET), bis-ethylhexyloxyphenol methoxyphenyl triazine (BEMT), diethylhexyl butamido triazone (DBT), butyl methoxydibenzoylmethane (BM), diethylamino hydroxybenzoyl hexyl benzoate (DHHB), methylene bisbenzotriazolyl tetramethylbutylphenol (MBBT), octocrylene (OC), 2-ethylhexyl salicylate (ES) in marine autotroph *Tetraselmis sp.* have shown severe acute toxicity potential of OUVFs. After a week of exposure (10, 100, and 1,000 μg/L), OC and HS were found to be more toxic to *Tetraselmis sp.* in terms of growth rate while BP-3, DHHB, and OC have negatively affected the metabolic activity followed by decreased esterase activity at 100 μg/L concentration, whereas ES and HS showed similar effects at 10 μg/L exposure concentration. These OUVFs exhibited a concentration-dependent decrement in growth rate and chlorophyll-a content during the 7-day treatment regime analyzed by flow cytometry. Contrary to this, neither the growth rate nor the chlorophyll-a content was significantly not affected upon MBBT, ET, BEMT, DBT, DHHB, and BM exposure in *Tetraselmis sp.* even up to 1,000 μg/L concentration (Thorel et al. 2020).

3.2.2 Effects on Consumers

At the consumer level, OUVFs have shown to induce a plethora of adverse effects in marine biota. A classical example is the impact of OUVFs on the coral reef ecosystem. Coral reef, an underwater ecosystem characterized by reef-building and occupies less than 0.1% of the world's ocean area. Exposure of coral reef, *Stylophora pistillata* to BP-3 for 4 h resulted in reef organizational deformities and eventual death. Studies revealed that exposure of *S. pistillata* larvae (planulae) and cultured primary cells to environmental concentrations (10 μL/L) of BP3 eventually inhibit the planulae growth *in vivo* and *in vitro* (Downs et al. 2016). The exposure of sunscreen has been shown to induce rapid bleaching in hard corals *via* mucus release and also promotes viral infection through a lytic phase in symbiotic zooxanthellae, *Acropora nubbins*. This leads to a dwindled symbiotic relationship and perturbed marine food chain that eventually threat marine biodiversity. The bleaching of coral was directly proportional to the OUVF concentration. TEM and epifluorescence analysis revealed loss of membrane integrity and photosynthetic pigments in the zooxanthellae released from the treated corals with approximately 30–98% damaged zooxanthellae appearing pale and transparent. Exposure to OUVF changed the cell morphology to swollen, vacuolated, devoid of the chloroplast, impaired thylakoids, dispersed cytoplasm, and membrane integrity. (Danovaro et al. 2008).

Scientific evidence has shown that OUVFs have the potential in affecting top-level consumers. For example, OC, an extensively used sunscreen formulation, was detected in the 26 liver samples (out of 56 analyzed) in Franciscana dolphin (*Pontoporia blainvillei*), from the Brazilian coastal areas, which is under special measures of conservation. An estimated concentration of 89–782 ng/g of lipid weight in liver tissue samples was detected (Gago-Ferrero et al. 2013b). It has been shown that the polyp (benthic) stage of the jellyfish population is more sensitive than the medusa stage (Bell et al. 2017). The sedentary lifestyle of mussels is prone to OUVFs with greater accumulation potential. One thousand mussels sampled along the Portuguese coastline showed the presence of tonalide (AHTN), EHMC, EHS (2-ethylhexyl salicylate) in 93% samples and among them, EHS and EHMC were the most prevalent with 73% detection frequency, followed by BP-3 (27%), IMC (Cinnamate derivatives), and 4-MBC (17%) (Castro et al. 2018). Analysis of personal care products including OUVFs in 62 commercial marine representative species that are used in seafood listed by European Union (EU) including plaice, octopus, perch, shrimp, mackerel, salmon, tuna, crab, cod, sea bream, and monkfish has shown the occurrence of 4-MBC with highest detection frequency (about 100%), although below the LOQ in most samples. The concentrations of BP-3 and BP-1 were found to be 0.0987 mg/Kg dw and 0.0989 mg/Kg dw, respectively, apart from 0.0562 mg/Kg dw concentration of 4-MBC in mussels. In seabream samples, the maximum concentration of OC was observed at 103.3 μg/Kg d.w due to its high log Kow value 6.9 (highly lipophilic) showing its maximum bioaccumulative property. The average highest concentration of OUVFs was found at 196 μg/kg dw in the seabream, whereas crab, octopus, and plaice had lower contamination, below than LOQ (Cunha et al. 2018). Besides, significant accumulation of OUVFs and musk fragrances was noticed in mussels (*Mytilus edulis* and *Mytilus galloprovincialis*), European flounder (*Platichthys flesus*), mullet (*Liza aurata*), clam (*Chamelea gallina*), and seaweeds (*Laminaria digitata* and *Saccharina latissima*) from European hotspots (Cunha et al. 2015b). Exposure of marine copepod, *Tigriopus japonicus* to 4-MBC (0.0005–0.01 mg/L) for quadruplet generations (F0-F3) showed a decreased hatching rate and significantly delayed development between nauplii-copepodite-adult (N-C-A) stages across F1–F3 generations even at the lowest exposure concentration (0.0005 mg/L). The up-regulation of the ecdysone receptor gene was induced by 4-MBC, which was consistent with the decrease of the N-C/N-A development stage. The transcription activity of *sod*, *gst,* and 8-oxoguanine DNA glycosylase (*ogg1*) genes in *T. japonicus* showed concentration-dependent uptrend to 4-MBC after 7 and 14 days of post-exposure. About twofold up-regulation of the *ogg1* gene, which is responsible for the repair of oxidative DNA damage, was noticed at the highest dose (0.01 mg/L) after 7-day exposure. Accordant with the up-regulation of oxidative stress-responsive genes and the ecdysone receptor gene (*EcR*), transcription of *vtg* remains unaffected after

4-MBC exposure (Chen et al. 2018a). In the case of the marine crustacean *A. salina,* the toxicity of 10 different OUVFs conducted for 48 h at Nauplii Instar II, III stages has found that at 2000 μg/L concentration OC, BM, and HS induced significant mortality at 88 ± 16%, 64 ± 19%, and 54 ± 16%, respectively, but at lower concentrations (from 0.02 to 200 μg/L), no significant toxic effects were recorded. It appears that OC was more toxic (LC_{50} of 600 μg/L) followed by BM (1800 μg/L) and HS (2,400 μg/L), whereas BP3, DHHB, MBBT, ET, BEMT, ES, and DBT exhibited no toxicity at the tested concentrations (Thorel et al. 2020).

Long-term dietary exposure of false clown anemonefish (*A. ocellaris*) to BP-3 for 90 days between 0 and 1,000 ng/g concentrations demonstrated no adverse effects on growth and survival whereas an increase in body weight has been noted in the treated group. The possible mechanism behind the increased body weight in treated fishes could be the compromised thyroid function during BP3 exposure (Wang et al. 2016), and hence the increased body size. Another plausible mechanism might have resulted from the thyroid disruption in HPT axis pathway (Chen et al. 2018b). Samples (muscle and stomach) collected from lionfish (*Pterois volitans*), near Grenada, West Indies were analyzed for 4-MBC, oxybenzone, OMC, and Padimate-O levels. The findings showed that 4-MBC and oxybenzone moieties were detected up to 12% and 35% in the lionfish muscle samples, whereas Padimate-O was not detected in any of the sample. Histopathological studies demonstrated no OUVFs mediated organismal toxicity in the lionfish (Horricks et al. 2019).

Available research shows that OUVFs have the potential to affect biota inhabiting respective niche in freshwater and coastal ecosystems. The umbrella of toxic effects noted ranges from growth inhibition in primary producers to endocrine disruption in consumers occupying the top-level food chain. Analogous to marine biota, effects of OUVFs on freshwater organisms are available but information for effective ecotoxicity risk assessment is not sufficient for both the ecosystems. Environmental risk assessment of OUVFs and potential adverse effects of OUVFs in aquatic organisms are given in Table 2 and 3.

4 Toxicity Mechanism of Organic UV Filters

OUVFs-induced toxicity mechanism is not clear. Since most of the OUVFs are lipophilic, they interact with plasma membrane components including cell surface receptors, ion channels, and trans-membrane proteins, etc. These chemical moieties enter into the cytoplasm and induce oxidative stress upon toxic insult. It has been shown that OUVFs primarily induce oxidative stress by ROS formation in a variety of organisms representing different trophic level such as algal species (*Desmodesmus subspicatus, Chlamydomonas reinhardtii), Daphnia magna, Sericostoma vittatum, Chironomus riparius, Chaetoceros gracilis, Danio rerio, Oryzias latipes, Tigriopus japonicus,* and coral reefs (Sieratowicz et al. 2011;

Table 2 Environmental risk assessment of OUVFs

S. No.	Name of the UV-filter	NOEC (mg/L)	PNEC (ng/l)	RQ	Degree of Risk	Organism	References
1.	BP-3	0.5[1, 4]	500[1]	2.60[2]	High risk[2]	*Isochrysis galbana*[3], *Mytilus galloprovinciallis*[3], *Paracentrotus lividus*[3], *Siriella armata*[3], *Daphnia magna*[4], *Danio rerio*[5]	Rodríguez et al. (2015)[1]; Sang and Leung (2016)[2]; Paredes et al. (2014)[3]; Sieratowicz et al. (2011)[4]; Kinnberg et al. (2015)[5]
		0.191[2]	13.87[2]				
		0.03[3], 0.03[3], 1.92[3], and 0.375[3]					
		0.191 (female fish)[5], 0.388 (male fish)[5]					
2.	4-MBC	0.1[1, 2, 4]	100[1, 2]	0.33[2]	Medium risk[2]	*Isochrysis galbana*[3], *Mytilus galloprovinciallis*[3], *Paracentrotus lividus*[3], *Siriella armata*[3], *Daphnia magna*[4]	Rodríguez et al. (2015)[1]; Sang and Leung (2016)[2]; Paredes et al. (2014)[3]; Sieratowicz et al. (2011)[4]
		0.018[3], 0.300[3], 0.300[3], and 0.03704[3]		>10[1]			
3.	EHMC	0.040[1, 2, 4]	40[1, 2]	3.29[2]	High risk[2]	*Isochrysis galbana*[3], *Mytilus galloprovinciallis*[3], *Paracentrotus lividus*[3], *Siriella armata*[3], *Daphnia magna*[4]	Rodríguez et al. (2015)[1]; Sang and Leung (2016)[2]; Paredes et al. (2014)[3]; Sieratowicz et al. (2011)[4]
		0.015[3], 0.500[3], 0.600[3], and 0.0625[3]					
4.	OC	10000[2]	1000000[2]	<0.001[2]	Low risk[2]	*Danio rerio*	Sang and Leung (2016)[2]
5.	2,4OH-BP	NA	800	0.007	No risk	Values derived from bulk water samples	Mao et al. (2018b)
6.	2,2′,4,4′OH-BP	NA	8,780	0.012	Low risk		
7.	2OH-4MeO-BP	NA	250	0.49	Medium risk		
8.	2,2′OH-4,4′MeO-BP	NA	14,000	0.002	No risk		

(continued)

Table 2 (continued)

S. No.	Name of the UV-filter	NOEC (mg/L)	PNEC (ng/l)	RQ	Degree of Risk	Organism	References
9.	2,2′OH-4MeO-BP	NA	4,400	0.002	No risk		
10.	4OH-BP	NA	7,100	0.002	No risk		
11.	4DHB	NA	12,300	0.002	No risk		

Superscript numbers indicate the derived PNEC, NOEC, and RQ values with respect to references and organisms

NOEC, No-Observed Effect Concentration; PNEC, Predicted No-Effect Concentration; RQ, Risk Quotient; BP-3, Benzophenone-3; 4-MBC, 4-methylbenzylidene camphor; EHMC, 2-ethyl-hexyl-4-trimethoxycinnamate; OC, Octocrylene; 2,4OH-BP, 2,4-dihydroxybenzophenone; 2,2′,4,4′OH-BP, 2,2′,4,4-′-tetrahydroxybenzophenone; 2OH-4MeO-BP, 2-hydroxy-4-methoxybenzophenone; 2,2′OH-4,4′MeO-BP, 2,2′-Dihydroxy-4,4′-dimethoxybenzophenone; 2,2′OH-4MeO-BP, 2,2′-Dihydroxy-4-methoxybenzophenone; 4OH-BP, 4-hydroxybenzophenone; 4DHB, 4,4′-dihyroxybenzophenone

Table 3 Toxic effects of OUVFs on aquatic organisms

Name of the UV filter/s	Model organism/s	Exposure concentration	Exposure duration	End points studied	Observation/s	References
Freshwater organisms						
BP-3	*Scenedesmus obliquus*	0.1–3 mg/L	10 days	Growth inhibition	Reduced growth inhibition up to 23% on day 10 of post-exposure. Decreased total chlorophyll content up to 10.4% Dry cell weight gradually declined.	Lee et al. (2020)
				Gene expression	Tas and ATPFoC genes over-expressed while Lhcb1 and HydA down-regulated.	
BP-1 + BP-3	*Chlamydomonas reinhardtii*	1.0, 2.0, 3.0, 4.0, and 5.0 mg/L	3 days	Photosynthetic pigments	Chlorophyll (chl-a) reduction 15–66% Two pigments chl-b and carotenoid were highly inhibited	Mao et al. (2018a)
				Growth inhibition	5–62% growth inhibition with increased concentration.	
BP-3, BP-4	*Chlorella vulgaris*	BP-3 (2.40 and 4.0 mg/L), BP-4 (200 and 220 mg/L)	96 h	Growth inhibition	Significant growth reduction by the end of 96-h post exposure with linear dose-response relationship BP-3 and BP-4 has antagonistic effect on *C. vulgaris*	Du et al. (2017)
BP-1, EHMC, and OD-PABA	*Raphidocelis subcapitata*	0.001 mg/L, 0.01 mg/L, and 100 mg/L	72 h	Growth inhibition	Algal species was most sensitive to OD-PABA	Molins-Delgado et al. (2016)
BP-3	*Desmodesmus subspicatus*	0.06, 0.13, 0.25, 0.5, 1.0 mg/L	72 h	Growth inhibition	Maximum 53.6% growth inhibition	Sieratowicz et al. (2011)
EHMC	*D. subspicatus*	0.015, 0.03, 0.06, 0.13, 0.25 mg/L	72 h		Maximum 23.9% growth inhibition	

(continued)

Table 3 (continued)

Name of the UV filter/s	Model organism/s	Exposure concentration	Exposure duration	End points studied	Observation/s	References
3-BC	*D. subspicatus*	0.5, 1.0, 2.0, 4.0, 8.0 mg/L	72 h		Maximum 45.6% growth inhibition	
4-MBC	*D. subspicatus*	0.5, 1.0, 2.0, 4.0, 8.0 mg/L	72 h		Maximum 46.8% growth inhibition	
BP-3, 4-MBC, and OC	*Chironomus riparius*	**For developmental toxicity–** BP-3: 0.75, 1.55, and 3.41 mg/kg, 4-MBC: 0.80, 2.05, and 4.17 mg/kg, OC: 0.53, 1.27, and 2.33 mg/kg	10 day	Developmental anomalies	Significantly reduced larval growth at conc. of 2.33, 2.05, and 0.75 mg/kg of OC, 4-MBC, and BP-3, respectively	Campos et al. (2017b)
				Larval emergence and weight	No significant weight reduction in OC exposed male and female at 8.47% and 16.63% observed, respectively; delayed larval emergence	
		For oxidative stress– BP-3: 0.23, 1.00, and 6.49 mg/kg, 4-MBC: 0.09, 1.12, and 14.13 mg/kg, OC: 0.23, 2.13, and 18.23 mg/kg	48 h	Oxidative stress	Change in oxidative stress marker including CAT, tGSH, GST, LPO and neurotoxic marker AChE	
				Energy consumption	Increased energy consumption	
BP-3	*Sericostoma vittatum*	BP-3: 0.89, 1.45, and 3.55 mg/kg,	6 days	Feeding behavior	Decreased carbohydrate level followed by the reduction in feeding rate by 54% and 69%, respectively.	Campos et al. (2017a)
4-MBC		4-MBC: 1.35, 2.57, and 6.95 mg/kg		Oxidative stress	BP-3 and 4-MBC induced increased total glutathione (tGSH). No significant change in CAT, LPO, GST, and ETS.	

BP-3, BP-4	*Daphnia magna*	BP-3 (0.25–3.00 mg/L) and BP-4 (22.50–52.50 mg/L)	48 h	Immobilization	BP-3 more toxic to water flea showing linear dose-response Mixture of BP-3 and BP-4 has antagonistic effect on water flea	Du et al. (2017)
BP1, BP3, 4MBC, and EHMC	*Daphnia magna*	Toxic units– 0.2, 0.5, and 0.8	72 h	Immobilization	Reduced immobilization Less toxic in mixture to *D. magna*	Molins-Delgado et al. (2016)
BP-3 EHMC 3-BC 4-MBC	*Daphnia magna*	**Acute:** BP-3: 0.63, 1.25, 2.5, 5, 10 mg/L EHMC: 0.08, 0.16, 0.31, 0.63, 1.25, 2.5 mg/L 3-BC: 0.8, 1.6, 3.2, 6.4, 12.8 mg/L 4-MBC: 0.4, 0.8, 1.6, 3.2, 6.4 mg/L **Chronic:** BP-3: 0.005, 0.013, 0.032, 0.08, 0.2, 0.5 mg/L EHMC: 0.005, 0.01, 0.02, 0.04, 0.08 mg/L 3-BC: 0.05, 0.1, 0.2, 0.4, 0.8 mg/L 4-MBC: 0.013, 0.025, 0.05, 0.1, 0.2 mg/L	72 h 21 days	Immobilization Off-springs production Parent daphnids length	Daphnids were more sensitive to EHMC followed by 4-MBC, BP-3 and 3-BC Reduced the mobility of daphnids BP-3 showed no effect on neonate production and length of adults EHMC at higher concentrations reduced the parent daphnids' length With increasing concentration, 3-BC reduced the parent daphnid's length, reduced the production of neonates per adult and also delayed the production of first brood Highest 4-MBC concentration reduced the parental length and neonates per adult	Sieratowicz et al. (2011)

(continued)

Table 3 (continued)

Name of the UV filter/s	Model organism/s	Exposure concentration	Exposure duration	End points studied	Observation/s	References
OMC, 2-EH, and 4-MBA	*Danio rerio* embryos	OMC: 6.2, 12.4, and 62 μg/mL 2-EH: 3.4, 6.8, and 34 μg/mL	96-h post-fertilization (hpf)	Developmental toxicity	Reduced hatching and heart rate. Caused pericardial edema, axial curvature, and deformed tail. Alteration at histopathological level	Nataraj et al. (2019)
		4-MBA: 0.35, 0.7, and 3.5 μg/mL		Oxidative stress	Induction and inhibition of SOD, CAT, GPx and GST activities and LPO was increased.	
4-MBC	*Danio rerio* embryos	0.08, 0.1, 0.15, 0.26, 0.44 and 0.77 mg/L	4 days	Developmental toxicity	Developmental deformation, axial curvature, reduced hatching rate.	Quintaneiro et al. (2019)
				Oxidative stress and gene expression	Induced oxidative stress (increased CAT and GST) and increased AChE activity. Perturbed gene expression of HPG-HPT axis.	
4-MBC	*Danio rerio* embryos	15 μM	72 h	Developmental abnormalities	Axial curvature and impaired motility Disorganized pattern of slow muscle fibers and axon path-finding errors during innervations of primary and secondary motor neurons	Li et al. (2016)
BP-3, BP-4	*Danio rerio*	BP-3 (1.50–5.50 mg/L), BP-4 (200.00–600.00 mg/L)	96 h	Mortality	Linear dose-response Mixture of BP-3 and BP-4 has antagonistic effect on *D. rerio* BP-3 had high toxicity than BP-4	Du et al. (2017)

OC	*Danio rerio*	28.61, 505.62, and 1248.70 μg/L	28 days	Gonado-somatic index Vitellogenesis Gene expression	Gonad-somatic index and percentage of vitellogenic oocytes were increased significantly in the ovaries At high accumulation level, down-regulation of *ar,* and significant up-regulation of *esr1* and *cyp19b* in gonads and vtg1 in liver in both sexes	Zhang et al. (2016)
BP-2	*Danio rerio* embryo	9.85 mg/L (40 μM)	5 days	Developmental defects	Induced facial deformation Adverse effects on segmentation process, blood circulation. Lipid accumulation in yolk sac via affecting lipid processing,.	Fong et al. (2016)
OC	*Danio rerio* (adult male)	22, 209, 383 μg/L	8–16 days	Organogenesis	Altered organ development	Blüthgen et al. (2014)
	Embryos	69, 293, 925 μg/L		Transcriptome analysis	At 383 μg/L exposure, alteration in 628 transcripts in brain and 136 transcripts in liver related to developmental processes and fat cell differentiation. Bioaccumulation of OC.	
BP-3	*Oryzias latipes*	0, 15, 50, 150, or 500 μg/L	14 days	Reproduction	Reduction in clutch size and egg formation	Kim et al. (2014)
				Endocrine disruption	Inhibition of sex steroids, down-regulation of *cyp11a, hsd17b3, cyp17, cyp19a,* and *star* genes.	
BP-3	*Danio rerio* embryos	84 μg/L	120 hpf	Embryo lethality	Highest mortality occurred within 24 hpf after that eleuthero-embryos displayed	Blüthgen et al. (2012)

(continued)

Table 3 (continued)

Name of the UV filter/s	Model organism/s	Exposure concentration	Exposure duration	End points studied	Observation/s	References
					normal swimming behavior at 120 hpf exposures In adult male, decreased expression of esr1, ar cyt-P450, and cyp19b in the brain.	
BP-3	*Danio rerio* (male)	10–600 μg/L	14 days	Sexual behavior	Impaired sexual behavior	Blüthgen et al. (2012)
				Gene expression	Down-regulation of hsd3b, *hsd17b3, hsd11b2,* and *cyp11b2* genes in testis.	
EHMC	*Danio rerio* (adult)	2.2 μg/L and 890 μg/L	14 days	Global gene expression	Altered 1,096 genes at 2.2 μg/L conc. and 1,137 genes at 890 μg/L conc. belonging to many pathways, EHMC slightly affecting the genes: *vtg1, esr1, esr2b, ar, cyp19b,* and *hsd17β3*.	Zucchi et al. (2011b)
BP-4	*Danio rerio*	3,000 μg/L	14 days	Gene expression	Shows estrogenic activity by up-regulation of estrogenic related genes: vtg-1,-3 and cyp196 in brain, in liver vtg-1,-3 down-regulated	Zucchi et al. (2011a)
BP-3	*Oryzias latipes*	10, 100, and 1,000 μg/L	14 days	Reproduction	Significant reduced egg production, reduced fertilized eggs followed by declined hatching rate Disrupt neuro-endocrine system and Vitellogenin (vtg1) induction	Coronado et al. (2008)

Marine organism						
10 organic UV filters (BP-3, HS, ET, BEMT, DBT, BM, DHHB, MBBT, OC and ES)	*Tetraselmis sp.*	10, 100, and 1,000 μg/L	7 days	Growth rate and metabolic activity	Reduced growth rate Decreased esterase activity, reduction in chlorophyll-a content, cell volume increased.	Thorel et al. (2020)
BP-3	*Tetraselmis suecica*	5, 10, 20, and 40 pg/cell	24 h	Growth rate and cell viability	Decreased cell volume 65% growth rate and metabolic activity increased No toxic effect on cell viability Increased in forward scatter signal and increase in cell autofluorescence (chlorophyll-a), esterase activity increased. Depolarization of cell was observed	Seoane et al. (2017)
BP-3, BP-4, 4-MBC and EHMC	*Isochrysis galbana*	0.5, 1.0, 1.5, 2.0, 2.5, 3.0, 3.5, 4.0, 4.5 μg/L	72 h	Growth inhibition	BP-3 most toxic to *I. galbana* followed by EHMC and 4-MBC	Paredes et al. (2014)
Sunscreen formulation	Phytoplankton (*Chaetoceros gracilis*)	0. 100, 200, and 300 mg/L	72 h	Growth inhibition	Reduced growth rate	Tovar-Sánchez et al. (2013)
10 organic UV filters (BP-3, HS, ET, BEMT, DBT, BM, DHHB, MBBT, OC and ES)	*A. salina*	0.02, 0.2, 2.0, 20, 200, and 2000 μg/L	48 h	Survival and mortality	Highest concentration of OC, BM, and HS negatively affects Nauplii survival with mortality at 88 ± 16%, 64 ± 19%, and 54 ± 16%, respectively.	Thorel et al. (2020)
				Reduced survival.	Concentration-dependent mortality observed.	
				Mortality	BP-3, ET, BEMT, DBT, DHHB, MBBT, and ES have no toxic effects at higher concentration.	

(continued)

Table 3 (continued)

Name of the UV filter/s	Model organism/s	Exposure concentration	Exposure duration	End points studied	Observation/s	References
4-MBC	*Tigriopus japonicas*	0.5, 1, 5, and 10 μg/l	7–14 days For 4 consecutive generations (F0-F3)	Hatching rate	Decreased hatching rate.	Chen et al. (2018a)
				Developmental stage defect	Developmental delay from the nauplii to copepodite (N-C) and from nauplii to adult (N-A)	
				Oxidative stress	Oxidative stress biomarker genes SOD, GST, and OGGT, EcR gene, increased in a concentration-dependent manner.	
				Gene expression	OGG1, gene up-regulated, vtg1 expression unaffected.	
BP-3	*Stylophora pistillata*	2.28, 22.8, 228, 2,280, 22,800, and 228,000 μg/l	4 h, 8 h, and 24 h Light and dark periods.	Coral survival Cell death Coral bleaching	Reduction in ciliary movement and deformed morphology (dewdrop) in planulae Planulae had less zooxanthellae Significantly relative chlorophyll fluorescence intensity decreased with concentration in both conditions Epidermal tissue layer, ciliated cell undergone cellular degradation DNA abasic sites increasing with concentration in dark (20-folds) and light (tenfolds) Mortality in *S. pistillata* increased with concentration and time	Downs et al. (2016)

					Mortality in calicoblast cells of *S. pistillata* with increasing concentration	
Sunscreen	*Acropora nubbins*	10 μL/L	18–96 h	Coral bleaching	Induced viral infection and released mucus from the corals Loss of membrane integrity and photosynthetic pigments	Danovaro et al. (2008)
BP-3	*Amphiprion ocellaris*	0–1,000 ng/g food	90 days	Growth and survival	Growth and survival not affected while body weight increased Intra-colonial behaviors were analyzed quantitatively	Chen et al. (2018b)
OC	*Pontoporia blainvillei*	Field study	NA	Accumulation and lipid content	Detected up to 89–782 ng/g of lipid weight in liver tissues of fish No correlation established between the OC and liver lipid content in the fish	Gago-Ferrero et al. (2013b)

BP-1, Benzophenone-1; BP-3, Benzophenone-3; BP-4, Benzophenone-4; 3-BC, 3-benzylidene camphor; OD-PABA, 2-ethylhexyl 4-(dimethylamino)benzoate; 4-MBC, 4-methylbenzylidene camphor; EHMC, 2-ethyl-hexyl-4-trimethoxycinnamate; OC, Octocrylene; OMC, Octyl methoxycinnamate; 2-EH, 2-ethylhexanol; 4-MBA, 4-methoxybenzaldehyde; BEMT, bis-ethylhexyloxyphenol methoxyphenyl triazine; BM, butyl methoxydibenzoylmethane; MBBT, methylene bis-benzotriazolyl tetramethylbutylphenol; ES, 2-ethylhexyl salicylate; DHHB, diethylaminohydroxybenzoyl hexyl benzoate; DBT, diethylhexyl butamido triazone; ET, ethylhexyl triazone; HS, homosalate

Mao et al. 2018a; Campos et al. 2017a, b; Tovar-Sánchez et al. 2013; Downs et al. 2016; Blüthgen et al. 2012; Coronado et al. 2008; Fong et al. 2016; Li et al. 2016).

OUVFs-induced oxidative stress is mediated by reactive oxygen species (ROS) formation from the molecular oxygen (O_2) including singlet oxygen, superoxide ions, hydrogen peroxide, hydroxyl radical, peroxides, and lipid peroxides. The defense mechanism operated against oxidative stress is characterized by the involvement of various antioxidant enzymes that have the role to minimize the reactive oxygen species inside the cell. For example, superoxide dismutase (SOD) plays a major role by catalyzing the dismutation/partitioning of the superoxide ($O^{-}{}_2$) radicals into hydrogen peroxide or molecular oxygen. SOD is the most important antioxidant defense enzyme nearly found in all living organisms. The resultant of SOD activity is hydrogen peroxide (H_2O_2) and degraded by another enzyme catalase (CAT), the first line of defense enzyme to reduce the oxidative stress. The involvement of CAT activity is to break down the H_2O_2 into water and oxygen (Chelikani et al. 2004). The induction or inhibition of SOD and CAT activity (lower or higher respectively) depends on the concentration of OUVFs and the extent of compromise in the exposed individuals. Glutathione peroxidase (GPx) is an enzyme with peroxidase activity that targets H_2O_2 to reduce it to water and thus reduces the lipid hydroperoxides to their corresponding alcohol (Muthukumar et al. 2011). On the other hand, glutathione *S*-transferase (GST) is a detoxification enzyme mainly involved in phase II metabolism and catalyzes the conjugation of reduced glutathione to hydrophobic compounds that are highly water soluble (Douglas 1987). Together with GST, glutathione (GSH) also has the antioxidant defense role and exists in reduced and oxidized form in organisms with an elevated level in the organisms in response to OUVFs exposure. This antioxidant is important to prevent cellular organelles from ROS induced oxidative damage (Pompella et al. 2003). These ROS also interact with the plasma membrane by lipid peroxidation (LPO) of the polyunsaturated fatty acids by stealing the electron from the lipid, resulting in cell damage. The lipid peroxidation has been observed with OUVFs toxicity with increasing concentration. The end products of LPO are reactive aldehydes including malondialdehyde (MDA), a major biomarker and MDA tends to react with deoxyadenosine and deoxyguanosine in DNA leading to DNA adducts formation that results in DNA damage (Marnett 1999) analyzed using conventional techniques such as chromosome aberration assay, micronucleus cytome assay, sister-chromatid exchange and single cell gel electrophoresis (comet) assay, etc.

Abcb4 is an ABC transport protein (P-glycoprotein) with an essential role in eliciting MXR activity, which is responsible for the efflux of the xenobiotics from the cell as biological defense mechanism. The ciliated protozoan *Tetrahymena thermophila* exhibits MXR activity upon 4-MBC exposure (Gao et al. 2016); however, there is no published evidence of this activity in vertebrate model systems. MXR activity has been considered as a tool for defense mechanism at cellular level imparting the role to protect the organisms from xenobiotics, such as OUVFs. The findings from the authors' laboratory (unpublished data) also corroborated the prevalence of oxidative stress-induced developmental deformities, neurotoxicity, MXR activity, and larval swimming behavior to environmental concentrations of

4-MBC exposure in vertebrate model system. These personal care products, to a certain extent, elicit neurotoxicity through the inhibition of acetylcholine esterase enzyme activity (Chen et al. 2018a, b; Quintaneiro et al. 2019; Campos et al. 2017a, b) and genotoxicity *via* oxidative DNA damage (Chen et al. 2018a, Zucchi et al. 2011b). At the molecular level, OUVFs affect the expression of genes involved in many developmental processes including reproduction and sexual behavior such as *ar, esr1, cyt-P450, cyp19b, vtg1, vtg3, hsd3b, hsd17b3, hsd11b2,* and *cyp11b2* (Zucchi et al. 2011a, b, Coronado et al. 2008, Blüthgen et al. 2012, Zhang et al. 2016), apoptosis *(bax, bcl-2, caspase-3, caspase-9),* lipid metabolism *(ras, wnt, and myc),* vitamin A metabolism *(scarb1, rbp2, bcmo1, abca1, lrat)*. Exposure of *Danio rerio* to 4-MBC eventually altered the axon target peripheral tissues in an unorganized manner *via* calcium homeostasis disruption in axon growth cone through nicotinic acetylcholine receptor (nAChR) inactivation, following the inhibition of acetylcholine esterase (Li et al. 2016). Moreover, it has been proposed that certain OUVFs have the ability to induce calcium influx and hence cardiac rhythmic alterations (Quintaneiro et al. 2019). Recently, altered gene transcripts of hypothalamic-pituitary-gonadal (HPG) axis in gonads and brain could be considered as a plausible mechanism for OUVFs-induced anti-androgenicity and estrogenicity in pisces (Liang et al. 2020). Besides the toxic effects noted *in vivo*, OUVFs have been shown to induce lipid peroxidation in the human trophoblast cells (HTR8/SVneo cells) (Yang et al. 2019). The proposed mechanism of action of OUVFs and graphical abstract showing OUVFs ecotoxicity is shown in Figs. 4 and 5.

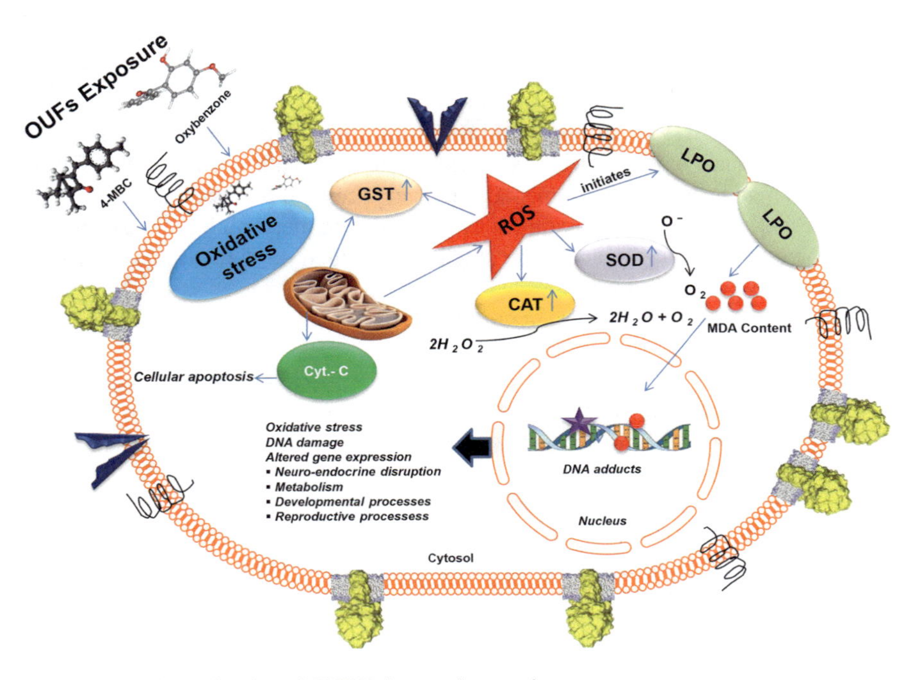

Fig. 4 Mechanism of action of OUVFs in aquatic organisms

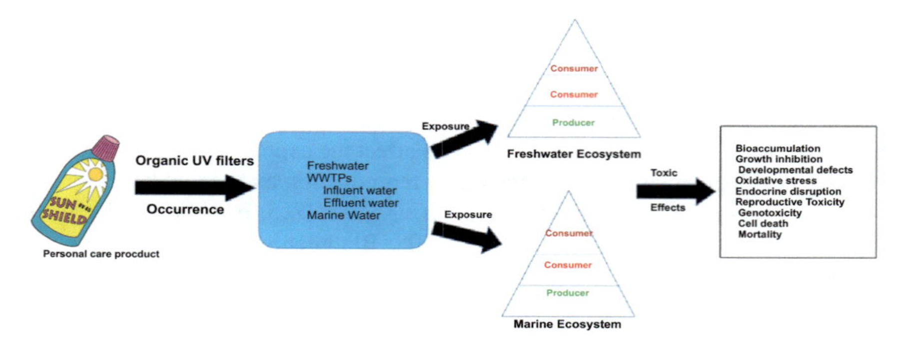

Fig. 5 Graphical abstract showing ecotoxicity of OUVFs

5 Knowledge Gaps and Future Research Directions

The existing information on OUVFs toxicity is not enough for precise environmental risk assessment. The following are the recommended knowledge gaps and future research directions:

- Systematic studies are required at a different level of biological organization to decipher the molecular mechanism of OUVFs, where information is not much available. Endocrine disrupting properties (anti-androgenic and estrogenicity) of OUVFs are not well characterized and more studies are warranted at receptor level interaction for molecular insights.
- Upon release, OUVFs kinetics and interaction with the organic matter may alter the organismal toxicity. Hence studies on the cross-talk between OUVFs and organic matter in the environmentally realistic scenario are essential.
- At present, very little information is available on the metabolite or transformation products of the OUVFs, and more studies are needed on the toxicity assessment of transformation products.
- Toxicokinetics of OUVFs is required to gather information on target organ toxicity.
- Epigenetic basis of mechanistic toxicity studies are imperative to decipher the mechanism between OUVFs and other multiple stressors (temperature, pH, dissolved oxygen) for developmental toxicity and deformed phenotypes across the generations.
- Since the behavioral response of aquatic organisms is widely used in environmental surveillance, toxicity data on OUVFs-induced behavioral effects at higher trophic levels are essential to monitor the proper functioning of the ecosystem at the population level.

6 Conclusion

In this review, we have systematically presented the information about occurrences of OUVFs in various environmental matrices including freshwater, marine water, and WWTPs (influents and effluents concentration), followed by potential adverse effects on freshwater and marine organisms from producer to consumer level along with its toxicity mechanism. The available scientific evidence suggests that OUVFs are a prime candidate for emerging environmental concern due to its potential toxic effects across the biota. Even though systematic studies are required for effective environmental risk assessment, it is parallelly the need of the hour to find out strategies to remediate these chemical entities from various environmental matrices. To the best of our knowledge, this is the first review on the detailed ecotoxicological effects of organic UV filters (OUVFs) on aquatic biota.

Acknowledgements The authors acknowledge Prof. Alok Dhawan, Director, CSIR-Indian Institute of Toxicology Research (IITR), Lucknow for his encouragement and support. The authors declare that they have no conflict of interest. No funding agency is involved. CSIR-IITR manuscript communication number: 3643.

Conflict of Interest The authors declare that they have no conflict of interest.

References

Apel C, Tang J, Ebinghaus R (2018a) Environmental occurrence and distribution of organic UV stabilizers and UV filters in the sediment of Chinese Bohai and Yellow Seas. Environ Pollut 235:85–94. https://doi.org/10.1016/j.envpol.2017.12.051

Apel C, Joerss H, Ebinghaus R (2018b) Environmental occurrence and hazard of organic UV stabilizers and UV filters in the sediment of European North and Baltic Seas. Chemosphere 212:254–261. https://doi.org/10.1016/j.chemosphere.2018.08.105

Balakrishna K, Rath A, Praveenkumarreddy Y, Guruge KS, Subedi B (2017) A review of the occurrence of pharmaceuticals and personal care products in Indian water bodies. Ecotoxicol Environ Saf 137:113–120. https://doi.org/10.1016/j.ecoenv.2016.11.014

Balmer ME, Buser HR, Müller MD, Poiger T (2005) Occurrence of some organic UV filters in wastewater, in surface waters, and in fish from Swiss lakes. Environ Sci Technol 39 (4):953–962. https://doi.org/10.1021/es040055r

Bell LJ, Ucharm G, Patris S, Diaz-Cruz MS, Roig MPS, Dawson MN (2017) Final report sunscreen pollution in Jellyfish Lake. Coral Reef Research Foundation, Palau

Blüthgen N, Zucchi S, Fent K (2012) Effects of the UV filter benzophenone-3 (oxybenzone) at low concentrations in zebrafish (Danio rerio). Toxicol Appl Pharmacol 263(2):184–194. https://doi.org/10.1016/j.taap.2012.06.008

Blüthgen N, Meili N, Chew G, Odermatt A, Fent K (2014) Accumulation and effects of the UV-filter octocrylene in adult and embryonic zebrafish (Danio rerio). Sci Total Environ 476:207–217. https://doi.org/10.1016/j.scitotenv.2014.01.015

Bratkovics S, Sapozhnikova Y (2011) Determination of seven commonly used organic UV filters in fresh and saline waters by liquid chromatography-tandem mass spectrometry. Anal Methods 3 (12):2943–2950. https://doi.org/10.1039/C1AY05390F

Bratkovics S, Wirth E, Sapozhnikova Y, Pennington P, Sanger D (2015) Baseline monitoring of organic sunscreen compounds along South Carolina's coastal marine environment. Mar Pollut Bull 101(1):370–377. https://doi.org/10.1016/j.marpolbul.2015.10.015

Brausch JM, Rand GM (2011) A review of personal care products in the aquatic environment: environmental concentrations and toxicity. Chemosphere 82(11):1518–1532. https://doi.org/10.1016/j.chemosphere.2010.11.018

Buser HR, Balmer ME, Schmid P, Kohler M (2006) Occurrence of UV filters 4-methylbenzylidene camphor and octocrylene in fish from various Swiss rivers with inputs from wastewater treatment plants. Environ Sci Technol 40(5):1427–1431. https://doi.org/10.1021/es052088s

Cadena-Aizaga MI, Montesdeoca-Esponda S, Torres-Padrón ME, Sosa-Ferrera Z, Santana-Rodríguez JJ (2019) Organic UV filters in marine environments: an update of analytical methodologies, occurrence and distribution. Trends Environ Anal Chem 25:e00079. https://doi.org/10.1016/j.teac.2019.e00079

Campos D, Gravato C, Fedorova G, Burkina V, Soares AM, Pestana JL (2017a) Ecotoxicity of two organic UV-filters to the freshwater caddisfly Sericostoma vittatum. Environ Pollut 228:370–377. https://doi.org/10.1016/j.envpol.2017.05.021

Campos D, Gravato C, Quintaneiro C, Golovko O, Žlábek V, Soares AM, Pestana JL (2017b) Toxicity of organic UV-filters to the aquatic midge Chironomus riparius. Ecotoxicol Environ Saf 143:210–216. https://doi.org/10.1016/j.ecoenv.2017.05.005

Castro M, Fernandes JO, Pena A, Cunha SC (2018) Occurrence, profile and spatial distribution of UV-filters and musk fragrances in mussels from Portuguese coastline. Mar Environ Res 138:110–118. https://doi.org/10.1016/j.marenvres.2018.04.005

Chelikani P, Fita I, Loewen PC (2004) Diversity of structures and properties among catalases. Cell Mol Life Sci 61(2):192–208. https://doi.org/10.1007/s00018-003-3206-5

Chen L, Li X, Hong H, Shi D (2018a) Multigenerational effects of 4-methylbenzylidene camphor (4-MBC) on the survival, development and reproduction of the marine copepod Tigriopus japonicus. Aquat Toxicol 194:94–102. https://doi.org/10.1016/j.aquatox.2017.11.008

Chen TH, Hsieh CY, Ko FC, Cheng JO (2018b) Effect of the UV-filter benzophenone-3 on intra-colonial social behaviors of the false clown anemonefish (Amphiprion ocellaris). Sci Total Environ 644:1625–1629. https://doi.org/10.1016/j.scitotenv.2018.07.203

Corinaldesi C (2001) Sunscreens and marine pollution: experimental study on sunscreen product effects on coastal marine environment. Thesis, University of Ancona, Ancona, Italy

Coronado M, De Haro H, Deng X, Rempel MA, Lavado R, Schlenk D (2008) Estrogenic activity and reproductive effects of the UV-filter oxybenzone (2-hydroxy-4-methoxyphenyl-methanone) in fish. Aquat Toxicol 90(3):182–187. https://doi.org/10.1016/j.aquatox.2008.08.018

Cunha SC, Pena A, Fernandes JO (2015a) Dispersive liquid–liquid microextraction followed by microwave-assisted silylation and gas chromatography-mass spectrometry analysis for simultaneous trace quantification of bisphenol A and 13 ultraviolet filters in wastewaters. J Chromatogr A 1414:10–21. https://doi.org/10.1016/j.chroma.2015.07.099

Cunha SC, Fernandes JO, Vallecillos L, Cano-Sancho G, Domingo JL, Pocurull E, Borrull F, Maulvault AL, Ferrari F, Fernandez-Tejedor M, Van den Heuvel F (2015b) Co-occurrence of musk fragrances and UV-filters in seafood and macroalgae collected in European hotspots. Environ Res 143:65–71. https://doi.org/10.1016/j.envres.2015.05.003

Cunha SC, Trabalón L, Jacobs S, Castro M, Fernandez-Tejedor M, Granby K, Verbeke W, Kwadijk C, Ferrari F, Robbens J, Sioen I (2018) UV-filters and musk fragrances in seafood commercialized in Europe Union: occurrence, risk and exposure assessment. Environ Res 161:399–408. https://doi.org/10.1016/j.envres.2017.11.015

Danovaro R, Bongiorni L, Corinaldesi C, Giovannelli D, Damiani E, Astolfi P, Greci L, Pusceddu A (2008) Sunscreens cause coral bleaching by promoting viral infections. Environ Health Perspect 116(4):441–447. https://doi.org/10.1289/ehp.10966

Daughton CG, Ternes TA (1999) Pharmaceuticals and personal care products in the environment: agents of subtle change? Environ Health Perspect 107(suppl 6):907–938. https://doi.org/10.1289/ehp.99107s6907

Díaz-Cruz MS, Llorca M, Barceló D (2008) Organic UV filters and their photodegradates, metabolites and disinfection by-products in the aquatic environment. TrAC Trends Anal Chem 27 (10):873–887. https://doi.org/10.1016/j.trac.2008.08.012

Diffey BL (1998) Ultraviolet radiation and human health. Clin Dermatol 16(1):83–89. https://doi.org/10.1016/S0738-081X(97)00172-7

Douglas KT (1987) Mechanism of action of glutathione-dependent enzymes. Adv Enzymol Relat Areas Mol Biol 59:103–167. https://doi.org/10.1002/9780470123058.ch3

Downs CA, Woodley CM, Fauth JE, Knutson S, Burtscher MM, May LA, Avadanei AR, Higgins JL, Ostrander GK (2011) A survey of environmental pollutants and cellular-stress markers of Porites astreoides at six sites in St. John, US Virgin Islands. Ecotoxicology 20(8):1914. https://doi.org/10.1007/s10646-011-0729-7

Downs CA, Kramarsky-Winter E, Segal R, Fauth J, Knutson S, Bronstein O, Ciner FR, Jeger R, Lichtenfeld Y, Woodley CM, Pennington P (2016) Toxicopathological effects of the sunscreen UV filter, oxybenzone (benzophenone-3), on coral planulae and cultured primary cells and its environmental contamination in Hawaii and the US Virgin Islands. Arch Environ Contam Toxicol 70(2):265–288. https://doi.org/10.1007/s00244-015-0227-7

Du Y, Wang WQ, Pei ZT, Ahmad F, Xu RR, Zhang YM, Sun LW (2017) Acute toxicity and ecological risk assessment of benzophenone-3 (BP-3) and benzophenone-4 (BP-4) in ultraviolet (UV)-filters. Int J Environ Res Public Health 14(11):1414. https://doi.org/10.3390/ijerph14111414

Fent K, Zenker A, Rapp M (2010) Widespread occurrence of estrogenic UV-filters in aquatic ecosystems in Switzerland. Environ Pollut 158(5):1817–1824. https://doi.org/10.1016/j.envpol.2009.11.005

Fong HC, Ho JC, Cheung AH, Lai KP, William KF (2016) Developmental toxicity of the common UV filter, benophenone-2, in zebrafish embryos. Chemosphere 164:413–420. https://doi.org/10.1016/j.chemosphere.2016.08.073

Frederick JE, Snell HE, Haywood EK (1989) Solar ultraviolet radiation at the earth's surface. Photochem Photobiol 50(4):443–450. https://doi.org/10.1111/j.1751-1097.1989.tb05548.x

Gago-Ferrero P, Diaz-Cruz MS, Barceló D (2012) An overview of UV-absorbing compounds (organic UV filters) in aquatic biota. Anal Bioanal Chem 404(9):2597–2610. https://doi.org/10.1007/s00216-012-6067-7

Gago-Ferrero P, Mastroianni N, Díaz-Cruz MS, Barceló D (2013a) Fully automated determination of nine ultraviolet filters and transformation products in natural waters and wastewaters by on-line solid phase extraction–liquid chromatography–tandem mass spectrometry. J Chromatogr A 1294:106–116. https://doi.org/10.1016/j.chroma.2013.04.037

Gago-Ferrero P, Alonso MB, Bertozzi CP, Marigo J, Barbosa L, Cremer M, Secchi ER, Azevedo A, Lailson-Brito J Jr, Torres JP, Malm O (2013b) First determination of UV filters in marine mammals. Octocrylene levels in Franciscana dolphins. Environ Sci Technol 47(11):5619–5625. https://doi.org/10.1021/es400675y

Gao L, Yuan T, Cheng P, Zhou C, Ao J, Wang W, Zhang H (2016) Organic UV filters inhibit multixenobiotic resistance (MXR) activity in Tetrahymena thermophila: investigations by the Rhodamine 123 accumulation assay and molecular docking. Ecotoxicology 25(7):1318–1326. https://doi.org/10.1007/s10646-016-1684-0

Ghazipura M, McGowan R, Arslan A, Hossain T (2017) Exposure to benzophenone-3 and reproductive toxicity: a systematic review of human and animal studies. Reprod Toxicol 73:175–183. https://doi.org/10.1016/j.reprotox.2017.08.015

Giokas DL, Salvador A, Chisvert A (2007) UV filters: from sunscreens to human body and the environment. TrAC Trends Anal Chem 26(5):360–374. https://doi.org/10.1016/j.trac.2007.02.012

Hagedorn-Leweke U, Lippold BC (1995) Absorption of sunscreens and other compounds through human skin invivo: derivation of a method to predict maximum fluxes. Pharm Res 12 (9):1354–1360. https://doi.org/10.1023/A:1016286026462

He K, Hain E, Timm A, Tarnowski M, Blaney L (2019) Occurrence of antibiotics, estrogenic hormones, and UV-filters in water, sediment, and oyster tissue from the Chesapeake Bay. Sci Total Environ 650:3101–3109. https://doi.org/10.1016/j.scitotenv.2018.10.021

Hopkins ZR, Blaney L (2016) An aggregate analysis of personal care products in the environment: identifying the distribution of environmentally-relevant concentrations. Environ Int 92:301–316. https://doi.org/10.1016/j.envint.2016.04.026

Horricks RA, Tabin SK, Edwards JJ, Lumsden JS, Marancik DP (2019) Organic ultraviolet filters in nearshore waters and in the invasive lionfish (Pterois volitans) in Grenada, West Indies. PLoS One 14(7):e0220280. https://doi.org/10.1371/journal.pone.0220280

Juliano C, Magrini GA (2017) Cosmetic ingredients as emerging pollutants of environmental and health concern. A mini-review. Cosmetics 4(2):11. https://doi.org/10.3390/cosmetics4020011

Kameda Y, Kimura K, Miyazaki M (2011) Occurrence and profiles of organic sun-blocking agents in surface waters and sediments in Japanese rivers and lakes. Environ Pollut 159(6):1570–1576. https://doi.org/10.1016/j.envpol.2011.02.055

Kasprzyk-Hordern B, Dinsdale RM, Guwy AJ (2009) The removal of pharmaceuticals, personal care products, endocrine disruptors and illicit drugs during wastewater treatment and its impact on the quality of receiving waters. Water Res 43(2):363–380. https://doi.org/10.1016/j.watres.2008.10.047

Kim S, Choi K (2014) Occurrences, toxicities, and ecological risks of benzophenone-3, a common component of organic sunscreen products: a mini-review. Environ Int 70:143–157. https://doi.org/10.1016/j.envint.2014.05.015

Kim S, Jung D, Kho Y, Choi K (2014) Effects of benzophenone-3 exposure on endocrine disruption and reproduction of Japanese medaka (Oryzias latipes)—a two generation exposure study. Aquat Toxicol 155:244–252. https://doi.org/10.1016/j.aquatox.2014.07.004

Kinnberg KL, Petersen GI, Albrektsen M, Minghlani M, Awad SM, Holbech BF, Green JW, Bjerregaard P, Holbech H (2015) Endocrine-disrupting effect of the ultraviolet filter benzophenone-3 in zebrafish, Danio rerio. Environ Toxicol Chem 34(12):2833–2840. https://doi.org/10.1002/etc.3129

Kumar S, Gupta RN (2013) Safety and regulatory issues on sunscreen products in India. Arch Appl Sci Res 5(2):145–153

Kupper T, Plagellat C, Brändli RC, De Alencastro LF, Grandjean D, Tarradellas J (2006) Fate and removal of polycyclic musks, UV filters and biocides during wastewater treatment. Water Res 40(14):2603–2612. https://doi.org/10.1016/j.watres.2006.04.012

Langford KH, Reid MJ, Fjeld E, Øxnevad S, Thomas KV (2015) Environmental occurrence and risk of organic UV filters and stabilizers in multiple matrices in Norway. Environ Int 80:1–7. https://doi.org/10.1016/j.envint.2015.03.012

Lee SH, Xiong JQ, Ru S, Patil SM, Kurade MB, Govindwar SP, Oh SE, Jeon BH (2020) Toxicity of benzophenone-3 and its biodegradation in a freshwater microalga Scenedesmus obliquus. J Hazard Mater 389:122149. https://doi.org/10.1016/j.jhazmat.2020.122149

Li W, Ma Y, Guo C, Hu W, Liu K, Wang Y, Zhu T (2007) Occurrence and behavior of four of the most used sunscreen UV filters in a wastewater reclamation plant. Water Res 41(15):3506–3512. https://doi.org/10.1016/j.watres.2007.05.039

Li VWT, Tsui MPM, Chen X, Hui MNY, Jin L, Lam RH, Yu RMK, Murphy MB, Cheng J, Lam PKS, Cheng SH (2016) Effects of 4-methylbenzylidene camphor (4-MBC) on neuronal and muscular development in zebrafish (Danio rerio) embryos. Environ Sci Pollut Res 23(9):8275–8285. https://doi.org/10.1007/s11356-016-6180-9

Liang M, Yan S, Chen R, Hong X, Zha J (2020) 3-(4-Methylbenzylidene) camphor induced reproduction toxicity and antiandrogenicity in Japanese medaka (Oryzias latipes). Chemosphere 249:126224. https://doi.org/10.1016/j.chemosphere.2020.126224

Liu YS, Ying GG, Shareef A, Kookana RS (2012) Occurrence and removal of benzotriazoles and ultraviolet filters in a municipal wastewater treatment plant. Environ Pollut 165:225–232. https://doi.org/10.1016/j.envpol.2011.10.009

Magi E, Scapolla C, Di Carro M, Rivaro P, Nguyen KTN (2013) Emerging pollutants in aquatic environments: monitoring of UV filters in urban wastewater treatment plants. Anal Methods 5(2):428–433. https://doi.org/10.1039/C2AY26163D

Mao F, He Y, Gin KYH (2018a) Evaluating the joint toxicity of two benzophenone-type UV filters on the green alga Chlamydomonas reinhardtii with response surface methodology. Toxics 6 (1):8. https://doi.org/10.3390/toxics6010008

Mao F, You L, Reinhard M, He Y, Gin KYH (2018b) Occurrence and fate of benzophenone-type UV filters in a tropical urban watershed. Environ Sci Technol 52(7):3960–3967. https://doi.org/10.1021/acs.est.7b05634

Mao F, He Y, Gin KYH (2019) Occurrence and fate of benzophenone-type UV filters in aquatic environments: a review. Environ Sci Water Res Technol 5(2):209–223. https://doi.org/10.1039/C8EW00539G

Marnett LJ (1999) Lipid peroxidation—DNA damage by malondialdehyde. Mutat Res 424 (1-2):83–95. https://doi.org/10.1016/S0027-5107(99)00010-X

McKenzie RL, Björn LO, Bais A, Ilyasd M (2003) Changes in biologically active ultraviolet radiation reaching the Earth's surface. Photochem Photobiol Sci 2(1):5–15. https://doi.org/10.1039/B211155C

Moeder M, Schrader S, Winkler U, Rodil R (2010) At-line microextraction by packed sorbent-gas chromatography–mass spectrometry for the determination of UV filter and polycyclic musk compounds in water samples. J Chromatogr A 1217(17):2925–2932. https://doi.org/10.1016/j.chroma.2010.02.057

Molins-Delgado D, Gago-Ferrero P, Díaz-Cruz MS, Barceló D (2016) Single and joint ecotoxicity data estimation of organic UV filters and nanomaterials toward selected aquatic organisms. Urban groundwater risk assessment. Environ Res 145:126–134. https://doi.org/10.1016/j.envres.2015.11.026

Muthukumar K, Rajakumar S, Sarkar MN, Nachiappan V (2011) Glutathione peroxidase3 of Saccharomyces cerevisiae protects phospholipids during cadmium-induced oxidative stress. Antonie Van Leeuwenhoek 99(4):761–771. https://doi.org/10.1007/s10482-011-9550-9

Nakata H, Sasaki H, Takemura A, Yoshioka M, Tanabe S, Kannan K (2007) Bioaccumulation, temporal trend, and geographical distribution of synthetic musks in the marine environment. Environ Sci Technol 41(7):2216–2222. https://doi.org/10.1021/es0623818

Nataraj B, Maharajan K, Hemalatha D, Rangasamy B, Arul N, Ramesh M (2019) Comparative toxicity of UV-filter Octyl methoxycinnamate and its photoproducts on zebrafish development. Sci Total Environ 718:134546. https://doi.org/10.1016/j.scitotenv.2019.134546

Negreira N, Rodríguez I, Ramil M, Rubí E, Cela R (2009) Sensitive determination of salicylate and benzophenone type UV filters in water samples using solid-phase microextraction, derivatization and gas chromatography tandem mass spectrometry. Anal Chim Acta 638(1):36–44. https://doi.org/10.1016/j.aca.2009.02.015

Paredes E, Perez S, Rodil R, Quintana JB, Beiras R (2014) Ecotoxicological evaluation of four UV filters using marine organisms from different trophic levels Isochrysis galbana, Mytilus galloprovincialis, Paracentrotus lividus, and Siriella armata. Chemosphere 104:44–50. https://doi.org/10.1016/j.chemosphere.2013.10.053

Pedrouzo M, Borrull F, Marcé RM, Pocurull E (2010) Stir-bar-sorptive extraction and ultra-high-performance liquid chromatography–tandem mass spectrometry for simultaneous analysis of UV filters and antimicrobial agents in water samples. Anal Bioanal Chem 397(7):2833–2839. https://doi.org/10.1007/s00216-010-3743-3

Plagellat C, Kupper T, Furrer R, De Alencastro LF, Grandjean D, Tarradellas J (2006) Concentrations and specific loads of UV filters in sewage sludge originating from a monitoring network in Switzerland. Chemosphere 62(6):915–925. https://doi.org/10.1016/j.chemosphere.2005.05.024

Poiger T, Buser HR, Balmer ME, Bergqvist PA, Müller MD (2004) Occurrence of UV filter compounds from sunscreens in surface waters: regional mass balance in two Swiss lakes. Chemosphere 55(7):951–963. https://doi.org/10.1016/j.chemosphere.2004.01.012

Pompella A, Visvikis A, Paolicchi A, De Tata V, Casini AF (2003) The changing faces of glutathione, a cellular protagonist. Biochem Pharmacol 66(8):1499–1503. https://doi.org/10.1016/S0006-2952(03)00504-5

Quintaneiro C, Teixeira B, Benedé JL, Chisvert A, Soares AM, Monteiro MS (2019) Toxicity effects of the organic UV-filter 4-Methylbenzylidene camphor in zebrafish embryos. Chemosphere 218:273–281. https://doi.org/10.1016/j.chemosphere.2018.11.096

Ramos S, Homem V, Alves A, Santos L (2015) Advances in analytical methods and occurrence of organic UV-filters in the environment—a review. Sci Total Environ 526:278–311. https://doi.org/10.1016/j.scitotenv.2015.04.055

Ramos S, Homem V, Alves A, Santos L (2016) A review of organic UV-filters in wastewater treatment plants. Environ Int 86:24–44. https://doi.org/10.1016/j.envint.2015.10.004

Rodil R, Moeder M (2008a) Development of a method for the determination of UV filters in water samples using stir bar sorptive extraction and thermal desorption–gas chromatography–mass spectrometry. J Chromatogr A 1179(2):81–88. https://doi.org/10.1016/j.chroma.2007.11.090

Rodil R, Moeder M (2008b) Development of a simultaneous pressurised-liquid extraction and clean-up procedure for the determination of UV filters in sediments. Anal Chim Acta 612 (2):152–159. https://doi.org/10.1016/j.aca.2008.02.030

Rodil R, Moeder M, Altenburger R, Schmitt-Jansen M (2009a) Photostability and phytotoxicity of selected sunscreen agents and their degradation mixtures in water. Anal Bioanal Chem 395 (5):1513–1524. https://doi.org/10.1007/s00216-009-3113-1

Rodil R, Schrader S, Moeder M (2009b) Non-porous membrane-assisted liquid–liquid extraction of UV filter compounds from water samples. J Chromatogr A 1216(24):4887–4894. https://doi.org/10.1016/j.chroma.2009.04.042

Rodil R, Quintana JB, López-Mahía P, Muniategui-Lorenzo S, Prada-Rodríguez D (2009c) Multi-residue analytical method for the determination of emerging pollutants in water by solid-phase extraction and liquid chromatography–tandem mass spectrometry. J Chromatogr A 1216 (14):2958–2969. https://doi.org/10.1016/j.chroma.2008.09.041

Rodríguez AS, Sanz MR, Rodríguez JB (2015) Occurrence of eight UV filters in beaches of Gran Canaria (Canary Islands). An approach to environmental risk assessment. Chemosphere 131:85–90. https://doi.org/10.1016/j.chemosphere.2015.02.054

Ruszkiewicz JA, Pinkas A, Ferrer B, Peres TV, Tsatsakis A, Aschner M (2017) Neurotoxic effect of active ingredients in sunscreen products, a contemporary review. Toxicol Rep 4:245–259. https://doi.org/10.1016/j.toxrep.2017.05.006

Sánchez-Quiles D, Tovar-Sánchez A (2015) Are sunscreens a new environmental risk associated with coastal tourism? Environ Int 83:158–170. https://doi.org/10.1016/j.envint.2015.06.007

Sang Z, Leung KSY (2016) Environmental occurrence and ecological risk assessment of organic UV filters in marine organisms from Hong Kong coastal waters. Sci Total Environ 566:489–498. https://doi.org/10.1016/j.scitotenv.2016.05.120

Schmid P, Kohler M, Gujer E, Zennegg M, Lanfranchi M (2007) Persistent organic pollutants, brominated flame retardants and synthetic musks in fish from remote alpine lakes in Switzerland. Chemosphere 67(9):S16–S21. https://doi.org/10.1016/j.chemosphere.2006.05.080

Schneider SL, Lim HW (2019) Review of environmental effects of oxybenzone and other sunscreen active ingredients. J Am Acad Dermatol 80(1):266–271. https://doi.org/10.1016/j.jaad.2018.06.033

Seoane M, Esperanza M, Rioboo C, Herrero C, Cid Á (2017) Flow cytometric assay to assess short-term effects of personal care products on the marine microalga Tetraselmis suecica. Chemosphere 171:339–347. https://doi.org/10.1016/j.chemosphere.2016.12.097

Sharifan H, Klein D, Morse AN (2016a) UV filters interaction in the chlorinated swimming pool, a new challenge for urbanization, a need for community scale investigations. Environ Res 148:273–276. https://doi.org/10.1016/j.envres.2016.04.002

Sharifan H, Klein D, Morse AN (2016b) UV filters are an environmental threat in the Gulf of Mexico: a case study of Texas coastal zones. Oceanologia 58(4):327–335. https://doi.org/10.1016/j.oceano.2016.07.002

Sheehan JM, Potten CS, Young AR (1998) Tanning in human skin types II and III offers modest photoprotection against erythema. Photochem Photobiol 68(4):588–592. https://doi.org/10.1111/j.1751-1097.1998.tb02518.x

Sieratowicz A, Kaiser D, Behr M, Oetken M, Oehlmann J (2011) Acute and chronic toxicity of four frequently used UV filter substances for Desmodesmus subspicatus and Daphnia magna. J Environ Sci Health A 46(12):1311–1319. https://doi.org/10.1080/10934529.2011.602936

Thorel E, Clergeaud F, Jaugeon L, Rodrigues A, Lucas J, Stien D, Lebaron P (2020) Effect of 10 UV filters on the brine shrimp Artemia salina and the marine microalga Tetraselmis sp. Toxics 8(2):29. https://doi.org/10.3390/toxics8020029

Tovar-Sánchez A, Sánchez-Quiles D, Basterretxea G, Benedé JL, Chisvert A, Salvador A, Moreno-Garrido I, Blasco J (2013) Sunscreen products as emerging pollutants to coastal waters. PLoS One 8(6):e65451. https://doi.org/10.1371/journal.pone.0065451

Tsui MM, Leung HW, Wai TC, Yamashita N, Taniyasu S, Liu W, Lam PK, Murphy MB (2014) Occurrence, distribution and ecological risk assessment of multiple classes of UV filters in surface waters from different countries. Water Res 67:55–65. https://doi.org/10.1016/j.watres.2014.09.013

Tsui MM, Lam JC, Ng TY, Ang PO, Murphy MB, Lam PK (2017) Occurrence, distribution, and fate of organic UV filters in coral communities. Environ Sci Technol 51(8):4182–4190. https://doi.org/10.1021/acs.est.6b05211

Wang J, Pan L, Wu S, Lu L, Xu Y, Zhu Y, Guo M, Zhuang S (2016) Recent advances on endocrine disrupting effects of UV filters. Int J Environ Res Public Health 13(8):782. https://doi.org/10.3390/ijerph13080782

Wick A, Fink G, Ternes TA (2010) Comparison of electrospray ionization and atmospheric pressure chemical ionization for multi-residue analysis of biocides, UV-filters and benzothiazoles in aqueous matrices and activated sludge by liquid chromatography–tandem mass spectrometry. J Chromatogr A 1217(14):2088–2103. https://doi.org/10.1016/j.chroma.2010.01.079

World Health Organization, World Meteorological Organization,United Nations Environment Programme, International Commission on Non-Ionizing Radiation Protection (2002) Global solar UV index: a practical guide. World Health Organization, Geneva. http://www.who.int/uv/publications/en/UVIGuide.pdf

Wu JW, Chen HC, Ding WH (2013) Ultrasound-assisted dispersive liquid–liquid microextraction plus simultaneous silylation for rapid determination of salicylate and benzophenone-type ultraviolet filters in aqueous samples. J Chromatogr A 1302:20–27. https://doi.org/10.1016/j.chroma.2013.06.017

Yan S, Liang M, Chen R, Hong X, Zha J (2020) Reproductive toxicity and estrogen activity in Japanese medaka (Oryzias latipes) exposed to environmentally relevant concentrations of octocrylene. Environ Pollut 261:114104. https://doi.org/10.1016/j.envpol.2020.114104

Yang C, Lim W, You S, Song G (2019) 4-Methylbenzylidene-camphor inhibits proliferation and induces reactive oxygen species-mediated apoptosis of human trophoblast cells. Reprod Toxicol 84:49–58. https://doi.org/10.1016/j.reprotox.2018.12.011

Zhang QY, Ma XY, Wang XC, Ngo HH (2016) Assessment of multiple hormone activities of a UV-filter (octocrylene) in zebrafish (Danio rerio). Chemosphere 159:433–441. https://doi.org/10.1016/j.chemosphere.2016.06.037

Zhou R, Lu G, Yan Z, Bao X, Zhang P, Jiang R (2019) Bioaccumulation and biochemical effects of ethylhexyl methoxy cinnamate and its main transformation products in zebrafish. Aquat Toxicol 214:105241. https://doi.org/10.1016/j.aquatox.2019.105241

Zucchi S, Blüthgen N, Ieronimo A, Fent K (2011a) The UV-absorber benzophenone-4 alters transcripts of genes involved in hormonal pathways in zebrafish (Danio rerio) eleuthero-embryos and adult males. Toxicol Appl Pharmacol 250(2):137–146. https://doi.org/10.1016/j.taap.2010.10.001

Zucchi S, Oggier DM, Fent K (2011b) Global gene expression profile induced by the UV-filter 2-ethyl-hexyl-4-trimethoxycinnamate (EHMC) in zebrafish (Danio rerio). Environ Pollut 159(10):3086–3096. https://doi.org/10.1016/j.envpol.2011.04.013

Zwiener C, Richardson SD, De Marini DM, Grummt T, Glauner T, Frimmel FH (2007) Drowning in disinfection byproducts? Assessing swimming pool water. Environ Sci Technol 41(2):363–372. https://doi.org/10.1021/es062367v

Micro and Nano-Plastics in the Environment: Research Priorities for the Near Future

Marco Vighi, Javier Bayo, Francisca Fernández-Piñas, Jesús Gago, May Gómez, Javier Hernández-Borges, Alicia Herrera, Junkal Landaburu, Soledad Muniategui-Lorenzo, Antonio-Román Muñoz, Andreu Rico, Cristina Romera-Castillo, Lucía Viñas, and Roberto Rosal

M. Vighi · J. Landaburu
IMDEA-Water Institute, Madrid, Spain
e-mail: marco.vighi@imdea.org; junkal.landaburu@imdea.org

J. Bayo
Department of Chemical and Environmental Engineering, Technical University of Cartagena, Cartagena, Spain
e-mail: javier-bayo@upct.es

F. Fernández-Piñas
Departamento de Biología, Facultad de Ciencias, Universidad Autónoma de Madrid, Madrid, Spain
e-mail: francisca.pina@uam.es

J. Gago · L. Viñas
Instituto Español de Oceanografía (IEO), Vigo, Spain
e-mail: jesus.gago@ieo.es; lucia.vinas@ieo.es

M. Gómez · A. Herrera
EOMAR: Marine Ecophysiology Group, IU-ECOAQUA, Universidad de Las Palmas de Gran Canaria, Las Palmas de Gran Canaria, Spain
e-mail: may.gomez@ulpgc.es; alicia.herrera@ulpgc.es

J. Hernández-Borges
Departamento de Química, Unidad Departamental de Química Analítica, Facultad de Ciencias, Universidad de La Laguna, San Cristóbal de La Laguna, Spain

Instituto Universitario de Enfermedades Tropicales y Salud Pública de Canarias, Universidad de La Laguna, San Cristóbal de La Laguna, Spain
e-mail: jhborges@ull.edu.es

S. Muniategui-Lorenzo
Grupo Química Analítica Aplicada, Instituto Universitario de Medio Ambiente (IUMA), Centro de Investigaciones Científicas Avanzadas (CICA), Facultade de Ciencias, Universidade da Coruña, A Coruña, Spain
e-mail: soledad.muniategui@udc.es

P. de Voogt (ed.), *Reviews of Environmental Contamination and Toxicology Volume 257*, Reviews of Environmental Contamination and Toxicology 257,
https://doi.org/10.1007/398_2021_69

Contents

1 Introduction 166
2 Definitions 169
3 Research Priorities 171
3.1 Environmental Sources 171
3.2 Sampling Procedures 174
3.3 Analytical Methods 181
3.4 Additives and Other Non-intentionally Added Substances 186
3.5 Sorption of Chemicals 188
3.6 Interaction with Microorganisms 189
3.7 Degradation and Fate of Microplastics 193
3.8 Direct Adverse Effects of Microplastics 194
3.9 Translocation and Transfer to the Food Web 196
3.10 Nanoplastics 199
4 Conclusions and Recommendations 200
References 204

Abstract Plastic litter dispersed in the different environmental compartments represents one of the most concerning problems associated with human activities. Specifically, plastic particles in the micro and nano size scale are ubiquitous and represent a threat to human health and the environment. In the last few decades, a huge amount of research has been devoted to evaluate several aspects of micro/nanoplastic contamination: origin and emissions, presence in different compartments, environmental fate, effects on human health and the environment, transfer in the food web and the role of associated chemicals and microorganisms. Nevertheless, despite the bulk of information produced, several knowledge gaps still exist. The objective of this paper is to highlight the most important of these knowledge gaps and to provide suggestions for the main research needs required to describe and understand the most controversial points to better orient the research efforts for the

A.-R. Muñoz
Departamento de Biología Animal, Facultad de Ciencias, Universidad de Málaga, Málaga, Spain
e-mail: roman@uma.es

A. Rico
IMDEA-Water Institute, Madrid, Spain

Cavanilles Institute of Biodiversity and Evolutionary Biology, University of Valencia, Valencia, Spain
e-mail: andreu.rico@imdea.org

C. Romera-Castillo
Department of Marine Biology and Oceanography, Institut de Ciències del Mar-CSIC, Barcelona, Spain
e-mail: crisrc@icm.csic.es

R. Rosal (✉)
Department of Chemical Engineering, University of Alcalá, Madrid, Spain
e-mail: roberto.rosal@uah.es

near future. Some of the major issues that need further efforts to improve our knowledge on the exposure, effects and risk of micro/nano-plastics are: harmonization of sampling procedures; development of more accurate, less expensive and less time-consuming analytical methods; assessment of degradation patterns and environmental fate of fragments; evaluating the capabilities for bioaccumulation and transfer to the food web; and evaluating the fate and the impact of chemicals and microorganisms associated with micro/nano-plastics. The major gaps in all sectors of our knowledge, from exposure to potentially harmful effects, refer to small size microplastics and, particularly, to the occurrence, fate and effects of nanoplastics.

Keywords Additives · Environmental risk · Internalization · Microbial colonization · Microplastics · Nanoplastics · Standardization

Abbreviations

AFM	Atomic force microscopy
ARB	Antibiotic resistant bacteria
ARG	Antibiotic resistance genes
CEC	Contaminant of emerging concern
DLS	Dynamic light scattering
DOC	Dissolved organic carbon
DSC	Differential scanning calorimetry
GIT	Gastrointestinal tract
HSI	Hyperspectral imaging
LC50	Lethal concentration for 50% of organisms
M/NPs	Micro- and nano-plastics
MBR	Membrane bioreactor
MP	Microplastic
NGS	Next-generation sequencing
NP	Nanoplastic
NTA	Nanoparticle tracking analysis
OCS	Operation clean sweep
PA	Polyamide
PCL	Polycaprolactone
PCP	Personal care product
PE	Polyethylene
PET	Polyethylene terephthalate
PHB	Polyhydroxybutyrate
PLA	Poly(lactic acid)
PP	Polypropylene
PS	Polystyrene
PU	Polyurethane
PVC	Poly(vinyl chloride)

Py-GC-MS	Pyrolysis gas chromatography-mass spectrometry
Py-GC-ToF-MS	Pyrolysis gas chromatography time of flight mass spectrometry
TGA	Thermogravimetry
WWTP	Wastewater treatment plant

1 Introduction

Although the first synthetic plastic polymer was discovered in the early 1900s (Andrady and Neal 2009), the presence of plastics in the world started to grow only in the 1950s eventually becoming ubiquitous and an integral part of modern life. Several reasons are explaining the enormous success of plastics. The extreme versatility of these materials, in terms of shapes, consistency, hardness and other properties, allows producing a practically endless variety of products. The possibility for manufacturing large series of items at low cost makes them the perfect material for producing disposable objects and all kinds of packaging. Plastics are almost chemically inert and may be easily sterilized, so they are excellent for containing food and for sanitary products. They are hardly altered and, therefore, the products made from them are long-lasting. However, this last property is also the main reason for the growing concern about plastics that has been raised worldwide: Plastics are highly persistent and, once introduced in the environment, it takes a very long time until they disappear.

Plastic manufacture represents about 6% of global oil consumption, and according to plastics usage projections, the plastics sector will account for 15% of the global emission of greenhouse gases by 2050 (WEF 2016). According to PlasticsEurope (2020), worldwide plastic production in 2019 amounted to 368 million tonnes. In Europe (EU plus UK, Norway and Switzerland), 29.1 million tonnes were collected as post-consumer waste through official schemes, equivalent to 47% of the amount of plastics produced in the same countries. Still, 25% of plastic post-consumer waste was sent to landfill and an undefined amount ended up in the environment (PlasticsEurope 2020). A large amount of these wastes ends up dispersed into the environment creating a worldwide pollution problem generally considered one of the major environmental issues associated with human activities (Baztan et al. 2017; GESAMP 2020; Koelmans et al. 2017a; UNEP 2016). Although most plastics come from land sources, the final receptor of plastic wastes are the oceans (Beaumont et al. 2019). Once there, they concentrate in particular areas due to the global cycle of currents, posing large risks for marine fauna (Kuhn et al. 2015; Lebreton et al. 2018; Thiel et al. 2018).

Once in the environment, plastic wastes suffer from several biotic and abiotic degradation processes. Abiotic mechanisms can be physical, which refers to erosion or fragmentation into smaller pieces, or chemical, due to the action of light and oxygen that lead to bond cleavage and the generation of molecules with new chemical moieties. The presence of light stabilizers or antioxidants, which are

Table 1 Potential risks associated with plastic particles according to their properties and environmental factors

Plastics properties	1. Transfer of additives used in the production of plastics – Sect. 3.4
	2. Release of unreacted monomers/oligomers – Sect. 3.4
	3. Physical impact on biota (higher size particles) – Sect. 3.8
	4. Translocation and transfer to the food web – Sect. 3.9
Environmental factors	5. Transfer of adsorbed environmental pollutants – Sect. 3.5
	6. Transport of non-indigenous species in the environment – Sect. 3.6

added to increase the service life of plastics, is another factor explaining the low environmental degradation rate of many plastics (Chamas et al. 2020). Biotic degradation generally follows abiotic fragmentation and takes place when microorganisms decompose break-down products under aerobic or anaerobic conditions to generate carbon dioxide, methane and biomass (Klein et al. 2018). The degradation of plastics also includes the leaching of additives included during compounding for a wide variety of purposes as well as non-intentionally added substances, which include impurities, catalysts, or polymerization by-products. All of them become eventually leached out from plastic materials during environmental degradation processes. However, the complete degradation to carbon dioxide and water hardly occurs in the environment, making plastic debris and smaller particles prone to travel long distances and/or to accumulate in most environmental compartments.

Irrespective of their origin, plastic particles are expected to pose a risk to the environment due to their inherent properties (i.e. molecular composition, additives, size and shape) or to environmental factors (Table 1). Risks associated with the inherent plastic properties can vary for the same polymer class due to differences in the manufacturing process. For example, a generic term for a class of plastic such as "*polyethylene*" (PE) includes many grades, differing in aspects like molecular weight, strength, crystallinity and even the detailed chemical structure, which leads to different monomer/oligomer release. Similarly, the additive composition can vary notably among polymers of the same class and result in different chemical leaching characteristics. The physical impact of large plastics on biota is essentially independent of polymer characteristics or toxic substances, as is associated with a physical harm, mainly related to their size and shape. Small debris may cause the blockage of the intestines of small animals and, for sufficiently small particles, there is the possibility of transfer through the food web and even translocate to tissues, thereby originating true toxic effects. The environmental factors are associated with the characteristics of the external medium, like the concentration of other pollutants, temperature, salinity or presence of potentially colonizing microorganisms. Similar microplastics (MPs) can behave differently in the environment depending on external variables providing some kind of "*en route*" signature (Leslie et al. 2017), which determines its capacity to disseminate microbial pathogens and transfer pollutants to living organisms. In addition, plastic debris could affect some aspects of the functioning of the ecosystem. For example, it has been hypothesized that they can contribute to decrease marine primary productivity and influence the carbon and nutrient cycles (Troost et al. 2018); however, further research is needed to provide a solid demonstration.

In the last few decades, huge research activity has been developed on the study of MPs and an enormous number of research papers and reviews have been published to quantify their presence in environmental compartments (Andrady 2011; Auta et al. 2017; Eerkes-Medrano et al. 2015; Li et al. 2018; Schell et al. 2020b); to evaluate their effects and risks for aquatic and terrestrial organisms (Burns and Boxall 2018; Chae and An 2018; de Sá et al. 2018); to assess their bioaccumulation and the effects of associated chemicals (Crawford and Quinn 2017; Verma et al. 2016; Wright et al. 2013); and to model their environmental behaviour (Everaert et al. 2018; Koelmans et al. 2016). Several international organizations and working groups have produced important technical reports and opinions (GESAMP 2015, 2016; SAM 2019; SAPEA 2019). A considerable number of international research projects have been funded in the last few years (notably, under JPI Oceans) and specific calls on this topic have been recently launched (H2020 under Food security and Environment Programmes). Despite this bulk of information, several knowledge gaps still exist that, in many cases, affect the relevance and the reliability of existing information. For example, the lack of harmonization and standardization of sampling and analytical methods makes it difficult to compare different studies. Therefore, even fundamental information like actual exposure in environmental compartments becomes difficult to judge.

Regulatory restrictions on MPs started in 2015 with the Microbead-Free Waters Act (USA) prohibiting the manufacturing and distribution of cosmetics containing plastic microbeads. A broader regulation came from a proposal from the European Chemicals Agency (ECHA) upon request of the European Commission to ban MPs intentionally added to a variety of goods including cosmetics, cleaning agents, paints and some industrial products. ECHA's restriction is currently under study in the European Parliament and the Council. In the meantime, some EU and non-EU countries, starting by the Netherlands, introduced different limitations in MP beads in cosmetic products. In 2019, the European Parliament voted the Directive 2019/904 on the reduction of the impact of certain plastic products on the environment, meaning single-use plastics and fishing gear containing plastic, which bans single-use plastic products by 2021, extends producer's responsibility schemes based on polluter-pays principle for items without available sustainable alternatives, and set the responsibility of Member States with marine waters for the collection of waste fishing gear containing plastic. As in the case of primary MPs, several countries already adopted or announced actions to limit the use of plastic starting by plastic bags and single-use items. Additional provisions have been included in the amended Waste Framework Directive while others are being considered by the EU Commission, US EPA and other Governments and agencies with different rate, extension and credibility.

The objective of this paper is not to provide an additional review on the presence and risks of microplastics and nanoplastics (M/NPs) in the environment, but to highlight and describe the major knowledge gaps and controversial points that researchers should deal with in the near future. A clear picture of the main research needs could be the fundamental basis for the coordination of future research efforts and for the development of specific project calls, at national and international level. This will allow developing proposals to cover these gaps and to improve our knowledge on the exposure, effect and risks of M/NPs, increasing our capability to develop risk mitigation measures to counteract one of the most important environmental problems in the start of the third millennium.

2 Definitions

Before presenting the main knowledge gaps and research priorities, a brief note on conventions and arbitrary definitions is needed. MPs are defined as fragments having <5 mm along its largest dimension (GESAMP 2019). Recently, Frias and Nash defined MPs as "synthetic solid particle or polymeric matrix, with regular or irregular shape and with size ranging from 1 μm to 5 mm, of either of primary or secondary manufacturing origin, which are insoluble in water" (Frias and Nash 2019). The definition should be broad enough to include natural polymers processed in such a way that they constitute anthropogenic litter, if spread into the environment (Hartmann et al. 2019). This definition, despite arbitrary and imprecise regarding its nomenclature (i.e. MP should include the μm range, and there is no reason to span to the mm range), has been adopted for the sake of harmonization after certain controversy (GESAMP 2019).

MPs are heterogeneous, exhibiting a range of shapes or morphologies from spherical beads to angular fragments and long fibres (Fig. 1). According to their origin, MPs are either primary or secondary. Primary MPs have been specifically manufactured with their size and include virgin plastic pellets used as raw materials for the fabrication of different products (Browne et al. 2011; Fendall and Sewell 2009; GESAMP 2019). According to GESAMP, secondary MPs "*result from wear and tear or fragmentation of larger objects*" (GESAMP 2019). The shape of plastic fragments is relevant because it determines drag, the viscous force exerted by a flowing fluid on any submerged particle that governs its terminal settling velocity and, therefore, the time a particle is being transported by water or air. Besides, and concerning the smaller sizes, particle shape influences suspension stability (Kim et al. 2015).

Concerning nanoplastics (NPs), the scientific literature used at least two different definitions: (1) Nano-sized plastic particles <1,000 nm (Andrady 2011; Cole et al. 2011); and (2) Nanoplastics <100 nm (in at least one of its dimensions) as defined for engineered nanoparticles (Bergami et al. 2016; Koelmans 2015). Lately, the first option that considers NPs as unintentionally produced plastic particles with colloidal behaviour and size range from 1 to 1,000 nm has gained popularity (Gigault et al. 2018). GESAMP also accepted this boundary, which must be understood as referred to the largest dimension by analogy with MPs (GESAMP 2019). The plain use of the 1,000 nm boundary without the limitation to unintentionally produced or secondary particles is less controversial (Hartmann et al. 2019).

Fig. 1 Photographs of different types of MPs collected in seawater in Ría de Vigo, NW Spain. Source: Instituto Español de Oceanografía

Although detected in essentially all ecosystems (Gago et al. 2018), the current debate on M/NPs tends to exclude fibres (Frias and Nash 2019; Henry et al. 2019). Polymeric fibres are produced by textile wearing, particularly during laundry (Napper and Thompson 2016). Before reaching conclusions on the impact assessment of fibres, two main methodological gaps need to be addressed. First, the lack of a proper definition of "*size*" in the case of fibres. Second, the definition of fibres of concern in the context of plastic pollution. Both aspects require clarification and standardization.

In the case of fibres, their largest dimension is particularly meaningless to establish cut-off among categories because large fibrous materials may pass through filters with smaller opening size, thereby complicating quantification. The behaviour of a fibre inside a fluid medium depends on its Stokes' or aerodynamic diameter. Accordingly, the relevant or characteristic dimension of a fibre is its equivalent diameter, which is generally a linear function of their physical diameter and depends less on fibre length. Besides, fibres are flexible. A clear definition of size cut-offs for fibres and in general for particles of low sphericity is lacking.

Besides, there is an issue concerning nomenclature. The term "*microfibre*" is common in many environmental studies as a synonym of "microplastic fibre" or synthetic fibre within the MP size range, but the denomination is controversial. There is a technical definition of microfibres to refer to a mass per unit length of thread, which can be conflicting. Some authors recommend avoiding the term "*microfibre*" and others suggest including a minimum length to diameter ratio in the definition (Liu et al. 2019b; Salvador-Cesa et al. 2017). The term "*nanofibre*" is also debated: the industry often considers "*nanofibres*" objects with diameters as large as 500 nm or 1,000 nm.

An additional issue concerning fibre composition is a need to include as anthropogenic debris not only those made of synthetic polymers, but also regenerated cellulose textiles (like rayon and lyocell). Both are included under the heading "man-made fibres" in ISO/TR 11827 Textiles – Composition testing – Identification of fibres. Besides, natural fibres that show evidence of industrial processing should be considered as a category of anthropogenic litter because they incorporate additives like bleaching agents, softening, or stiffening additives, synthetic dyes, light stabilizers, and flame retardants among others (Darbra et al. 2012). Clearly, fibres made of synthetic polymers, regenerated cellulose, or processed natural materials are generally sampled together. Moreover, the textile industry is moving towards the production of a wide range of hybrid natural/synthetic fabrics.

3 Research Priorities

3.1 Environmental Sources

3.1.1 State of the Art

Intentionally manufactured MPs, or primary MPs, are used with different purposes in many products. These include scrubbing phase in personal care products (PCPs), encapsulating agent for fragrances in detergents and softeners, or with several technical functions in fertilizers and plant protection products for agriculture, paints, coatings, inks, medical products and devices or food supplements. Most of these primary MPs are extremely persistent materials whose exposure could result in adverse effects nowadays or in the future due to continued use and the difficulty of being removed once in the environment. Therefore, there is a need in the industry for a transition to more suitable alternatives like natural products in PCPs or biodegradable polymers for other technical functions. In the USA, the Microbead-Free Waters Act of 2015 banned the manufacturing and distribution of cosmetics containing rinse-off plastic microbeads. In Europe, a wider restriction has been proposed by the European Chemicals Agency (ECHA 2019). The restriction affects a range of products in different sectors, including domestic and industrial uses. Several EU Member States have already introduced partial bans for MPs in specific products. Some exemptions are considered like MPs for use at industrial sites and in medical products for human or veterinary use, among others.

Secondary MPs have several different origins. Wastewater treatment plants (WWTP) have been identified as one of the main point sources of MPs in freshwater (Carr et al. 2016). Most studies indicated that primary and secondary wastewater treatments remove most MPs. Murphy et al. reported 98% of MPs removal from a conventional secondary WWTP plant located in Scotland (Murphy et al. 2016). Talvitie et al. observed 99% MP removal from a secondary WWTP, the primary treatment already removing 97.4–98.4% (Talvitie et al. 2017b). The same group evaluated four different wastewater treatment technologies (disc filters, rapid sand filters, dissolved air floatation and membrane bioreactor). They concluded that membrane bioreactor (MBR) was the most efficient technology with 99.9% removal capacity (Talvitie et al. 2017a). However, despite the high removal ability of current wastewater treatment technologies, and due to the high volume of treated wastewater continuously emitted to the environment, there is still a considerable emission of MPs from WWTPs to rivers. Edo et al. (2020) reported a release of 300 million MP particles (>25 μm) per day from a Spanish WWTP to the Henares River representing an approximate load of MPs of 350 particles/m^3 (Edo et al. 2020). One major contribution to MPs reaching WWTP is the wearing of synthetic clothes in domestic washing machines, notably those made of polyester and acrylic fibres (Napper and Thompson 2016). Additionally, industrially processed natural fibres, which contain potentially harmful additives, reach the environment in the same way (Edo et al. 2020). Fragments and other secondary plastic debris are also usual in the effluents of WWTP, which constitute a vehicle for them to reach freshwater and seawater environments (Bayo et al. 2020).

Stormwater runoff from urban and agricultural soils has been shown to represent an important source of MP pollution. Commercial and industrial areas are major contributors while synthetic rubber particles attributed to car tyre wear mostly appear in sediments due to road runoff (Liu et al. 2019a; Ziajahromi et al. 2020). Besides, plastic debris from materials used in the construction of wetlands, rests from agricultural plastics and many other secondary MPs reach natural environments driven by wind (Zhang et al. 2019b). A precise estimation of MPs emissions due to water runoff and atmospheric transport is difficult due to limited data available.

The atmosphere is the less studied environmental compartment concerning the occurrence and transport of MPs. The occurrence of airborne MPs has been documented in studies at ground or near-to-ground level (Brahney et al. 2020; Klein and Fischer 2019). Recently, and for the first time, direct evidence of the presence of MPs at high altitudes has been provided that demonstrated their presence even beyond the planetary boundary layer (González-Pleiter et al. 2021). The available data are difficult to interpret due to the rapid atmospheric mixing and the occurrence of random deposition events, but generally suggest that the source of most airborne MPs is urban due to the higher concentrations detected near populated areas (Wright et al. 2020). However, the mobility of airborne MPs is high, and they can be transported to areas far from any source of pollution (Bergmann et al. 2019; Bullard et al. 2021). The literature reports concentrations in the order of a few MPs per cubic metre and deposition rates reaching values up to the order of hundreds of MPs per square metre and day (Abbasi et al. 2019; Dris et al. 2016).

The use of plastic packaging in the food sector clearly proved to be a vehicle for MPs release to packaged food (Fadare et al. 2020; Kedzierski et al. 2020). Trays made from extruded polystyrene have been deemed responsible for food transfer of

MPs in levels ranging from 4.0 to 18.7 MP/kg of packaged food (Kedzierski et al. 2020). The occurrence of MPs in drinking water, both tap and bottled, has also been studied with results showing concentrations in the order of tens of MPs per litre (Schymanski et al. 2018; Shruti et al. 2020). Food plastic packaging enhances storage, transport, protection, and preservation, but contributes to human exposure to MPs in products intended for human consumption. The presence of MPs in food is a topic widely covered in the literature with estimations of annual MP intake in order of tens of thousands of particles (Cox et al. 2019). Teabags packaging were shown to release billions (10^9) of M/NP particles (polyacrylate and polyethylene terephthalate, PET) into a single cup of beverage (Hernandez et al. 2019). MPs in a wide array of seafood products have been detected due to the pollution of seas (Sun et al. 2019).

3.1.2 Knowledge Gaps and Research Needs

In view of the increasing regulatory restrictions affecting intentionally manufactured MPs, it is foreseen that they will represent a minor cause for concern in the future. Additionally, it has been observed that <10% of the MPs found in the effluent of WWTPs are pellets, which can be classified as primary MPs (Dyachenko et al. 2017). Therefore, it can be concluded that secondary MPs represent a bigger threat than intentionally manufactured MPs. This includes a better management of plastic litter, which, in the form of larger debris (mesoplastics, 5–25 mm or macroplastics, >25 mm) is an important source of M/NPs due to fragmentation, and the limitation of unnecessary plastic items like plastic packaging materials. Most of the original research efforts have been conducted on the marine environment. More studies are needed about MPs in freshwaters, which proved to be receiving bodies comparable to the marine environment (Li et al. 2018). An important limitation for assessing the fate of M/NPs is the limited data available for assessing the sources and the origin (e.g. primary or secondary MNPs) and the mass balance in the different environmental compartments (water, air, soil) of the smaller sizes of MPs and of NPs (Schell et al. 2020b).

A deeper insight into WWTP processes is required to avoid MPs emissions. The understanding of physical, chemical and biological mechanisms affecting MPs in WWTP is a related need (Bayo et al. 2020). Even if removal rates are high in conventional WWTP and most MPs are recovered with sludge, they find a way to go back to the environment via sludge use in agriculture as fertilizer. Several studies revealed a concentration of MPs in sewage sludge ranging from a few to several hundred particles per gram of dry sludge (Edo et al. 2020; Magni et al. 2019). Accordingly, synthetic polymers mainly consisting of fibres can be detected in agricultural soils even years after sludge application (Zubris and Richards 2005). Therefore, ways of managing WWTP sludge that ensures its safety and avoids the spreading of MPs into the environment need to be urgently developed ensuring a safe use by source separation, composting or risk assessment.

The generation of secondary M/NPs from food contact materials and human exposure to them is a major cause for concern nowadays, even in the absence of evidence about their risk for human health. Quantitative data are needed on the presence of MPs in food from plastic containers and the contribution of food packaging to the global emissions of M/NPs. There are very limited data on the smaller sizes of MPs and on NPs, which are the size ranges of higher concern.

Overall, limited data exist on the occurrence and transport of MPs in important environmental compartments. There are only a few studies addressing the atmospheric deposition rates of MPs and no data truly reporting their occurrence in the atmosphere. Data from soil and sediments are also scarce and fragmentary and little is known about the role played by sea bottoms as an ultimate sink. The relevance of the different transport pathways for the environmental distribution of M/NPs is challenged by the limited amount of data. The data required for modelling are globally insufficient and, in some regions, rely on extrapolations made with too many uncertainties.

3.2 Sampling Procedures

3.2.1 State of the Art

Sampling campaigns in surface water (particularly in the sea) are usually performed using manta trawl nets with a mesh size usually in the 200–500 μm range (Fig. 2 and Table 2). Data are usually reported as MPs/km^2. The use of flow meters is highly recommended to report data in MP/m^3. Sampling should be carried out under optimal sea conditions, with a Beaufort scale between 0 and 2. Otherwise, it is necessary to do the calculations of wind correction factor as proposed by Kukulka et al. (2012) who showed that under strong wind conditions neuston nets tend to collect fewer plastic particles due to vertical wind-induced mixing (Kukulka et al. 2012). In some cases, water is pumped into the nets (Fig. 3).

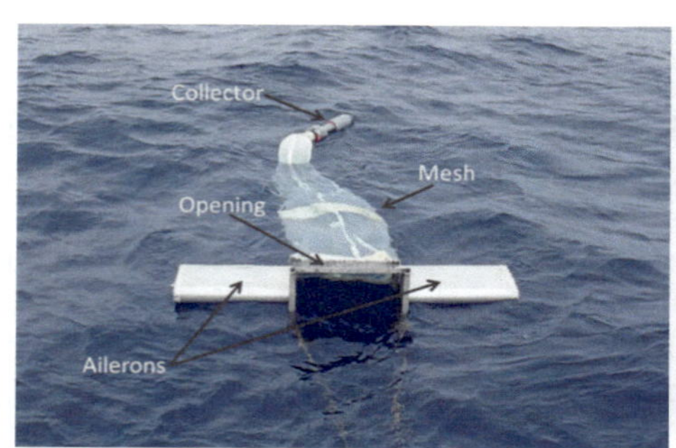

Fig. 2 Manta net picture and its parts (left): stainless steel structure with a front opening of 60 × 40 cm and rear opening of 60 × 25 cm. On the sides, two stainless steel ailerons. Some modifications are used such as the use of buoys (right). Source: EOMAR-Universidad de Las Palmas de Gran Canaria

Table 2 Net types and information collected from manta trawl studies. Comparison of different types of nets used, MP colour categories and units used to report the data; FTIR analysis; and if zooplankton abundance and MP/zooplankton ratio was reported. Adapted from Herrera et al. (2020)

Region	Net/size (μm)	MP colour	MP type	Sample sieving	FTIR analysis	Zooplankton	Report Items/km^2	Report Items/m^3	Ratio MP/Zoo	Reference
North Pacific Central Gyre	Manta 330	No	Yes	Yes	No	Yes	Yes	Yes	Yes	(Moore et al. 2001)
Southern California coastal waters	Manta 333	No	Yes	Yes	No	Yes	No	Yes	No	(Moore et al. 2002)
Santa Monica Bay, California	Manta 333	No	Yes	Yes	No	Yes	No	Yes	No	(Lattin et al. 2004)
Southern California Current, Pacific Ocean	Manta 505	No	No	No	No	No	No	Yes	No	(Gilfillan et al. 2009)
Northeast Bering Sea, Pacific Ocean	Sameoto 505	No	Yes	Yes	Yes	No	No	Yes	No	(Doyle et al. 2011)
Southern California, Pacific Ocean	Manta 505	No	Yes	Yes	Yes	No	No	Yes	No	(Doyle et al. 2011)
North Western Mediterranean Sea	Manta 333	No	No	No	No	Yes	Yes	No	No	(Collignon et al. 2012)
South Pacific subtropical gyre	Manta 333	No	Yes	Yes	No	No	Yes	No	No	(Eriksen et al. 2013)
Bay of Calvi, Mediterranean Sea	Manta 333	No	No	Yes	No	Yes	Yes	No	Yes	(Collignon et al. 2014)
Sardinian Sea, Western Mediterranean	Manta 500	No	No	No	No	No	No	Yes	No	(de Lucia et al. 2014)
Portuguese coastal waters	Neuston 280–335	No	No	No	Yes	Yes	No	Yes	Yes	(Frias et al. 2014)
Western Tropical Atlantic Ocean	Manta 300	Yes	Yes	No	No	No	No	Yes	No	(Ivar do Sul et al. 2014)
Eastern Pacific Ocean	Neuston 333	No	No	No	No	No	Yes	No	No	(Law et al. 2014)

(continued)

Table 2 (continued)

Region	Net/size (μm)	MP colour	MP type	Sample sieving	FTIR analysis	Zooplankton	Report Items/km^2	Report Items/m^3	Ratio MP/Zoo	Reference
Goiana Estuary, Northeast coast of Brazil	Plankton 300	Yes	Yes	No	No	Yes	No	Yes	Yes	(Lima et al. 2014)
Northeast Atlantic Ocean	Pump 250	Yes	Yes	No	No, Raman	No	No	Yes	No	(Lusher et al. 2014)
Southern coast Korea	Manta 333	No	No	Yes	Yes	No	No	Yes	No	(Song et al. 2014)
Western Mediterranean Sea	Manta 334	No	Yes	No	No	Yes	Yes	No	No	(Faure et al. 2015)
Spanish Northwest coast	Manta 335	No	No	No	No	No	Yes	No	No	(Gago et al. 2015)
East Asian seas, Japan	Neuston 350	No	No	No	Yes	No	No	Yes	No	(Isobe et al. 2015)
South East Sea of Korea	Manta 330	No	Yes	Yes	Yes	No	No	Yes	No	(Kang et al. 2015)
Arctic waters, Norway	Manta 333	Yes	Yes	No	Yes	Yes	Yes	Yes	No	(Lusher et al. 2015)
Arctic waters, Norway	Pump 250	Yes	Yes	No	Yes	Yes	No	Yes	No	(Lusher et al. 2015)
Mediterranean Sea	Manta 200	No	Yes	No	No	No	Yes	No	No	(Cózar et al. 2015)
Black Sea	WP2 200	No	Yes	No	No	Yes	No	Yes	Yes	(Aytan et al. 2016)
Mediterranean Sea, near coast	Manta 333	No	No	No	Yes	Yes, zooscan	Yes	No	Yes	(Pedrotti et al. 2016)
Central and Western Mediterranean Sea	Manta 333	No	No	No	No	No	Yes	No	No	(Ruiz-Orejón et al. 2016)

Mediterranean Sea	Neuston 200	No	Yes	Yes	Yes	No	Yes	Yes	No	(Suaria et al. 2016)
Northern Gulf of Mexico	Neuston 335	No	Yes	No	Yes	Yes, zooscan	No	Yes	Yes	(Di Mauro et al. 2017)
Pelagos Sanctuary, Western Mediterranean Sea	Manta 333	Yes	No	Yes	Yes	No	Yes	No	No	(Fossi et al. 2017)
Bay of Brest, France	Manta 335	No	Yes	Yes	No, Raman	No	No	Yes	No	(Frère et al. 2017)
Stockholm Archipelago, Baltic Sea	Manta 335	Yes	Yes	No	Yes	No	Yes	Yes	No	(Gewert et al. 2017)
Southern Ocean, Antarctica	Manta 350	No	No	No	Yes	No	Yes	Yes	No	(Isobe et al. 2017)
Atlantic Ocean	Pump 250	Yes	Yes	No	Yes	No	No	Yes	No	(Kanhai et al. 2017)
North-East Atlantic	Manta 333	Yes	Yes	Yes	No	No	Yes	Yes	No	(Maes et al. 2017)
Mediterranean Sea, > 10 km from coast	Manta 333	No	No	No	Yes	Yes, zooscan	Yes	No	Yes	(Pedrotti et al. 2016)
Israeli Mediterranean coast	Manta 333	Yes	Yes	No	No	No	Yes	Yes	No	(van der Hal et al. 2017)
Bornholm Basin, Baltic Sea	Bongo 150	Yes	Yes	No	No	No	No	Yes	No	(Beer et al. 2018)
Pearl River estuary, Hong Kong waters	Manta 333	No	Yes	Yes	Yes	No	Yes	Yes	No	(Cheung et al. 2018)
Guanabara Bay, South-eastern Brazil	Neuston 64	Yes	Yes	No	Yes	Yes	Yes	Yes	No	(Figueiredo and Vianna 2018)
North Western Australia, Indian Ocean	Manta 355	No	No	No	Yes	No	No	Yes	No	(Kroon et al. 2018)
North Western Australia, Indian Ocean	Plankton (subsurf)	No	No	No	Yes	No	No	Yes	No	(Kroon et al. 2018)

(continued)

Table 2 (continued)

Region	Net/size (μm)	MP colour	MP type	Sample sieving	FTIR analysis	Zooplankton	Report Items/km^2	Report Items/m^3	Ratio MP/Zoo	Reference
Chabahar Bay, Gulf of Oman	Neuston 333	Yes	Yes	Yes	Yes	No	No	Yes	No	(Aliabad et al. 2019)
North Atlantic Ocean, Azores	Bongo 200	No	Yes	Yes	No	Yes, zooscan	Yes	Yes	Yes	(Herrera et al. 2020)
North Atlantic Ocean, Madeira and Canary Islands	Manta 200	No	Yes	Yes	No	Yes, zooscan	Yes	No	Yes	(Herrera et al. 2020)
Persian Gulf	Neuston 300	Yes	Yes	No	Yes	No	Yes	No	No	(Kor and Mehdinia 2020)
Baltic Sea and Gulf of Bothnia	Manta 333	No	No	Yes	NIR-hyperspectral imaging	No	No	Yes	No	(Schönlau et al. 2020)
Baltic Sea and Gulf of Bothnia	Pump ≥300	No	No	No	NIR-hyperspectral imaging	No	No	Yes	No	(Schönlau et al. 2020)
Baltic Sea and Gulf of Bothnia	Pump 50	No	No	No	NIR-hyperspectral imaging	No	No	Yes	No	(Schönlau et al. 2020)
Nuup Kangerlua, West Greenland	Pump 10	No	Yes	No	Yes	No	No	Yes	No	(Rist et al. 2020)
Nuup Kangerlua, West Greenland	Bongo 300	No	Yes	No	Yes	No	No	Yes	No	(Rist et al. 2020)

Fig. 3 Example of sampling in a river with a battery of nets with different mesh size fed by a pump with known water flow. Source: IMDEA Water (Madrid Institute for Advanced Studies on Water, Spain, left). In situ pump built by KC Denmark (Silkeborg, Denmark) and EU CleanSea project; (**a**) Faulhaber Swiss motor system, 3863H0224CR – 24 V, GB 150 nm with a planetary gearing 44/1–4,8:1 16 nm; (**b**) Water inlet with rotating blades to pull in the surrounding water; (**c**) stack of filters with inserted 500 μm, 300 μm, and 50 μm filters; (**d**) water outlet and flow meter (right). Source: Schönlau et al. (2020)

It is also important to mention the recent use of new systems (see Fig. 3), like in situ pumps, fractionated cascade filtration and other devices, suitable to collect suspended microplastics in the surface and also in the water column (Abeynayaka et al. 2020; Karlsson et al. 2020; Rist et al. 2020; Schönlau et al. 2020; Setälä et al. 2016). The main improvement of these systems is the more accurate measurement of the water volume filtered. Sediment sampling can be done in intertidal or subtidal areas. Intertidal sampling is generally performed seasonally to account for tide variability. Usually, the sediments are sampled along transects (approx. 100 m) set in parallel to the water edge and defined by GPS position. Sampling sites can be distributed along the high tide line, along the low tide line and/or in between. This aspect must be taken into consideration when comparing results. The sampling unit (e.g. a square of 50 × 50 cm) must be replicated along the transect collecting the first centimetres of sediment. Subtidal sampling is generally performed using grabs or corers being Van Veen grab and Box corers the most usual devices. A Box corer can be used for sediments allowing sampling the first centimetres with minimal impact on sediment integrity. In this case, a replicate (up to six per site) is recommended to check the homogeneity of sampling sites.

3.2.2 Knowledge Gaps and Research Needs

The methods for MPs sampling vary considerably among studies making it difficult to compare data from different studies (GESAMP 2020). Table 2, adapted from Herrera et al. (2020), presents the different nets and data categorization from recently published studies (Herrera et al. 2020). The different sampling methods, MPs categories and units used make impossible the comparison among literature data. A harmonized approach should be established to get relevant and comparable information from sampling campaigns. Data sharing protocols and data platforms at regional level (using the Regional Sea Conventions) and worldwide (like the UN initiative; Global partnership on Marine Litter platform) must include global standards for sampling, identification and quantification (GESAMP 2020).

One major knowledge gap refers to the need for obtaining information about smaller size ranges. Most studies in the marine environment use mesh sizes above 200 μm. However, the available data indicated that MPs below current sampling limits might be dominant. Cai et al. (2018) found that >80% of the MPs collected from surface waters of South China Sea were <200 μm (average 145 μm). Enders et al. (2015) sampled MPs down to sizes of 10 μm and concluded that small MPs are ubiquitous in the ocean surface layer, the majority (64%) of particles being <40 μm. Technical developments are necessary to implement standardized procedures for sampling MPs with such small sizes. In some sampling campaigns batteries of plankton nets in series, with mesh size as small as 20 μm are already in use, in addition a new plastic-free pump-filter system has recently been successfully used that allows the collection of MPs down to 10 μm (Rist et al. 2020). Concerning sizes in the higher range, standardized cut-offs should be established. Specifically, for marine pollution studies, sampling size ranging from 200 μm to >1 mm is ideal for comparing samples taken with different types of mesh. It may include 1,000, 500, 330 and 200 μm opening sizes.

Besides, MPs should be consistently classified according to typology and colour. Typology should include the following categories or others that can be reduced to them: fragments, fibres, lines, pellets, films, foams. RGB colour system can be determined from images using software like ImageJ or similar. Both colour and typology can help to determine the source of pollution or can indicate whether predation is being selective or not. The following nine colour categories are suggested: no colour/white/clear, yellow, orange/brown, red/pink, green, blue, purple, grey/silver, black/dark.

Plankton count is important, as microplastics/plankton ratio is an indicator of the probability of plastic entering the food web through filter feeders. Zooplankton can be quantified using a stereomicroscope and software like Zooprocess (https://sites.google.com/view/piqv/). The identification and counting of the different groups of organisms can be carried out with the ECOTAXA web application (https://ecotaxa.obs-vlfr.fr/). Reports should include the zooplankton abundance in items/km^2 and, preferably, in items/m^3.

There is an urgent need for information about the fate of smaller size MPs, i.e. those falling below 100 or 200 μm and are currently outside systematic sampling, except for in situ pumping systems. For all sizes, there is a need for agreeing standardized methods to allow comparability from different sources. Standardization is needed in all environmental compartments and in large scale monitoring campaigns. There is a need to record cross-cutting data like plankton abundance, proper georeferencing and environmental conditions. The use of adequate procedures for sample processing and storage should be documented in all studies. This includes the mention to the use (or not) of clean air conditions, the thorough description of controls and any other details that can be relevant for comparing data. A more comprehensive monitoring of freshwater ecosystems is needed as they received far less attention than marine ones, despite the evidence that most plastic litter comes from land sources.

3.3 Analytical Methods

3.3.1 State of the Art

Large MPs (1–5 mm) are usually identified by optical microscopes (sizes usually >100 μm) or using the naked eye detection for differentiation from non-plastic materials. This approach allows evaluating colour, shape, size and number of plastic particles and, accordingly, several guidelines have been reported for harmonizing the visual identification of MP particles (Lv et al. 2020). Dyes (e.g. fluorescent dye Nile Red) are used sometimes to improve discrimination. However, visual identification is not usually accurate enough for scientific and monitoring purposes and other techniques are required. High-throughput alternatives based on specific equipment used for zooplankton (ZooScan or FlowCam), flow cytometry or by a high-resolution scanner in combination with automatic image analysis or computer vision have been recently developed to count and classify MPs into different visual classes, thereby reducing analysis time and cost (e.g. SMACC, which is freely distributed by EOMAR group from University of Las Palmas de Gran Canaria) (Lorenzo-Navarro et al. 2020). In any case, and even if high-throughput techniques are used, the analysis of plastic pollutants is complex, expensive, and time-consuming. Therefore, there is a need for establishing the size of subsample to be analysed based on robust statistical criteria, which is a caution very rarely addressed in the literature. Relatively simple statistics can be used to assess the accuracy of results within a certain error margin (Kedzierski et al. 2019).

Analytical methods based on spectroscopic techniques such as Fourier transform infrared spectroscopy (FTIR) or Raman spectroscopy allow non-destructive chemical characterization and are commonly used to accurately identify plastic polymers (and sometimes additives) over a wide range of particle sizes (Xu et al. 2019). These techniques require a small amount of sample although they normally involve careful spectra gathering and long analysis time. Larger particles can be analysed by attenuated total reflectance (ATR) FTIR spectroscopy with high speed and accuracy (200–5,000 μm). Depending on the set-up of the application, small particles can also be measured only down to the range of 20 μm (reflectance or transmittance modes) due to the diffraction limit (Lv et al. 2020). Recently, the combination of FTIR with an IR microscope (single point, focal plane array or linear array) has emerged as micro-FTIR for the characterization of samples with sizes down to 10 μm (Löder et al. 2015). Raman spectroscopy can detect small plastic particles down to 1 μm and using micro-Raman even smaller sizes could be achieved, although limited by fluorescence from some polymers or from biogenic materials. In addition, aged and weathered plastics must be included in the spectral libraries used for identifying environmental samples. An important drawback is that micro-spectroscopic identification is a very time-consuming task, especially when analysing the entire sample and smaller particle size ranges in complex environmental samples.

To cope with these limitations, automatic image software based on library search and chemometric analyses have been developed that reduce working time and cost sometimes using freeware software tools (e.g. siMPle, developed by Aalborg University and Alfred Wegener Institute) (Meyns et al. 2019). Lately, hyperspectral imaging (HSI) is used to characterize larger MPs (>200–300 μm). It produces 3D hyperspectral image hypercube, which contains spatial and spectral information such as morphological features and chemical characteristics of the analyte. The main drawback of HSI is data processing complexity because users must develop customized algorithms and models to extract information (Fu et al. 2020).

In contrast to spectroscopic techniques, the thermal analysis is being increasingly used for MP characterization, which includes pyrolysis gas chromatography-mass spectrometry (Py-GC-MS), thermogravimetric analysis (TGA), hyphenated TGA such as TGA-mass spectrometry (TGA-MS), TGA-thermal desorption-gas chromatography-mass spectrometry (TGA-TD-GC-MS), TGA-differential scanning calorimetry (DSC) (Peñalver et al. 2020). These methods give information about chemical composition and mass-based quantification but not about size, shape, or number of MPs in each given sample. Besides, they are destructive techniques. One of the main advantages of Py-GC-MS is the possibility of the chemical characterization of polymer and organic additives in the same analysis (Fischer and Scholz-Böttcher 2017; Fries et al. 2013). Py-GC-MS does not usually require any pretreatment and only needs a very small amount of sample (in the low mg or μg range and even only one particle). X-ray fluorescence can assess additives or adsorbed metals, while scanning electron microscopy reveals information on morphology and composition of MPs. These are complementary techniques with generally high cost (Fries et al. 2013).

3.3.2 Knowledge Gaps and Research Needs

There are still no harmonized analytical methods for quantifying and determining the occurrence and composition of M/NPs in environmental samples. For most environmental applications, the methods currently applied for the detection, characterization, and identification of M/NPs are complex, tedious, time-consuming, and difficult to automate. All of them suffer from matrix effects and require controversial sample pretreatments. There are issues relating to the use of standard metrics, pretreatment and separation methods and there is also an urgent need to improve rapid and reliable analytical methods, particularly for small size MPs as quantitative data for particles with a size smaller than c.a. 50 μm are scarce in the literature. There is an important difficulty derived from the huge variety of polymers and additives that can be included in MPs such as plasticizers, flame retardants, pigments, stabilizers and many others that have been used to modify their properties and characteristics (Hermabessiere et al. 2017). Additionally, the weathering of plastics can modify their composition or some of their characteristics making it difficult to detect them in environmental matrices (Fernández-González et al. 2021; Jahnke et al. 2012).

A first non-solved issue is the metrics used to report "plastics", which is closely related to the analytical methods required. One of the most usual units found in published articles is items of plastic per unit of volume, weight or similar of the environmental matrix (i.e. items/kg sediment). This approach requires methods that individually separate and identify every single item. This type of methodology is extremely time-consuming, impossible to apply to NPs and, to some extent, dependent on the analytical operator and, for brittle materials, a possible source of errors (if the items break in two pieces the result would be the double number of items). Another approach, more in line with what is normally done in the analytical quantification of pollutants in the environment is to report the results as the weight of plastic per weight or volume of the environmental matrix (i.e. mg/kg sediment). This approach is applicable to M/NPs and would be probably more accurate and lead to more comparable results with the drawback of the difficulty to separately quantify the different plastic components of environmental mixtures. The need to clearly define what to measure and how to report data is critical if the results are to be used for risk assessment and included regulatory frameworks such as the Marine Strategy Framework Directive or, in the future, the Water Framework Directive.

Depending on the sample to be analysed, a suitable pretreatment might be necessary. At some stage, floatation, as a density fractionation method, is generally required followed by suitable filtration. However, there is a wide variability in the type of solutions in which floatation can be developed: saturated solutions of NaCl (the cheapest and most common), NaI, NaBr, $ZnCl_2$, $ZnBr_2$, $CaCl_2$, sodium or lithium metatungstate, among others (Gong and Xie 2020; Li et al. 2020; Miller et al. 2017; Prata et al. 2019a; Silva et al. 2018). There is a clear need to establish a floatation protocol, since the use of a certain type of solution determines the plastics that can be separated. For this purpose, apart from the density of the floating solution and polymers, the toxicity of the salts, their cost as well as their possible interaction with specific materials (NaI reacts with cellulose filters) should be considered. Besides, for MPs of very small sizes (typically <10 μm) floatation may not be suitable, while the separation of fibres by floatation is also difficult (Miller et al. 2017). An important issue regarding floatation is that its automation remains a challenge.

The removal of organic matter is another important step in samples from soils, sediments, sewage sludge, biosamples and many others. Organic matter removal can be achieved with acid (HNO_3, H_2SO_4, $HClO_4$ or mixtures of them) or basic (i.e. NaOH, KOH) treatments, oxidizing agents (H_2O_2 with or without Fe(II) as catalyser-Fenton's reagent) or enzymatic digestion (using proteolytic enzymes like trypsin, papain, pepsin or collagenase) at different temperatures and times, without a clear harmonized procedure (Bretas Alvim et al. 2020; Miller et al. 2017; Prata et al. 2019a; Primpke et al. 2020). It should be considered that a complete elimination of organic matter might not always be possible. Besides, chemical and structural integrity of the polymer, which is an organic compound itself, may be affected, a fact that should be carefully evaluated (Munno et al. 2018; Prata et al. 2019a). Aggressive pretreatments can be strongly influenced by the ageing of plastics specimens, as well as their composition and size.

Concerning fibres, and, in general, MPs of small size (i.e. <300 μm), it is also frequent to incubate them in a dye solution like Nile Red (which is also fluorescent) or Rose Bengal (Prata et al. 2019a, b), once they are separated from the sample matrix. This is normally done by immersing the filtration membranes into the dye solution (Primpke et al. 2020). Staining facilitates the visual identification in different ways. For example, Nile Red will improve their observation by fluorescence or any other imaging technique while Rose Bengal will not normally stain MPs but natural particles (Bretas Alvim et al. 2020; Primpke et al. 2020). Despite the benefits achieved by dye staining, this method alone is a non-specific approach that may yield false positives. Another important issue regarding sample pretreatment is the composition of the filters to be used for the recovery of MPs from liquid samples or from the supernatant of the density separation, especially if they are directly used for further spectroscopic analysis since their compatibility should be considered (i.e. with FTIR or Raman).

There is a lack of relevant information on the instrumental set-up required to replicate the environmental studies undergone (e.g. up to 25% of the published papers do not reported relevant instrumental operational details); therefore, some MPs identifications might be compromised (Andrade et al. 2020). In addition, statistical assessment of sample-associated errors should be systematically addressed. The subsample derived to micro-FTIR or other characterization techniques must not be arbitrary. The accuracy of estimations, representing half-width of the confidence interval, should be routinely reported. Standardized analytical procedures and more efficient analysis workflow of environmental samples should be carried out regarding sizes, shapes and material identity focussing on the development of automatized systems to avoid biases in plastic identification and providing a reliable estimation of environmental contamination from MPs (Campanale et al. 2020).

Appropriate quality assurance/quality control (QA/QC) procedures are required to improve data reliability. Thus, cross-contamination/procedural blanks should be routinely performed during all steps of the analytical procedure, especially when measuring small fragments and fibres to assess the representativity of results. Besides, the results of procedural blanks should be reported. The recovery rates of the analytical procedure using spiked samples are also relevant to assess the accuracy of the selected methodology, avoiding the risk of under- or overestimation of the reported MPs. The validation of the analytical methods for measurements of MPs is hampered by a general lack of standards and reference materials (Seghers et al. 2021). In fact, it is particularly challenging to prepare reference materials able to mimic the MPs found in environmental samples. However, there is an urgent need to develop such standards to achieve reliable monitoring of MP contamination. In addition, interlaboratory comparison exercises are also required to detect potential biases, uncertainties and other sources of error and to demonstrate proficiency and competence. As recent examples, the European Commission's Joint Research Centre (JRC) has been involved in the preparation of a reference material for MPs (PET) in water and proficiency tests on MPs in water (PET) and in sediments (PE). QUASI MEME/NORMAN organized an international laboratory intercomparison exercise to determine the polymer type and number or mass of polymer particles in different samples, which revealed an urgent need for harmonization (van Mourik et al. 2021). These challenging analytical progresses will contribute to improve the reliability of MP analysis to support monitoring programmes, research and decision-making.

An even greater challenge is the identification of smaller-sized M/NPs, especially for complex matrices and if particles are affected by plastic weathering that may cause misclassification. All techniques generally available have a particle size limit of a few micrometres; consequently, new methods must be developed to cover the smaller sizes of MPs (<10 μm) and the nanometre range (<1,000 nm). Recently, promising techniques widely used for characterizing nanomaterials have been applied to small MPs. Field flow fractionation (single-particle mode of inductively coupled plasma-mass spectrometry) allow active particle separations. Hydrodynamic chromatography, a solution-phase liquid chromatographic separation method, is advantageous for particle size determination in the range from 10 nm to 1 μm. For physicochemical quantification, dynamic light scattering (DLS) and nanoparticle tracking analysis (NTA) is useful for the hydrodynamic size and zeta potential measurements of M/NPs. Recently, atomic force microscopy (AFM) is an emerging nanoscale characterization technique of materials. Besides morphological information, chemical properties are also achieved in combination with spectroscopic IR or Raman IR techniques (Dominguez et al. 2014). So far, there have been a very limited number of studies using hybrid AFM techniques (AFM/IR or AFM/Raman) to detect and characterize M/NPs (Fu et al. 2020). Raman spectroscopy can also be combined with SEM allowing a spatial resolution down to several hundred of nanometres (Zhang et al. 2020). However, the techniques for detecting NPs are still complex and difficult to apply to environmental samples. There is an urgent need to develop and implement more precise, more reliable and less time-consuming methodologies for the identification and quantification of MPs (particularly small size MPs) and NPs in environmental matrices.

3.4 Additives and Other Non-intentionally Added Substances

3.4.1 State of the Art

Commercial plastics are not pure polymers. They include many additives to improve their processability and their properties that include a wide series of different chemicals and materials: fillers, plasticizers, colourants, stabilizers, flame retardants, compatibilizers, among others, which are found in different proportions in the formulation of plastic materials (Ambrogi et al. 2017). Most additives are included at levels of very low per cent by weight, although some of them, like flame retardants or plasticizers, may reach much higher values (Hahladakis et al. 2018). It is well documented that the additives found in plastics have the potential to contaminate the environment. Inorganic substances like metals become easily leached, while organic compounds are released directly or as degradation products after photochemical reaction (Bandow et al. 2017). In contact with water, additives migrate to the aquatic media (Koelmans et al. 2014; Mato et al. 2001; Romera-Castillo et al. 2018). The migration of additives in food contact plastic materials poses an additional issue to human health and food quality if transferred beyond certain limits (Bhunia et al. 2013). Noteworthy, some additives, like antioxidants, ultraviolet (UV) absorbers and biological preservatives, are responsible for enhanced persistence of many plastics that degrade very slowly under environmental conditions (Hahladakis et al. 2018). Besides intentionally added chemicals, other substances like mono- and oligomers from the plastic structure can be released threatening the environment (Amamiya et al. 2019; Saido et al. 2014). As an example, unreacted styrene from polystyrene (PS) packaging materials has been detected in different matrices (Arvanitoyannis and Bosnea 2004). The transfer of PET oligomers to drinking water has also been described (Hoppe et al. 2017). This issue has been dealt within the context of food safety, but, as non-intentionally added substances, there is also a concern associated with their leaching to the environment (Hoppe et al. 2016).

3.4.2 Knowledge Gaps and Research Needs

The risk assessment of additives faces an important problem of lack of information about the chemical nature of the additives themselves and their concentrations in plastic materials because of the secrecy associated with proprietary formulations. Some information is available, but much more is needed about additives and their potential toxicity when incorporated to plastic products. Significant efforts have been made in the field of plastic packaging. A database has been created that includes near 1,000 chemicals plus several more thousands possibly associated with plastic packaging. Some of them are known by posing a significant risk to human health and to the environment according to ECHA, including endocrine-disrupting chemicals and persistent and bioaccumulative compounds (Groh et al. 2019). The issue is not only the enormous number of different chemicals in use but the lack of transparency and incompleteness of publicly available information on the use of many substances. The fact that many plastic objects are produced in countries with limited access to the information does not help. Obviously, their long-term toxicity and possible mixture effects are essentially unknown except for some of the most hazardous chemicals.

The environmental conditions affect the intensity of chemical migration from the plastic to the aquatic media. Turbulence was found to enhance plastic leaching especially of those additives less soluble in water (e.g. phthalates or Irgafos® 168) in comparison with those with higher solubility (e.g. BPA) (Suhrhoff and Scholz-Böttcher 2016). Other variables, such as salinity, affect differently the intensity of the plastic leaching depending on the intrinsic nature of the additive (Suhrhoff and Scholz-Böttcher 2016). The effect of photochemical ageing on the release of additives and depolymerization fragments has been studied but results are not yet conclusive (Lee et al. 2020; Romera-Castillo et al. 2018; Suhrhoff and Scholz-Böttcher 2016; Zhu et al. 2020). The degradation stage of the plastic can also affect the leaching rate and its toxicity to marine fauna (Bejgarn et al. 2015; Saido et al. 2014).

There is also a considerable lack of knowledge about the possible degradation products originated from additives under environmental conditions. The analyses required tracing the huge amount of oxidation, photodegradation and biotransformation products that can be originated from plastic additives are highly challenging. The identification of additives and their degradation products is a very complex issue due to the large number of different types of chemicals used, the relatively high molecular weight of many additives, their presence in mixtures and their inclusion in complex matrices, particularly when dealing with food transfer chemicals (Blázquez-Blázquez et al. 2020). As for the environmental fate of plastic debris, additives received much less attention compared with food contact materials. There is a need to characterize the additives associated with the MPs detected in environmental samples. Non-target screening analysis and toxicity studies are required to assess the risk of additives leaching from plastic debris. The transfer of additives to the food chain upon MPs ingestion and their effect on freshwater and marine organisms are largely unknown. The possible formation of toxic degradation products from additives upon oxidation or photochemical processes is another complex issue that requires attention. The degradation pattern and leaching of dissolved organic carbon from biodegradable polymers should be clarified, as up to date, the environmental impact of their degradation products is poorly understood. In addition, there is scarce information about the processes that such additives suffer when recycling polymers or the additives released in the environment from recycled polymers. All of these are key issues to promote more sustainable and non-toxic reusable products reducing the impact of plastics on the environment.

3.5 Sorption of Chemicals

3.5.1 State of the Art

Plastic has the capacity to absorb and desorb organic compounds, such as persistent or emerging organic pollutants (Rodrigues et al. 2019; Wang et al. 2018). The available data suggest that certain hydrophobic pollutants adsorb onto the surface of MPs reaching a concentration much higher than in surrounding water (Mato et al. 2001). Field campaigns detected aromatic hydrocarbons, polychlorinated biphenyls, organochlorine pesticides, brominated diphenyl ethers and organophosphorus flame retardants among other pollutants in MPs from marine environments and sediments (Camacho et al. 2019). Despite the ample evidence that MPs accumulate persistent organic pollutants, this does not result in MPs being important for their global dispersion and there is scant evidence that MPs are an important transfer vector for bioaccumulative chemicals (Lohmann 2017). Physicochemical studies suggested that the adsorption of hydrophobic pollutants would be insufficient to increase the exposure to toxic substances in the marine environment (Koelmans et al. 2016). Gouin et al. used thermodynamic calculations to show that the importance of MPs as a carrier of hydrophobic substances is probably limited in marine environments because the partition of pollutants among plastic, air and water would not result in significant adsorbed amounts even for volumes of plastic orders of magnitude above current values (Gouin et al. 2011). Recent experimental work confirmed the theoretical hypotheses showing that the presence of MPs (polyethylene particles 150 μm diameter) does not increase the bioconcentration of lipophilic chemicals in fish (*Danio rerio*) or marine plankton (Beiras et al. 2019; Schell et al. 2020a). However, limited studies are available to evaluate possible different patterns in organisms with different physiological characteristics (e.g. invertebrates) and in different environmental matrices (e.g. soil or sediments). Therefore, the role of plastics as a vector of organic pollutants to biological organisms is still unclear. The concentration of certain chemicals such as contaminants of emerging concern (CEC) or persistent pollutants is particularly high in inland waters, close to zones of intensive agriculture, industrial placements or near the discharge of wastewater treatment plants. Adsorbed chemicals might be released when the conditions of the external medium change, such as inside the animal guts (e.g. pH conditions). Compounds eventually translocated to organs or tissues may induce damage and move through the trophic web until humans.

3.5.2 Knowledge Gaps and Research Needs

The relative importance of plastics as a vector in the transport of chemical contaminants to biota is influenced by several factors like polymer characteristics, material ageing, chemical environment, and residence time in the organism when ingested, among others. Although experimental and modelling evidence suggests that plastics would not represent an important pathway for the transfer of sorbed chemicals, more work is needed to assess this fact across a wider range of organisms (Bakir et al. 2016). It is a fact that plastics can sorb pollutants from the environment relatively quickly and their concentration on their surface or dissolved in the glassy phase can become orders of magnitude higher than in the surrounding aquatic environment (León et al. 2018; Mato et al. 2001). If sorbed chemicals desorb upon ingestion, this could provide a route for transferring pollutants to biota (Teuten et al. 2007). The relative importance of this pathway has yet to be fully evaluated in freshwater, soils and sediments. Plastics from intensive agriculture or in contact with wastewater discharges may be exposed to high concentration of toxics and, therefore, cause a higher carrier effect than that of marine debris.

Sediments and soil have been recognized as a major sink of MPs, probably one order of magnitude higher than oceans (Tourinho et al. 2019). There is a need to extend studies to them as well as including relevant organisms such as soil invertebrates. The effect of MP ageing should also be considered. The interaction of pollutants and microplastics changes from hydrophobic interaction to hydrogen bonding as hydrophilic moieties appear upon oxidation and photo-oxidation. The interaction with polar and semipolar compounds should be emphasized including antibiotics, new pesticides and other (CEC).

3.6 Interaction with Microorganisms

3.6.1 State of the Art

Once in natural environments, plastics are easily colonized by different types of microorganisms, including bacteria, archaea and eukaryotes such as fungi and protists (Kettner et al. 2017; McCormick et al. 2016). The term "plastisphere" first coined by Zettler et al. identifies plastic as a new niche or habitat for microorganisms (Amaral-Zettler et al. 2020; Zettler et al. 2013). Advanced DNA sequencing protocols (metabarcoding analyses as well as shotgun metagenomics; generally known as next-generation sequencing, NGS, techniques) facilitated the knowledge of the diversity of microorganisms forming biofilms on different types of polymers and have allowed comparison with those free-living in the water column or attached to the sediments (Jacquin et al. 2019). Most studies about plastic-colonizing microorganisms have been made in marine environments and only a few have targeted freshwater environments (Hoellein et al. 2014; McCormick et al. 2016). The experimental approach of reported studies is highly variable. Some are based on the in situ sampling of plastics or MPs in aquatic environments (Bryant et al. 2016; De Tender et al. 2017; Lee et al. 2014), while others selected different types of artificial polymers and sizes and incubated them under controlled experimental conditions using microcosms (Ogonowski et al. 2018). Biodegradable plastics have also been found more easily colonizable than non-biodegradable ones (Dussud et al. 2018). Most studies describe that location (in situ environment), rather than polymer type, determines microbial community on plastic biofilms, but data are still scarce to provide general trends.

Biofilm formation in any matrix (including plastics) involves a series of phases from initial colonization to maturation. Early attachment by pioneer microorganisms is usually facilitated by the formation of a surface organic layer on the substrate. At this stage, physical properties of the material such as roughness, charge, density, mechanical stability or hydrophobicity may play crucial roles. Subsequent production of extracellular polymeric substances determines the capacity of biofilms to grow and to establish cell–cell interactions, which is an evolutionary strategy to survive in unfavourable environments (Flemming et al. 2007). Usually, during the maturation phase, a succession of new settlers occurs and finally the biofilm disperses, and the free microbes look for new niches to be established. Several studies have followed the dynamics of the formation of biofilms on plastics showing that biofilm formation was stable enough to reconstruct temporal dynamics allowing the identification of indicator species of the different stages of biofilm formation (De Tender et al. 2017). Some results reported significant changes in microbial diversity depending on polymer type (De Tender et al. 2017; Ogonowski et al. 2018; Webb et al. 2009). It was suggested that differences between substrates may be stronger during early stages of biofilm formation (Oberbeckmann et al. 2016). In this context, most studies have examined microbial colonization in mid- to long-term experiments, while early microbial colonization has seldom been studied although it is a critical phase for biofilm conditioning. A recent study characterized bacterial communities in the early stage of biofilm formation on seven different types of MPs (including biodegradable as well as non-biodegradable ones) deployed in two different WWTP effluents (Martínez-Campos et al. 2021). An early colonization phase MPs-core microbiome was identified. Furthermore, linear discriminant analysis effect size analyses (LEfSe) allowed identifying core microbiomes specific for each type of polymer suggesting that each type might select early attachment of bacteria.

An important issue of microbial colonization is that plastics may be first colonized by taxa that can degrade plastic polymers to some extent. For example, different species of *Pseudomonas* which have been found on plastics have been associated with the degradation of PE, polypropylene (PP) or poly(vinyl chloride) (PVC) (McCormick et al. 2014, 2016). Oberbeckmann et al. (2016) also found PET-colonizing organisms able to degrade polymers such as taxa belonging to the family Rhodobacteraceae. One of its members, *Rhodococcues ruber*, has also been reported to degrade PE in biofilms (Oberbeckmann et al. 2016). Martinez-Campos et al. reported in their study of early bacterial colonizers on MPs deployed in WWTP effluents that genera *Pseudomonas*, *Variovorax*, *Aquabacterium* or *Acidovorax*, which have species able to metabolize recalcitrant substances including plastics, were dominant in MPs (Martínez-Campos et al. 2021). It is interesting that biodegradable MPs were also enriched on these potential degrading taxa (*Aquabacterium* and *Pseudomonas* in PHB and *Variovorax* in PCL). In fact, taxa specialized in complex carbon degradation (including some recalcitrant compounds) have also been found (Bryant et al. 2016). This raises the question as to whether these colonizers might be involved in plastic or other organic compounds degradation as a source of carbon for their growth and metabolism. It is important to note that conventional plastics have been designed to be intrinsically persistent, not only because of the nature of their polymeric backbone, but also because they are blended with stabilizers that limit oxidative or photochemical degradation. Therefore, the biodegradation of "non-biodegradable" plastics is difficult, which is the reason they constitute a group of persistent organic pollutants. Oxo-degradable plastics, which contain additives that accelerate oxidation processes, have been recently banned in the EU (Directive 2019/904).

NGS analyses have underpinned that MPs might host pathogens within attached microorganisms and thus, might act as vectors to distribute them in their movement through aquatic ecosystems. Sequences belonging to potential pathogens such as *Vibrio* and *Arcobacter* spp. have been identified in MPs in marine as well as freshwater environments (Amaral-Zettler et al. 2015; Harrison et al. 2014; Martínez-Campos et al. 2021). Campylobacteraceae, a family that includes several taxa associated with human gastrointestinal infections, and potential fish pathogens like *Aeromonas*, have been identified on MPs in an urban river (McCormick et al. 2014). It is a remarkable finding that biodegradable MPs such as PLA, PHB and PCL showed a significant abundance of genera with potential pathogenic members such as *Pseudomonas*, *Comamonas*, *Aeromonas* and *Vibrio*; this might be of concern since the capacity of the MPs to act as vector of potentially pathogenic taxa may be facilitated by their biodegradability (Martínez-Campos et al. 2021). In addition, viruses, such as the SARS-CoV-2, which is the coronavirus responsible for the COVID-19 pandemic, are of concern given that the virus may remain active during several days on plastics (van Doremalen et al. 2020). The materials used to be protected from the pandemic, in particular facemasks, are usually made of PP and PA and are being disposed carelessly in the environment. Thus, they are becoming a real environmental problem because the virus has already been found in wastewaters (Chavarria-Miró et al. 2020) and in many countries.

There is also a growing concern that MPs and plastics in general may be reservoirs of antibiotic resistance bacteria (ARB) and cognate antibiotic resistance genes (ARG) (Laganà et al. 2019; Wang et al. 2020; Yang et al. 2019). ARB may survive in the presence of one or more antibiotics and that might be a potential threat for human health. In addition, ARGs are carried usually on broad-host range plasmids or other mobile elements that may be potentially transferred by horizontal gene transfer to nearby receptors, which may contribute to global spread of antibiotic resistance (Sultan et al. 2018; Wang et al. 2020). It was recently reported that MPs could concentrate ARGs such as *sulI*, *tetA*, *tetC*, *tetX*, *ermE* and *ermF* from the surrounding water (Martínez-Campos et al. 2021; Wang et al. 2020).

3.6.2 Knowledge Gaps and Research Needs

The colonization of plastic substrates by microorganisms is still largely unknown. Specific in situ environments might select the indicator species and early-stage development of plastisphere communities should be studied not only in marine but also in soil and freshwater habitats. An important issue seldom tackled is whether there are changes in community composition along with the transport from WWTP to rivers and to the ocean. Several studies have found that geography and season are the main factors in shaping microbial communities in plastics (Lee et al. 2014; Oberbeckmann et al. 2014). The transport of non-indigenous species or pathogens (like bacteria or viruses) using plastic debris as transport mechanism should be compared to natural materials. The stability and physical properties of plastics may favour the attachment and transport of mobile and sessile species to new areas. In this context, MPs have been found in remote regions such as in the Arctic in deep-sea sediments (Bergmann et al. 2017), seawater (Cincinelli et al. 2017; Cózar et al. 2017; Lusher et al. 2015) and sea ice (Obbard et al. 2014; Peeken et al. 2018), and recently in a freshwater lake (González-Pleiter et al. 2020). The colonization of MPs from remote regions has not been addressed yet. An important knowledge gap exists about the potential of MPs to act as vectors of ARBs/ARGs in remote locations because they would shed light on the global issue of antibiotic resistance (Hendriksen et al. 2019; Pärnänen et al. 2019).

Only a few studies about microbial colonization of plastic wastes have been carried out on soil environments, albeit recent evidence indicates that plastics are abundant in soils. The recent study of Puglisi et al. confirms the novel hypothesis that different plastics host different bacterial communities, and that their structure can be correlated with the physicochemical properties of the plastics, particularly their degradation degree (Puglisi et al. 2019). The most degraded polyethylene films were found to host a bacterial community similar to the surrounding soil. Meanwhile the study of Zhang et al. concluded that the bacterial communities colonizing microplastics were significantly different in structure from those in the surrounding soil, plant litter and macroplastics (Zhang et al. 2019a).

Overall, the role of MPs as a new niche for microorganisms is not well understood. It is a new microbial habitat that might already be performing a role at the ecological level. MPs may contribute to disperse microorganisms even to remote areas or to host pathogens. This could include ARBs/ARGs possibly posing an important issue to human health and the environment. Colonized MPs may alter the feeding behaviour of many aquatic organisms that may feed on them. Further research is needed on the potential transfer of pathogens or ARGs to the aquatic trophic web. Attention must be paid to microbial assemblages that may be involved in polymer degradation and metabolism of xenobiotics, including biodegradable plastics. This merits further research as new degradation pathways may be discovered.

3.7 *Degradation and Fate of Microplastics*

3.7.1 State of the Art

The breakdown of plastic is known to be triggered by environmental factors like light, oxygen, temperature and mechanical erosion. Recently, it has been demonstrated that some aquatic invertebrates can also contribute to MP breakdown (Mateos-Cárdenas et al. 2020). Mineralization as the final stage of polymer degradation takes a very long time. Polyolefins, which are the most abundant polymers in marine samples, may persist even hundreds of years exposed to hydrolysis and photo-oxidation conditions (Barnes et al. 2009). Several studies showed how the abiotic ageing of polymers leads to their fragmentation in smaller pieces (Gewert et al. 2015; Kalogerakis et al. 2017). Surface cracking makes the rest of polymeric material more prone to degradation, while the mechanical properties related to the fabrication process might play a major role in the fragmentation propagation of cracks and eventually in the disintegration of specimens. The data indicate that fragments are generated when cracking lines converge, so cracking is key to predict the number and size of the fragments produced from a given material (Julienne et al. 2019). The biodegradation of "non-biodegradable" MPs does exist, but it is very slow as most plastics are very resistant to microorganisms because high molecular weight and hydrophobic surfaces make them inaccessible to microbial enzymes. Some studies identified strains capable of certain biodegradation of conventional polymers, but at a very slow rate (Skariyachan et al. 2017). Multiomics and synthetic microbial communities have been explored to enhance plastic biodegradation with a certain degree of success (Jaiswal et al. 2020). Many fungi are able to degrade complex carbon polymers such as lignin, which might imply that they can also degrade plastics. Lignin-degrading enzymes such as oxidases, laccases and peroxidases have been reported as responsible for the degradation of plastic polymers by fungi (Shah et al. 2007). Most reports describe fungi able to degrade polyurethane (PU) by extracellular polyurethanases (Russell et al. 2011). Brunner et al. found several fungi growing on plastic debris floating in the shoreline, which were able to degrade PU (Brunner et al. 2018). The capacity of fungi to degrade PE is controversial although that capability has been reported (Ojha et al. 2017). Regarding biodegradable plastics, fungal depolymerases have been found capable to degrade PHB films (Panagiotidou et al. 2014).

Most polymeric molecules are chemically (and toxicologically) inert, but the same does not stand for their degradation products. Plastics can be broken down into smaller pieces and the smaller they are, the higher their surface to volume ratio, with more plastic surface potentially leaching. There are more than 250,000 tonnes of plastics floating in the ocean exposed to UV radiation and oxygen degradation, which are the main abiotic factors responsible for plastic degradation and leaching (Andrady 2011; Eriksen et al. 2014). Up to 23,600 tonnes of dissolved organic carbon (DOC) can be released from plastic reaching the ocean every year (Romera-Castillo et al. 2018). About 7% of the plastic weight can be lost in form of DOC under UV irradiation (Zhu et al. 2020). The leached material mainly consists of low molecular weight compounds (<350 Da) and its release is enhanced by UV radiation (Lee et al. 2020). It has been shown that leached compounds may alter the marine food web by stimulating marine bacterial growth (Romera-Castillo et al. 2018; Zhu et al. 2020). It has also been shown that they can impair the photosynthetic capacity and growth of cyanobacteria (Tetu et al. 2019).

3.7.2 Knowledge Gaps and Research Needs

Fracture behaviour and the propagation of cracks in aged plastics require further studies. The parameters governing plastic fragmentation have not been completely identified and there are great difficulties to monitor fragmentation patterns in real environments. Therefore, artificial weathering protocols should be developed to clarify fragmentation kinetics (Andrade et al. 2019; Julienne et al. 2019). It is necessary to gain information about the influence of abiotic and biotic factors in the fragmentation process of plastic debris to model the number of small pieces of MPs and NPs in environmental compartments, which is a major question still open in M/NPs research (Koelmans et al. 2017b).

The degradation patterns (time, intermediate products) of different types of polymers in different environmental conditions are not well understood in terms of DOC generation and should be better investigated. The biodegradation of conventional plastics by microorganisms is poorly known, in particular regarding the isolation of depolymerization enzymes and their mechanism of action. Another poorly known but relevant issue is the fate of non-traditional plastic polymers, the so-called biodegradable polymers, which include their subclass of compostable plastics. Their actual degradation patterns, as well as the environmental impact of intermediate products, should be thoroughly investigated. The available data point towards a non-negligible environmental impact of MPs derived from the biodegradation of biodegradable plastics (González-Pleiter et al. 2019). This is an aspect of high economic and social relevance and should be thoroughly investigated. Plastic leachates can also have consequences in human health after interacting with other chemicals. For instance, the chlorine often applied in WWTP as a disinfectant can react with the organic compounds migrated from plastic to form toxic disinfection by-products such as trihalomethanes (Lee et al. 2020).

3.8 Direct Adverse Effects of Microplastics

3.8.1 State of the Art

One of the major properties of most plastic polymers is their lack of chemical reactivity. Moreover, MPs in the size range of a few tens of μm up to a few mm are too large to be capable to cross cell membranes and enter cells. Therefore, the adverse effects of MPs on living organisms, if only the effects of particles are considered, excluding chemical additives that may be present in some formulations (Sect. 3.4), cannot be considered toxic effects. Indeed, toxicity is the reaction, inside of the cell, of a chemical substance with a chemical cellular receptor (specific toxicity) or with the general chemical environment of the cell (non-specific or narcotic-type toxicity) (Verhaar et al. 1992).

MPs ingested by higher organisms pass through the digestive tract and can be eliminated through faeces. Possibly, a different pathway may occur with very small size MPs (a few μm) or NPs (Sects. 3.9 and 3.10). The ingestion of MPs may produce physical injuries, inducing inflammation and stress, or it may result in a blockage of the gut and subsequent reduced energy intake or respiration. MPs may also produce behavioural effects such as reduction of feeding efficiency (Besseling et al. 2017; Cole et al. 2015; de Sá et al. 2015).

In recent years, a huge amount of information has been produced on the adverse effects of MPs on aquatic (freshwater and seawater) and terrestrial organisms. Several types of organisms have been tested covering various taxonomic groups (e.g. crustaceans, insects, molluscs, annelids, fish), ecological role and feed habit (e.g. filter feeders, grazers, predators), habitats (e.g. planktonic, benthonic, sediment-dwelling, soil organisms). Short term (e.g. mortality) and long-term (e.g. growth, reproduction) endpoints have been addressed. Tests have been performed using various types of test materials (microbeads, fibres, tyre debris) of different size, also in relation to the size of tested animals, shape and chemical composition (de Sá et al. 2018) (Kögel et al. 2020). In most studies tests have been designed to provide (LC_{50} results) with concentrations that are beyond the environmentally realistic range (Lenz et al. 2016). However, in some cases, the concentration range used spans from environmentally realistic levels up to orders of magnitude higher (Lusher 2015). An important problem is that the particles used in different tests span through a wide range of ranges and shapes. A rescaling method has been recently proposed to adjust data from sources using different types of MPs when determining species sensitivity distributions (Koelmans et al. 2020).

3.8.2 Knowledge Gaps and Research Needs

Several comprehensive reviews have been published collecting the bulk of the information available and trying to perform hazard and risk assessments (Adam et al. 2019; Burns and Boxall 2018; Kögel et al. 2020). The main conclusion that may be derived is that no adverse effects have been observed at concentrations comparable to the upper range of the distribution of the levels that have been measured in natural environments. Usually, adverse effects under laboratory conditions have been observed at levels that are orders of magnitude higher than environmentally realistic levels. A comparison between a safe concentration, estimated on the basis of available data on adverse effects, and projection of the concentration of global marine microplastics in a "*business as usual*" scenario, indicates that by the end of this century, the concentration of floating MPs will reach a level about two orders of magnitude lower than a threshold of risk, while a potential risk level will be reached only by the concentration of MPs that wash ashore in the marine environment (Everaert et al. 2018). Based on the available information, it may be concluded that the adverse effects of MPs do not represent a priority for further research, although some specific issues may be further investigated. For example, some details on the types of effects and modes of action remain to be clarified. Besides, the environmental relevance of the plastic material is rarely considered for impact assessment studies. More information is needed on the different test materials representing plastics from real consumer products and on their environmental behaviour when exposed to ageing conditions similar to natural environments.

3.9 Translocation and Transfer to the Food Web

3.9.1 State of the Art

The uptake of chemicals by living organisms is a process by which a contaminant is stored in the tissues to a level higher than the surrounding environment (Gobas and Morrison 2000). These processes may describe uptake patterns: (1) Bioconcentration is the accumulation of a chemical in the tissues of an organism because of direct exposure to the surrounding medium (e.g. water). The bioconcentration factor is the ratio of a contaminant concentration in biota to its concentration in the surrounding medium once equilibrium is reached. (2) Bioaccumulation is the accumulation of chemicals in the tissue of organisms through any route, including respiration, food ingestion or direct contact. (3) Biomagnification is the increase of internal chemical concentration from lower to higher levels of the food chain, which depends not only on the physicochemical properties of the chemical but also on the trophic relations (Solomon et al. 2013). These processes may not occur with insoluble particulate materials, such as MPs, that cannot cross cellular membranes and enter the cells by passive diffusion regulated by partitioning mechanisms, as soluble chemicals do (Devito 2000; Schultz 1976). Therefore, they cannot accumulate in tissues unless other types of active processes such as endocytosis-related mechanisms occur (Felix et al. 2017). The cellular uptake process of insoluble particulates is completely different from the passive diffusion process of bioaccumulative chemical compounds, and the assessment of their potential to bioaccumulate may require the development of new test systems, models and mechanistic understanding (ECETOC 2019; Handy et al. 2018; Petersen et al. 2019; Roch et al. 2020).

The ingestion of MPs by aquatic and terrestrial animals has been widely documented in the literature (Wesch et al. 2016). According to Gouin, a huge number of individual organisms (about 87,000) belonging to more than 800 different species have been analyzed and MPs have been found in the gastrointestinal tract (GIT) of more than 20% of them, with an average of 4 MP items per individual (Gouin 2020). However, the presence of MPs in the GIT does not mean that a bioaccumulation process is occurring and does not necessarily represent a transfer to the food web comparable to biomagnification processes occurring with bioaccumulative chemicals. This consideration refers to MPs of medium-large size, i.e. those that are easily detected and analysed. For small size MPs, smaller than some tens of microns, and particularly for NPs, smaller than 1 μm, the problem is much more complex. Indeed, these particles cannot be detected, measured and counted with the conventional analytical procedures usually applied to detect MPs. Due to technical limitations, the concentrations of NPs in the environment are currently unknown (GESAMP 2020). To date, there are no examples in the literature demonstrating the ingestion of NPs by free-living organisms, though some laboratory studies have attempted to investigate ingestion of NPs using labelled particles and environmentally unrealistic exposure levels (Skjolding et al. 2017).

3.9.2 Knowledge Gaps and Research Needs

The capability of M/NPs to cross cell membranes and enter tissues is highly controversial. Large size MPs cannot, and the problem is limited to very small MPs and NPs. However, we lack information to determine to what extent small MPs and NPs may cross cell membranes. The potential for translocation of small size MPs and NPs from GIT to internal tissues, at least at very high levels, has been demonstrated (Triebskorn et al. 2019). Most of these results have been obtained using fluorescence-labelled NPs (Fig. 4). However, the reliability of results based on fluorescence-labelled MP-NPs is highly controversial, as it has been recently demonstrated that these results may be biased by experimental artefacts due to lipid accumulation of the leached fraction of hydrophobic fluorescent dye (Schür et al. 2019). An additional cause for concern is represented by a general lack of reproducibility of studies on the translocation of M/NPs into living organisms (Burns and

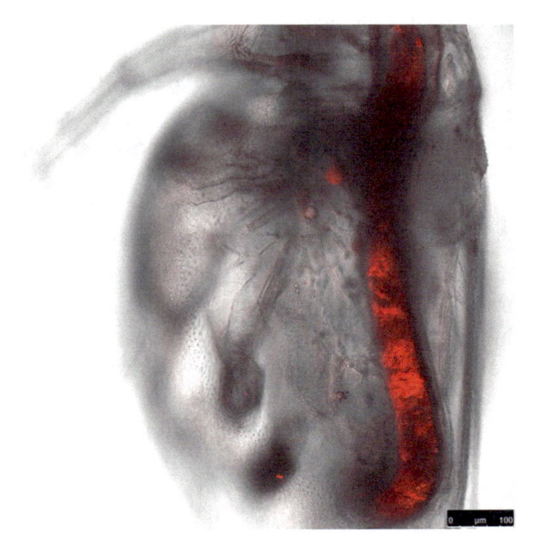

Fig. 4 Rhodamine B labelled NPs in the gastrointestinal trait of *Daphnia magna. Daphnia* was exposed for 24 h to PE NPs (about 100 nm) at a concentration of 10 mg/mL, corresponding to about 1.7×10^{10} NP/mL. The experiment was performed at IMDEA Water (Madrid Institute for Advanced Studies on Water, Spain)

Boxall 2018). It is also important to note that nano-sized plastics exhibit strong sorption affinities for toxic compounds (Mattsson et al. 2015). Therefore, their role as vector of fluorescent dies and other pollutants could be higher for NPs than for MPs.

It is reasonable to suppose that there exists a size threshold below which particles can cross cell membranes and may be transferred into body tissues through passive diffusion mechanisms. Above this threshold, insoluble particles would enter cells only by endocytosis-related mechanisms or membrane disruption. Previous research performed with engineered nanoparticles showed that particle size is a key factor influencing passive permeation, and only particles below 10–50 nm tend to penetrate. Moreover, this is a complex process that tends to be regulated by several other factors such as hydrophobicity/hydrophilicity and surface modification of the particle, as well as environmental factors (pH, osmotic pressure, ionic strength; Nakamura and Watano 2018). As for plastic particles, the first two are not expected to have a large influence given the hydrophobicity and non-reactivity of the constituent polymers, but the environmental factors can probably contribute to different cell permeation rates. Besides, the shape of the particles may affect their uptake into the cells, so that the permeability of tubular or irregular NPs may be different from that of rather globular particles.

Given the above, the major research priorities on the issue of bioaccumulation, biomagnification and transfer in the food web of M/NPs are: (1) developing, improving and calibrating standardized methods to extract, isolate and identify MPs in organisms, particularly focused on the measurements of small-sized MPs and NPs; (2) development of methods, test procedures and modelling approaches to assess, measure and predict the possible occurrence of cellular uptake of M/NPs through endocytosis-related mechanisms and permeation; and (3) assessing the capability of NPs (several tens of nanometres or less) to cross cellular membranes and to enter cells and tissues through passive, diffusion-based mechanisms and quantifying the actual dimensional threshold and factors that influence this process. Experiments should be made using NPs of different sizes labelled with procedures different than fluorescence. A possibility is to use metal-doped NPs, which can be detected by ICP-MS thanks to the metal incorporated (Mitrano et al. 2019). Another option is the use of ^{14}C labelled polymers, and recently latexes of PS nanoparticles with size as low as 20 nm have been prepared using radiolabelled styrene that can be used to study the in vivo uptake of NPs in simulated environmental conditions (Al-Sid-Cheikh et al. 2020, 2018). However, metal and ^{14}C labelled NPs are valid only for specific polymers, difficult to prepare and/or very expensive. Additionally, there are very few data on the presence of inhaled MPs in lungs either in animals or humans. Only a few studies have addressed this issue showing the presence of polymeric fibres in human lung tissues of people exposed to airborne microplastics (Pauly et al. 1998; Vianello et al. 2019). This is an important research topic that should be addressed in future studies.

3.10 Nanoplastics

3.10.1 State of the Art

NPs have already been mentioned in several sections above. The reasons for concern are clear. Just one ethylene-vinyl acetate commercial buoy (14.8 cm × 9.5 cm, 110 g) may produce >10^{20} fragments of 100 nm NPs (own calculation). However, they represent the major knowledge gap in the general topic of plastics in the environment. It is reasonable to hypothesize that NPs may represent a major concern for the environment and human health. Nevertheless, to date, the information required to support this hypothesis is insufficient. According to Koelmans et al. "*Nanoplastics is probably the least known area of marine litter but potentially also the most hazardous*" (Koelmans et al. 2015). Specifically, there is no adequate information to assess the exposure and effects of nanoplastics and, therefore, to characterize the risk for the environment and human health.

3.10.2 Knowledge Gaps and Research Needs

The first key issue is the problem of sources and origins of NPs. As for MPs, NPs may be emitted as primary (i.e. intentionally produced) or secondary (i.e. non-intentionally produced) particles. Primary nanometre-sized particles may be produced using known synthesis processes and may be used for several applications (Rao and Geckeler 2011; Stephens et al. 2013). Secondary NPs can be produced in specific processes like the thermal cutting of PS foam (Zhang et al. 2012). Another possible source is the fragmentation of MPs to smaller-sized particles eventually reaching the nanoscale (Andrady 2011). However, to date, no precise data exist on these processes and any quantitative estimate of the possible emissions of primary and secondary NPs is largely hypothetical (Koelmans et al. 2015; Lehner et al. 2019). It is reasonable to suppose that NPs may be present in large amounts in all environmental matrices, but this is speculative and must be supported by experimental data.

A key issue is the lack of suitable and reliable methods for sampling and analysing NPs. The problem was already highlighted for small-sized MPs (e.g. <20 μm) in previous sections and it is even more complex for NPs. For the sampling in the water environment, suitable methods for filtering in field large volumes of water, up to the nano-level are not realistically applicable. Therefore, a possibility could be taking water samples and processing in the lab with suitable approaches (e.g. ultra-filtration). This cannot be accomplished with large volume samples as for those needed for MP sampling (typically some cubic metres). Smaller volumes may be enough if NP concentration would be higher than those of MPs. This is probable, but it is just a speculative hypothesis because even the order of magnitude of NP concentration in natural waters is unknown. Comparable and even higher problems may arise for sampling in other environmental matrices like soil, sediments or biota. The methods for analysis and detection of NPs are still in an early stage of development. Currently, there are no routine methods available that permit detection of NPs in any environmental matrices, including biota and food (Lehner et al. 2019; Peng et al. 2020). As mentioned before, thermo-analytical methods coupled with mass spectrometry offer the possibility of detecting polymer particles with sizes outside current analytical possibilities. The detection limits are expected to reach the nanogram per litre range with minimum sample pretreatment (Yakovenko et al. 2020). Recently, pyrolysis gas chromatography time of flight mass spectrometry (Py-GC-ToF-MS) has been proposed for the detection of marker ions from the compounds generated in the pyrolysis of M/NPs (Sullivan et al. 2020).

In addition, several studies have demonstrated that nanoparticle toxicity is extremely complex, and that the biological activity of nanoparticles will depend on a variety of physicochemical properties such as particle size, shape, agglomeration state, crystal structure, chemical composition, surface area and surface properties (Hofmann-Amtenbrink et al. 2015). Moreover, the concept of toxicity itself is unclear for insoluble particulate materials. As stated above, toxicity is a reaction between a chemical substance and a chemical structure of a living organism. Unlike medium-to-large MPs, NPs have higher possibilities to cross cell membranes and to enter cells (Sect. 3.9). Small size increases reactivity and possible breakdown in monomers (Lehner et al. 2019). Therefore, the possibility for true toxic effects increases. Experimental tests on the adverse effects of NPs on the environment and human health have been performed using in vitro assays (Lehner et al. 2019) and in vivo tests, mainly on aquatic organisms (Kögel et al. 2020; Liu et al. 2020, 2021). However, the information available is scarce compared to that recovered for MPs. In general, adverse effects have been observed at relatively high concentrations, but the lack of information on environmentally realistic concentrations in natural ecosystems makes impossible any characterization of risk. Moreover, the types of adverse effects possible and the modes of action are largely unknown.

The main knowledge gaps and needs for research on NPs may be summarized as follows: (1) better defining and quantitatively estimating emission sources; (2) developing suitable sampling procedures in different environmental media; (3) developing suitable and reliable analytical methods to quantify their environmental occurrence; (4) assessing their capability to cross cell membranes and bioaccumulation potential; (5) assessing the possible modes of action and quantifying their adverse effects; and (6) characterizing the risks for different environmental compartments and for and human health.

4 Conclusions and Recommendations

In the last few years, a large amount of resources has been allocated to conduct research in the field of plastic pollution and its effects. This has contributed to some extent to improve our knowledge on the occurrence of large MPs in different environmental compartments and to perform preliminary hazard and risk assessments, which generally indicate low risks for living organisms. We have also learned that this new research area needs further technological developments and cannot always rely on the approaches traditionally implemented in the risk assessment of chemical contaminants (sampling and analytical methods, toxicological risk assessments, etc.). Moreover, the development of techniques and concepts urgently requires coordination and harmonization among different researchers and stakeholders. In general, many researchers share the idea that the results obtained were less than expected in relation to the efforts devoted, and that there are difficulties in the implementation and comparison of available scientific results. In addition, it seems that the focus on some key aspects needed to produce a suitable risk assessment has been frequently missed. Through this paper we have proposed a list of scientific issues that require to be better defined, clarified or studied to advance the field. These have been discussed through the text and schematically listed in Table 3. We sincerely hope that the list of research needs helps to optimize the use of human and economic resources dedicated to improving the risk assessment of M/NPs and that aids in the development of effective mitigation strategies to counteract these risks in the near future.

Table 3 Synthesis of the main knowledge gaps that need to be considered for future research in the field of M/NPs

Knowledge gap	Why it should be considered
Definitions	
M/NPs shape is generally ignored	– Shape determines the viscous force and transport in fluid media – Shape influences colloidal stability – Shape can also influence uptake by living organisms
Current size cut-off is inadequate to fibres	– Filter cut-off and size definition based on largest dimension do not match – Fibre length does not account for aerodynamic or hydrodynamic behaviour
Other anthropogenic fibres are rarely considered	– Non-plastic industrially processed fibres bear the same additives and appear together with those made of synthetic polymers
Environmental sources	
The mechanisms of generation of secondary M/NPs in freshwater environments are poorly known	– The understanding of physical, chemical and biological mechanisms affecting M/NPs fate in inland environments is needed
Wastewater treatment processes need improvements to limit the emission of M/NPs	– Wastewater treatment process and regulations (including sludge management) should be developed to avoid current emission of M/NPs
It is unclear if secondary M/NPs from food packaging may be a risk for human health	– Risk assessment of M/NPs ingestion via food transfer cannot be performed with currently available data
Available data are insufficient for modelling	– More data are needed to feed models that can predict the accumulation and fate of M/NPs

(continued)

Table 3 (continued)

Knowledge gap	Why it should be considered
Sampling procedures	
Standardization of sampling methods	– Standardization of sampling procedures is needed to allow comparing data from monitoring campaigns – Shape, colour and cross-cutting data about zooplankton and environmental conditions should be systematically reported
Scarcity of data for the lower size M/NPs	– Limited data are available for M/NPs below 100 μm, and primarily in the nanometre range, which is the size range with risk to translocate to biological tissues
Limited attention to freshwater, soil and air compartments	– Additional monitoring efforts should be oriented to inland ecosystems as most plastic litter is originated from land sources
Analytical methods	
Non-comparable metrics	– Metrics is a critical need to include M/NPs monitoring in regulatory frameworks
Standardization of pretreatment methods	– Pretreatments aimed at removing organic matter or separation from non-plastic particles affect the quality and comparability of results from different sources
Scarcity of data for the smaller size M/NPs	– As for sampling methods, analytical methods for small size MPs and or NPs are poorly reliable, inadequate or fully lacking
Poor statistics	– The fraction of sample analysed by micro-FTIR or other techniques should not be arbitrary, and accuracy must be reported
Insufficient information about cross-contamination/procedural blanks and recovery rates	– The results of procedural blanks should be reported, and recovery rates should be evaluated
Additives and other non-intentional substances	
Lack of information about additives in marketed plastics	– It is difficult to obtain information about substances included in marketed materials
Limited information for modelling or risk assessment	– The impact of environmental variables on the leaching of additives under realistic conditions is required for risk assessment
Lack of information about by-products or metabolites of additives	– The possible formation of toxic degradation products from additives upon oxidation or photochemical processes requires attention
Sorption of chemicals	
Different groups of living organisms (e.g. Invertebrates) and environmental compartments (e.g. Sediments, soil) should be considered	– To date, major attention has been devoted to fish and the aquatic compartment. The role of species with different biological traits in the compartments that are the major sinks of MPs and chemical contaminants is still unclear

(continued)

Table 3 (continued)

Knowledge gap	Why it should be considered
The sorption of semipolar pollutants in aged plastics received less attention	– The effect of ageing should also be considered because of the different interaction of pollutants with hydrophilic moieties, which may interact with polar and semipolar compounds including antibiotics and other CEC
Interaction with microorganisms	
The role of MPs as niche for microorganisms is not well understood.	– MPs are a new microbial habitat that might already be performing a role at the ecological level
MPs may contribute to disperse microorganisms even to remote areas	– Potentially transported microbes include pathogens (bacteria and viruses such as SARS-CoV-2) and ARB possibly threatening human health and the environment
Colonized MPs area a source of food for some aquatic organisms	– Further research is needed to shed light on the potential transfer of pathogens or ARB/ARG to these organisms and to the aquatic trophic web
Microbes involved in polymer degradation might be detected in biofilms	– Microbial assemblages on MPs may be a source of organisms involved in polymer degradation
Degradation and fate of microplastics	
Little is known about smaller size MPs and no data exist about NPs	– The generation of small fragments of M/NPs must be modelled in order to perform a risk assessment
Aged M/NPs may be relatively reactive	– Plastic leachates can interact with certain chemicals with consequences for human health and the ecosystems
Limited information on biodegradable polymers	– The degradation pattern and leaching of dissolved organic carbon from biodegradable polymers are poorly known
Translocation and transfer to the food web	
Limited data exist about the capacity of M/NPs to cross cell membranes	– There is a need for developing standardized methods to identify MPs in organisms, particularly small-sized MPs and NPs including the mechanisms of cellular uptake
The importance of size for M/NPs uptake needs to be determined	– The influence of different parameters, notably size, on the efficiency of internalization must be addressed in order to prioritize risk
Nanoplastics	
There is almost no information about the presence, fate and effects of NPs in the environment	– No risk assessment is possible without reliable data about NPs concentration in environmental compartments – Reference materials are needed to further develop toxicity tests

Acknowledgements The authors wish to thank the support provided by the Spanish Ministry of Science through the Thematic Network of Micro- and Nanoplastics in the Environment (RED2018-102345-T, EnviroPlaNet). Andreu Rico is supported by the Talented Researcher Support Programme - Plan GenT (CIDEGENT/2020/043) of the Generalitat Valenciana.

Conflict of Interest The authors declare that they have no conflict of interest.

References

Abbasi S, Keshavarzi B, Moore F, Turner A, Kelly FJ, Dominguez AO, Jaafarzadeh N (2019) Distribution and potential health impacts of microplastics and microrubbers in air and street dusts from Asaluyeh County. Iran Environ Pollut 244:153–164

Abeynayaka A, Kojima F, Miwa Y, Ito N, Nihei Y, Fukunaga Y, Yashima Y, Itsubo N (2020) Rapid sampling of suspended and floating microplastics in challenging riverine and coastal water environments in Japan. Water 12:1903

Adam V, Yang T, Nowack B (2019) Toward an ecotoxicological risk assessment of microplastics: comparison of available hazard and exposure data in freshwaters. Environ Toxicol Chem 38:436–447

Aliabad MK, Nassiri M, Kor K (2019) Microplastics in the surface seawaters of Chabahar Bay, Gulf of Oman (Makran Coasts). Mar Pollut Bull 143:125–133

Al-Sid-Cheikh M, Rowland SJ, Stevenson K, Rouleau C, Henry TB, Thompson RC (2018) Uptake, whole-body distribution, and depuration of nanoplastics by the scallop *Pecten maximus* at environmentally realistic concentrations. Environ Sci Technol 52:14480–14486

Al-Sid-Cheikh M, Rowland SJ, Kaegi R, Henry TB, Cormier M-A, Thompson RC (2020) Synthesis of ^{14}C-labelled polystyrene nanoplastics for environmental studies. Commun Mater 1:97

Amamiya K, Saido K, Chung S-Y, Hiaki T, Lee DS, Kwon BG (2019) Evidence of transport of styrene oligomers originated from polystyrene plastic to oceans by runoff. Sci Total Environ 667:57–63

Amaral-Zettler LA, Zettler ER, Slikas B, Boyd GD, Melvin DW, Morrall CE, Proskurowski G, Mincer TJ (2015) The biogeography of the plastisphere: implications for policy. Front Ecol Environ 13:541–546

Amaral-Zettler LA, Zettler ER, Mincer TJ (2020) Ecology of the plastisphere. Nat Rev Microbiol 18:139–151

Ambrogi V, Carfagna C, Cerruti P, Marturano V (2017) Additives in polymers. In: Jasso-Gastinel CF, Kenny JM (eds) Modification of polymer properties. William Andrew Publishing, pp 87–108

Andrade J, Fernández-González V, López-Mahía P, Muniategui S (2019) A low-cost system to simulate environmental microplastic weathering. Mar Pollut Bull 149:110663

Andrade JM, Ferreiro B, López-Mahía P, Muniategui-Lorenzo S (2020) Standardization of the minimum information for publication of infrared-related data when microplastics are characterized. Mar Pollut Bull 154:111035

Andrady AL (2011) Microplastics in the marine environment. Mar Pollut Bull 62:1596–1605

Andrady AL, Neal MA (2009) Applications and societal benefits of plastics. Philos Trans R Soc Lond B 364:1977–1984

Arvanitoyannis IS, Bosnea L (2004) Migration of substances from food packaging materials to foods. Crit Rev Food Sci Nutr 44:63–76

Auta HS, Emenike CU, Fauziah SH (2017) Distribution and importance of microplastics in the marine environment: a review of the sources, fate, effects, and potential solutions. Environ Int 102:165–176

Aytan U, Valente A, Senturk Y, Usta R, Esensoy Sahin FB, Mazlum RE, Agirbas E (2016) First evaluation of neustonic microplastics in Black Sea waters. Mar Environ Res 119:22–30

Bakir A, O'Connor IA, Rowland SJ, Hendriks AJ, Thompson RC (2016) Relative importance of microplastics as a pathway for the transfer of hydrophobic organic chemicals to marine life. Environ Pollut 219:56–65

Bandow N, Will V, Wachtendorf V, Simon F-G (2017) Contaminant release from aged microplastic. Environ Chem 14:394–405

Barnes DKA, Galgani F, Thompson RC, Barlaz M (2009) Accumulation and fragmentation of plastic debris in global environments. Philos Trans R Soc Lond B 364:1985–1998

Bayo J, Olmos S, López-Castellanos J (2020) Microplastics in an urban wastewater treatment plant: the influence of physicochemical parameters and environmental factors. Chemosphere 238:124593

Baztan J, Bergmann M, Booth A, Broglio E, Carrasco A, Chouinard O, Clüsener-Godt M, Cordier M, Cozar A, Devrieses L, Enevoldsen H, Ernsteins R, Ferreira-da-Costa M, Fossi MC, Gago J, Galgani F, Garrabou J, Gerdts G, Gomez M, Gómez-Parra A, Gutow L, Herrera A, Herring C, Huck T, Huvet A, Ivar do Sul JA, Jorgensen B, Krzan A, Lagarde F, Liria A, Lusher A, Miguelez A, Packard T, Pahl S, Paul-Pont I, Peeters D, Robbens J, Ruiz-Fernández AC, Runge J, Sánchez-Arcilla A, Soudant P, Surette C, Thompson RC, Valdés L, Vanderlinden JP, Wallace N (2017) Breaking down the plastic age. In: Baztan J, Jorgensen B, Pahl S, Thompson RC, Vanderlinden J-P (eds) Fate and impact of microplastics in marine ecosystems. Elsevier, pp. 177–181

Beaumont NJ, Aanesen M, Austen MC, Börger T, Clark JR, Cole M, Hooper T, Lindeque PK, Pascoe C, Wyles KJ (2019) Global ecological, social and economic impacts of marine plastic. Mar Pollut Bull 142:189–195

Beer S, Garm A, Huwer B, Dierking J, Nielsen TG (2018) No increase in marine microplastic concentration over the last three decades – a case study from the Baltic Sea. Sci Total Environ 621:1272–1279

Beiras R, Muniategui-Lorenzo S, Rodil R, Tato T, Montes R, López-Ibáñez S, Concha-Graña E, Campoy-López P, Salgueiro-González N, Quintana JB (2019) Polyethylene microplastics do not increase bioaccumulation or toxicity of nonylphenol and 4-MBC to marine zooplankton. Sci Total Environ 692:1–9

Bejgarn S, MacLeod M, Bogdal C, Breitholtz M (2015) Toxicity of leachate from weathering plastics: An exploratory screening study with *Nitocra spinipes*. Chemosphere 132:114–119

Bergami E, Bocci E, Vannuccini ML, Monopoli M, Salvati A, Dawson KA, Corsi I (2016) Nano-sized polystyrene affects feeding, behavior and physiology of brine shrimp *Artemia franciscana* larvae. Ecotox Environ Saf 123:18–25

Bergmann M, Wirzberger V, Krumpen T, Lorenz C, Primpke S, Tekman MB, Gerdts G (2017) High quantities of microplastic in Arctic deep-sea sediments from the HAUSGARTEN observatory. Environ Sci Technol 51:11000–11010

Bergmann M, Mützel S, Primpke S, Tekman MB, Trachsel J, Gerdts G (2019) White and wonderful? Microplastics prevail in snow from the Alps to the Arctic. Sci Adv 5:eaax1157

Besseling E, Foekema EM, van den Heuvel-Greve MJ, Koelmans AA (2017) The effect of microplastic on the uptake of chemicals by the lugworm *Arenicola marina* (L.) under environmentally relevant exposure conditions. Environ Sci Technol 51:8795–8804

Bhunia K, Sablani SS, Tang J, Rasco B (2013) Migration of chemical compounds from packaging polymers during microwave, conventional heat treatment, and storage. Compr Rev Food Sci Food Saf 12:523–545

Blázquez-Blázquez E, Cerrada ML, Benavente R, Pérez E (2020) Identification of additives in polypropylene and their degradation under solar exposure studied by gas dhromatography–mass spectrometry. ACS Omega 5:9055–9063

Brahney J, Hallerud M, Heim E, Hahnenberger M, Sukumaran S (2020) Plastic rain in protected areas of the United States. Science 368:1257–1260

Bretas Alvim C, Mendoza-Roca JA, Bes-Piá A (2020) Wastewater treatment plant as microplastics release source – quantification and identification techniques. J Environ Manag 255:109739

Browne MA, Crump P, Niven SJ, Teuten E, Tonkin A, Galloway T, Thompson R (2011) Accumulation of microplastic on shorelines woldwide: sources and sinks. Environ Sci Technol 45:9175–9179

Brunner I, Fischer M, Rüthi J, Stierli B, Frey B (2018) Ability of fungi isolated from plastic debris floating in the shoreline of a lake to degrade plastics. PLoS One 13:e0202047

Bryant JA, Clemente TM, Viviani DA, Fong AA, Thomas KA, Kemp P, Karl DM, White AE, DeLong EF (2016) Diversity and activity of communities inhabiting plastic debris in the North Pacific gyre. MSystems 1:e00024-00016

Bullard JE, Ockelford A, O'Brien P, McKenna Neuman C (2021) Preferential transport of microplastics by wind. Atmos Environ 245:118038

Burns EE, Boxall ABA (2018) Microplastics in the aquatic environment: evidence for or against adverse impacts and major knowledge gaps. Environ Toxicol Chem 37:2776–2796

Cai M, He H, Liu M, Li S, Tang G, Wang W, Huang P, Wei G, Lin Y, Chen B, Hu J, Cen Z (2018) Lost but can't be neglected: huge quantities of small microplastics hide in the South China Sea. Sci Total Environ 633:1206–1216

Camacho M, Herrera A, Gómez M, Acosta-Dacal A, Martínez I, Henríquez-Hernández LA, Luzardo OP (2019) Organic pollutants in marine plastic debris from Canary Islands beaches. Sci Total Environ 662:22–31

Campanale C, Savino I, Pojar I, Massarelli C, Uricchio VF (2020) A practical overview of methodologies for sampling and analysis of microplastics in riverine environments. Sustainability 12:6755

Carr SA, Liu J, Tesoro AG (2016) Transport and fate of microplastic particles in wastewater treatment plants. Water Res 91:174–182

Chae Y, An Y-J (2018) Current research trends on plastic pollution and ecological impacts on the soil ecosystem: a review. Environ Pollut 240:387–395

Chamas A, Moon H, Zheng J, Qiu Y, Tabassum T, Jang JH, Abu-Omar M, Scott SL, Suh S (2020) Degradation rates of plastics in the environment. ACS Sustain Chem Eng 8:3494–3511

Chavarria-Miró G, Anfruns-Estrada E, Guix S, Paraira M, Galofré B, Sáanchez G, Pintó R, Bosch A (2020) Sentinel surveillance of SARS-CoV-2 in wastewater anticipates the occurrence of COVID-19 cases. MedRxiv. 2020.2006.2013.20129627

Cheung PK, Fok L, Hung PL, Cheung LTO (2018) Spatio-temporal comparison of neustonic microplastic density in Hong Kong waters under the influence of the Pearl River Estuary. Sci Total Environ 628-629:731–739

Cincinelli A, Scopetani C, Chelazzi D, Lombardini E, Martellini T, Katsoyiannis A, Fossi MC, Corsolini S (2017) Microplastic in the surface waters of the Ross Sea (Antarctica): occurrence, distribution and characterization by FTIR. Chemosphere 175:391–400

Cole M, Lindeque P, Halsband C, Galloway TS (2011) Microplastics as contaminants in the marine environment: a review. Mar Pollut Bull 62:2588–2597

Cole M, Lindeque P, Fileman E, Halsband C, Galloway TS (2015) The impact of polystyrene microplastics on feeding, function and fecundity in the marine copepod *Calanus helgolandicus*. Environ Sci Technol 49:1130–1137

Collignon A, Hecq J-H, Glagani F, Voisin P, Collard F, Goffart A (2012) Neustonic microplastic and zooplankton in the North Western Mediterranean Sea. Mar Pollut Bull 64:861–864

Collignon A, Hecq J-H, Galgani F, Collard F, Goffart A (2014) Annual variation in neustonic micro- and meso-plastic particles and zooplankton in the Bay of Calvi (Mediterranean–Corsica). Mar Pollut Bull 79:293–298

Cox KD, Covernton GA, Davies HL, Dower JF, Juanes F, Dudas SE (2019) Human consumption of microplastics. Environ Sci Technol 53:7068–7074

Cózar A, Sanz-Martín M, Martí E, González-Gordillo JI, Ubeda B, Gálvez JÁ, Irigoien X, Duarte CM (2015) Plastic accumulation in the Mediterranean Sea. PLoS One 10:e0121762

Cózar A, Martí E, Duarte CM, García-de-Lomas J, van Sebille E, Ballatore TJ, Eguíluz VM, González-Gordillo JI, Pedrotti ML, Echevarría F, Troublè R, Irigoien X (2017) The Arctic Ocean as a dead end for floating plastics in the North Atlantic branch of the Thermohaline Circulation. Sci Adv 3:e1600582

Crawford CB, Quinn B (2017) 6 – The interactions of microplastics and chemical pollutants. In: Crawford CB, Quinn B (eds) Microplastic pollutants. Elsevier, pp 131–157

Darbra RM, Dan JRG, Casal J, Àgueda A, Capri E, Fait G, Schuhmacher M, Nadal M, Rovira J, Grundmann V, Barceló D, Ginebreda A, Guillén D (2012) Additives in the textile industry. In: Bilitewski B, Darbra RM, Barceló D (eds) Global risk-based management of chemical additives I: production, usage and environmental occurrence. Springer, Berlin, Heidelberg, pp 83–107

de Lucia GA, Caliani I, Marra S, Camedda A, Coppa S, Alcaro L, Campani T, Giannetti M, Coppola D, Cicero AM, Panti C, Baini M, Guerranti C, Marsili L, Massaro G, Fossi MC, Matiddi M (2014) Amount and distribution of neustonic micro-plastic off the western Sardinian coast (Central-Western Mediterranean Sea). Mar Environ Res 100:10–16

de Sá LC, Luís LG, Guilhermino L (2015) Effects of microplastics on juveniles of the common goby (*Pomatoschistus microps*): confusion with prey, reduction of the predatory performance and efficiency, and possible influence of developmental conditions. Environ Pollut 196:359–362

de Sá LC, Oliveira M, Ribeiro F, Rocha TL, Futter MN (2018) Studies of the effects of microplastics on aquatic organisms: what do we know and where should we focus our efforts in the future? Sci Total Environ 645:1029–1039

De Tender C, Devriese LI, Haegeman A, Maes S, Vangeyte J, Cattrijsse A, Dawyndt P, Ruttink T (2017) Temporal dynamics of bacterial and fungal colonization on plastic debris in the North Sea. Environ Sci Technol 51:7350–7360

Devito SC (2000) Absorption through cellular membranes. Handbook of property estimation methods for chemicals. In: Boethling RS, Mackay D (eds) Handbook of property estimation methods for chemicals: environmental and health sciences. CRC Press-Lewis, Boca Raton, pp 261–278

Di Mauro R, Kupchik MJ, Benfield MC (2017) Abundant plankton-sized microplastic particles in shelf waters of the northern Gulf of Mexico. Environ Pollut 230:798–809

Dominguez G, McLeod AS, Gainsforth Z, Kelly P, Bechtel HA, Keilmann F, Westphal A, Thiemens M, Basov DN (2014) Nanoscale infrared spectroscopy as a non-destructive probe of extraterrestrial samples. Nat Commun 5:5445

Doyle MJ, Watson W, Bowlin NM, Sheavly SB (2011) Plastic particles in coastal pelagic ecosystems of the Northeast Pacific Ocean. Mar Environ Res 71:41–52

Dris R, Gasperi J, Saad M, Mirande C, Tassin B (2016) Synthetic fibers in atmospheric fallout: a source of microplastics in the environment? Mar Pollut Bull 104:290–293

Dussud C, Hudec C, George M, Fabre P, Higgs P, Bruzaud S, Delort A-M, Eyheraguibel B, Meistertzheim A-L, Jacquin J, Cheng J, Callac N, Odobel C, Rabouille S, Ghiglione J-F (2018) Colonization of non-biodegradable and biodegradable plastics by marine microorganisms. Front Microbiol 9:1571–1571

Dyachenko A, Mitchell J, Arsem N (2017) Extraction and identification of microplastic particles from secondary wastewater treatment plant (WWTP) effluent. Anal Methods 9:1412–1418

ECETOC (2019) An evaluation of the challenges and limitations associated with aquatic toxicity and bioaccumulation studies for sparingly soluble and manufactured particulate substances. An evaluation of the challenges and limitations associated with aquatic toxicity and bioaccumulation studies for sparingly soluble and manufactured particulate substances, Technical Report No 132. ECETOC, Brussels

ECHA (2019) Annex XV Restriction Report. Proposal for a restriction of intentionally added microplastics. Version 1.1. 20th/March/2019. Helsinki

Edo C, González-Pleiter M, Leganés F, Fernández-Piñas F, Rosal R (2020) Fate of microplastics in wastewater treatment plants and their environmental dispersion with effluent and sludge. Environ Pollut 259:113837

Eerkes-Medrano D, Thompson RC, Aldridge DC (2015) Microplastics in freshwater systems: a review of the emerging threats, identification of knowledge gaps and prioritisation of research needs. Water Res 75:63–82

Enders K, Lenz R, Stedmon CA, Nielsen TG (2015) Abundance, size and polymer composition of marine microplastics ≥10μm in the Atlantic Ocean and their modelled vertical distribution. Mar Pollut Bull 100:70–81

Eriksen M, Maximenko N, Thiel M, Cummins A, Lattin G, Wilson S, Hafner J, Zellers A, Rifman S (2013) Plastic pollution in the South Pacific subtropical gyre. Mar Pollut Bull 68:71–76

Eriksen M, Lebreton LCM, Carson HS, Thiel M, Moore CJ, Borerro JC, Galgani F, Ryan PG, Reisser J (2014) Plastic pollution in the world's oceans: more than 5 trillion plastic pieces weighing over 250,000 tons afloat at sea. PLoS One 9:e111913

Everaert G, Van Cauwenberghe L, De Rijcke M, Koelmans AA, Mees J, Vandegehuchte M, Janssen CR (2018) Risk assessment of microplastics in the ocean: modelling approach and first conclusions. Environ Pollut 242:1930–1938

Fadare OO, Wan B, Guo L-H, Zhao L (2020) Microplastics from consumer plastic food containers: are we consuming it? Chemosphere 253:126787

Faure F, Demars C, Wieser O, Kunz M, de Alencastro LF (2015) Plastic pollution in Swiss surface waters: nature and concentrations, interaction with pollutants. Environ Chem 12:582–591

Felix LC, Ortega VA, Goss GG (2017) Cellular uptake and intracellular localization of poly (acrylic acid) nanoparticles in a rainbow trout (*Oncorhynchus mykiss*) gill epithelial cell line, RTgill-W1. Aquat Toxicol 192:58–68

Fendall LS, Sewell MA (2009) Contributing to marine pollution by washing your face: microplastics in facial cleansers. Mar Pollut Bull 58:1225–1228

Fernández-González V, Andrade-Garda JM, López-Mahía P, Muniategui-Lorenzo S (2021) Impact of weathering on the chemical identification of microplastics from usual packaging polymers in the marine environment. Anal Chim Acta 1142:179–188

Figueiredo GM, Vianna TMP (2018) Suspended microplastics in a highly polluted bay: abundance, size, and availability for mesozooplankton. Mar Pollut Bull 135:256–265

Fischer M, Scholz-Böttcher BM (2017) Simultaneous trace identification and quantification of common types of microplastics in environmental samples by pyrolysis-gas chromatography–mass spectrometry. Environ Sci Technol 51:5052–5060

Flemming H-C, Neu TR, Wozniak DJ (2007) The EPS matrix: the "house of biofilm cells". J Bacteriol 189:7945–7947

Fossi MC, Baini M, Panti C, Galli M, Jiménez B, Muñoz-Arnanz J, Marsili L, Finoia MG, Ramírez-Macías D (2017) Are whale sharks exposed to persistent organic pollutants and plastic pollution in the Gulf of California (Mexico)? First ecotoxicological investigation using skin biopsies. Comp Biochem Physiol C Toxicol Pharmacol 199:48–58

Frère L, Paul-Pont I, Rinnert E, Petton S, Jaffré J, Bihannic I, Soudant P, Lambert C, Huvet A (2017) Influence of environmental and anthropogenic factors on the composition, concentration and spatial distribution of microplastics: a case study of the Bay of Brest (Brittany, France). Environ Pollut 225:211–222

Frias JPGL, Nash R (2019) Microplastics: finding a consensus on the definition. Mar Pollut Bull 138:145–147

Frias JPGL, Otero V, Sobral P (2014) Evidence of microplastics in samples of zooplankton from Portuguese coastal waters. Mar Environ Res 95:89–95

Fries E, Dekiff JH, Willmeyer J, Nuelle M-T, Ebert M, Remy D (2013) Identification of polymer types and additives in marine microplastic particles using pyrolysis-GC/MS and scanning electron microscopy. Environ Sci Process Impacts 15:1949–1956

Fu W, Min J, Jiang W, Li Y, Zhang W (2020) Separation, characterization and identification of microplastics and nanoplastics in the environment. Sci Total Environ 721:137561

Gago J, Henry M, Galgani F (2015) First observation on neustonic plastics in waters off NW Spain (spring 2013 and 2014). Mar Environ Res 111:27–33

Gago J, Carretero O, Filgueiras AV, Viñas L (2018) Synthetic microfibers in the marine environment: a review on their occurrence in seawater and sediments. Mar Pollut Bull 127:365–376

GESAMP (2015) Sources, fate and effects of microplastic in the marine environment: a global assessment. In: Kershaw PJ (ed) Sources, fate and effects of microplastic in the marine environment: a global assessment, Rep. Stud. GESAMP No. 93, p 220

GESAMP (2016) Sources, fate and effects of microplastic in the marine environment: part two of a global assessment. In: Kershaw PJ (ed) Sources, fate and effects of microplastic in the marine environment: a global assessment, Rep. Stud. GESAMP No. 93, p 220

GESAMP (2019) Guidelines for the monitoring and assessment of plastic litter in the ocean. In: Kershaw PJ, Turra A, Galgani F (eds) Guidelines for the monitoring and assessment of plastic litter in the ocean, Rep. Stud. GESAMP No. 99, p 130

GESAMP (2020) Proceedings of the GESAMP International Workshop on assessing the risks associated with plastics and microplastics in the marine environment. In: Kershaw PJ, Carney B, Villarrubia-Gómez P, Koelmans AA, Gouin T (eds) Proceedings of the GESAMP International Workshop on assessing the risks associated with plastics and microplastics in the marine environment. Reports to GESAMP No. 103, p 68

Gewert B, Plassmann MM, MacLeod M (2015) Pathways for degradation of plastic polymers floating in the marine environment. Environ Sci Process Impacts 17:1513–1521

Gewert B, Ogonowski M, Barth A, MacLeod M (2017) Abundance and composition of near surface microplastics and plastic debris in the Stockholm Archipelago, Baltic Sea. Mar Pollut Bull 120:292–302

Gigault J, Halle A, Baudrimont M, Pascal P-Y, Gauffre F, Phi T-L, El Hadri H, Grassl B, Reynaud S (2018) Current opinion: what is a nanoplastic? Environ Pollut 235:1030–1034

Gilfillan LR, Ohman MD, Doyle MJ, Watson W (2009) Occurrence of plastic micro-debris in the Southern California current system. In: Reports CCOFI (ed) Occurrence of plastic micro-debris in the Southern California current system, vol 50, pp 123–133

Gobas FAFP, Morrison HA (2000) Bioconcentration and biomagnification in the aquatic environment. In: Boethling RS, Mackay D (eds) Handbook of property estimation methods for chemicals: environmental and health sciences. CRC Press-Lewis, Boca Rtaon, pp 189–231

Gong J, Xie P (2020) Research progress in sources, analytical methods, eco-environmental effects, and control measures of microplastics. Chemosphere 254:126790

González-Pleiter M, Tamayo-Belda M, Pulido-Reyes G, Amariei G, Leganés F, Rosal R, Fernández-Piñas F (2019) Secondary nanoplastics released from a biodegradable microplastic severely impact freshwater environments. Environ Sci Nano 6:1382–1392

González-Pleiter M, Velázquez D, Edo C, Carretero O, Gago J, Barón-Sola Á, Hernández LE, Yousef I, Quesada A, Leganés F, Rosal R, Fernández-Piñas F (2020) Fibers spreading worldwide: microplastics and other anthropogenic litter in an Arctic freshwater lake. Sci Total Environ 722:137904

González-Pleiter M, Edo C, Aguilera Á, Viúdez-Moreiras D, Pulido-Reyes G, González-Toril E, Osuna S, de Diego-Castilla G, Leganés F, Fernández-Piñas F, Rosal R (2021) Occurrence and transport of microplastics sampled within and above the planetary boundary layer. Sci Total Environ 761:143213

Gouin T (2020) Towards improved understanding of the ingestion and trophic transfer of microplastic particles – critical review and implications for future research. Environ Toxicol Chem 39:1119–1137

Gouin T, Roche N, Lohmann R, Hodges G (2011) A thermodynamic approach for assessing the environmental exposure of chemicals absorbed to microplastic. Environ Sci Technol 45:1466–1472

Groh KJ, Backhaus T, Carney-Almroth B, Geueke B, Inostroza PA, Lennquist A, Leslie HA, Maffini M, Slunge D, Trasande L, Warhurst AM, Muncke J (2019) Overview of known plastic packaging-associated chemicals and their hazards. Sci Total Environ 651:3253–3268

Hahladakis JN, Velis CA, Weber R, Iacovidou E, Purnell P (2018) An overview of chemical additives present in plastics: migration, release, fate and environmental impact during their use, disposal and recycling. J Hazard Mater 344:179–199

Handy RD, Ahtiainen J, Navas JM, Goss G, Bleeker EAJ, von der Kammer F (2018) Proposal for a tiered dietary bioaccumulation testing strategy for engineered nanomaterials using fish. Environ Sci Nano 5:2030–2046

Harrison JP, Schratzberger M, Sapp M, Osborn AM (2014) Rapid bacterial colonization of low-density polyethylene microplastics in coastal sediment microcosms. BMC Microbiol 14:232

Hartmann NB, Hüffer T, Thompson RC, Hassellöv M, Verschoor A, Daugaard AE, Rist S, Karlsson T, Brennholt N, Cole M, Herrling MP, Hess MC, Ivleva NP, Lusher AL, Wagner M (2019) Are we speaking the same language? Recommendations for a definition and categorization framework for plastic debris. Environ Sci Technol 53:1039–1047

Hendriksen RS, Munk P, Njage P, van Bunnik B, McNally L, Lukjancenko O, Röder T, Nieuwenhuijse D, Pedersen SK, Kjeldgaard J, Kaas RS, Clausen PTLC, Vogt JK, Leekitcharoenphon P, van de Schans MGM, Zuidema T, de Roda Husman AM, Rasmussen S, Petersen B, Bego A, Rees C, Cassar S, Coventry K, Collignon P, Allerberger F, Rahube TO, Oliveira G, Ivanov I, Vuthy Y, Sopheak T, Yost CK, Ke C, Zheng H, Baisheng L, Jiao X, Donado-Godoy P, Coulibaly KJ, Jergović M, Hrenovic J, Karpíšková R, Villacis JE, Legesse M, Eguale T, Heikinheimo A, Malania L, Nitsche A, Brinkmann A, Saba CKS, Kocsis B, Solymosi N, Thorsteinsdottir TR, Hatha AM, Alebouyeh M, Morris D, Cormican M, O'Connor L, Moran-Gilad J, Alba P, Battisti A, Shakenova Z, Kiiyukia C, Ng'eno E, Raka L, Avsejenko J, Bērziņš A, Bartkevics V, Penny C, Rajandas H, Parimannan S, Haber MV, Pal P, Jeunen G-J, Gemmell N, Fashae K, Holmstad R, Hasan R, Shakoor S, Rojas MLZ, Wasyl D, Bosevska G, Kochubovski M, Radu C, Gassama A, Radosavljevic V, Wuertz S, Zuniga-Montanez R, Tay MYF, Gavačová D, Pastuchova K, Truska P, Trkov M, Esterhuyse K, Keddy K, Cerdà-Cuéllar M, Pathirage S, Norrgren L, Örn S, Larsson DGJ, Heijden TV, Kumburu HH, Sanneh B, Bidjada P, Njanpop-Lafourcade B-M, Nikiema-Pessinaba SC, Levent B, Meschke JS, Beck NK, Van CD, Phuc ND, Tran DMN, Kwenda G, Tabo D-a, Wester AL, Cuadros-Orellana S, Amid C, Cochrane G, Sicheritz-Ponten T, Schmitt H, JRM A, Aidara-Kane A, Pamp SJ, Lund O, Hald T, Woolhouse M, Koopmans MP, Vigre H, Petersen TN, Aarestrup FM, The Global Sewage Surveillance project c (2019) Global monitoring of antimicrobial resistance based on metagenomics analyses of urban sewage. Nat Commun 10:1124

Henry B, Laitala K, Klepp IG (2019) Microfibres from apparel and home textiles: prospects for including microplastics in environmental sustainability assessment. Sci Total Environ 652:483–494

Hermabessiere L, Dehaut A, Paul-Pont I, Lacroix C, Jezequel R, Soudant P, Duflos G (2017) Occurrence and effects of plastic additives on marine environments and organisms: a review. Chemosphere 182:781–793

Hernandez LM, Xu EG, Larsson HCE, Tahara R, Maisuria VB, Tufenkji N (2019) Plastic teabags release billions of microparticles and nanoparticles into tea. Environ Sci Technol 53:12300–12310

Herrera A, Raymond E, Martínez I, Álvarez S, Canning-Clode J, Gestoso I, Pham CK, Ríos N, Rodríguez Y, Gómez M (2020) First evaluation of neustonic microplastics in the Macaronesian region, NE Atlantic. Mar Pollut Bull 153:110999

Hoellein T, Rojas M, Pink A, Gasior J, Kelly J (2014) Anthropogenic litter in urban freshwater ecosystems: distribution and microbial interactions. PLoS One 9:e98485

Hofmann-Amtenbrink M, Grainger DW, Hofmann H (2015) Nanoparticles in medicine: current challenges facing inorganic nanoparticle toxicity assessments and standardizations. Nanomed Nanotechnol Biol Med 11:1689–1694

Hoppe M, de Voogt P, Franz R (2016) Identification and quantification of oligomers as potential migrants in plastics food contact materials with a focus in polycondensates – a review. Trends Food Sci Technol 50:118–130

Hoppe M, Fornari R, de Voogt P, Franz R (2017) Migration of oligomers from PET: determination of diffusion coefficients and comparison of experimental versus modelled migration. Food Addit Contam A 34:1251–1260

Isobe A, Uchida K, Tokai T, Iwasaki S (2015) East Asian seas: a hot spot of pelagic microplastics. Mar Pollut Bull 101:618–623

Isobe A, Uchiyama-Matsumoto K, Uchida K, Tokai T (2017) Microplastics in the Southern Ocean. Mar Pollut Bull 114:623–626

Ivar do Sul JA, Costa MF, Fillmann G (2014) Microplastics in the pelagic environment around oceanic islands of the Western Tropical Atlantic Ocean. Water Air Soil Pollut 225:2004

Jacquin J, Cheng J, Odobel C, Pandin C, Conan P, Pujo-Pay M, Barbe V, Meistertzheim A-L, Ghiglione J-F (2019) Microbial ecotoxicology of marine plastic debris: a review on colonization and biodegradation by the "Plastisphere". Front Microbiol 10:865–865

Jahnke A, Mayer P, McLachlan MS (2012) Sensitive equilibrium sampling to study polychlorinated biphenyl disposition in Baltic Sea sediment. Environ Sci Technol 46:10114–10122

Jaiswal S, Sharma B, Shukla P (2020) Integrated approaches in microbial degradation of plastics. Environ Technol Innov 17:100567

Julienne F, Delorme N, Lagarde F (2019) From macroplastics to microplastics: role of water in the fragmentation of polyethylene. Chemosphere 236:124409

Kalogerakis N, Karkanorachaki K, Kalogerakis GC, Triantafyllidi EI, Gotsis AD, Partsinevelos P, Fava F (2017) Microplastics generation: onset of fragmentation of polyethylene films in marine environment mesocosms. Front Mar Sci 4:84

Kang J-H, Kwon OY, Lee K-W, Song YK, Shim WJ (2015) Marine neustonic microplastics around the southeastern coast of Korea. Mar Pollut Bull 96:304–312

Kanhai LDK, Officer R, Lyashevska O, Thompson RC, O'Connor I (2017) Microplastic abundance, distribution and composition along a latitudinal gradient in the Atlantic Ocean. Mar Pollut Bull 115:307–314

Karlsson TM, Kärrman A, Rotander A, Hassellöv M (2020) Comparison between manta trawl and in situ pump filtration methods, and guidance for visual identification of microplastics in surface waters. Environ Sci Pollut Res 27:5559–5571

Kedzierski M, Villain J, Falcou-Préfol M, Kerros ME, Henry M, Pedrotti ML, Bruzaud S (2019) Microplastics in Mediterranean Sea: a protocol to robustly assess contamination characteristics. PLoS One 14:e0212088

Kedzierski M, Lechat B, Sire O, Le Maguer G, Le Tilly V, Bruzaud S (2020) Microplastic contamination of packaged meat: occurrence and associated risks. Food Pack Shelf Life 24:100489

Kettner MT, Rojas-Jimenez K, Oberbeckmann S, Labrenz M, Grossart H-P (2017) Microplastics alter composition of fungal communities in aquatic ecosystems. Environ Microbiol 19:4447–4459

Kim HJ, Lee S-H, Lee J-H, Jang SP (2015) Effect of particle shape on suspension stability and thermal conductivities of water-based bohemite alumina nanofluids. Energy 90:1290–1297

Klein M, Fischer EK (2019) Microplastic abundance in atmospheric deposition within the Metropolitan area of Hamburg, Germany. Sci Total Environ 685:96–103

Klein S, Dimzon IK, Eubeler J, Knepper TP (2018) Analysis, occurrence, and degradation of microplastics in the aqueous environment. In: Wagner M, Lambert S (eds) Freshwater microplastics. Emerging environmental contaminants? Springer, pp 51–67

Koelmans AA (2015) Modeling the role of microplastics in bioaccumulation of organic chemicals to marine aquatic organisms. A critical review. In: Bergmann M, Gutow L, Klages M (eds) Marine anthropogenic litter. Springer, Cham, pp 309–324

Koelmans AA, Besseling E, Foekema EM (2014) Leaching of plastic additives to marine organisms. Environ Pollut 187:49–54

Koelmans AA, Besseling E, Shim WJ (2015) Nanoplastics in the aquatic environment. Critical review. In: Bergmann M, Gutow L, Klages M (eds) Marine anthropogenic litter. Springer, Cham, pp 325–340

Koelmans AA, Bakir A, Burton GA, Janssen CR (2016) Microplastic as a vector for chemicals in the aquatic environment: critical review and model-supported reinterpretation of empirical studies. Environ Sci Technol 50:3315–3326

Koelmans AA, Besseling E, Foekema E, Kooi M, Mintenig S, Ossendorp BC, Redondo-Hasselerharm PE, Verschoor A, van Wezel AP, Scheffer M (2017a) Risks of plastic debris: unravelling fact, opinion, perception, and belief. Environ Sci Technol 51:11513–11519

Koelmans AA, Kooi M, Law KL, Van Sebille E (2017b) All is not lost: deriving a top-down mass budget of plastic at sea. Environ Res Lett 12:114028

Koelmans AA, Redondo-Hasselerharm PE, Mohamed Nor NH, Kooi M (2020) Solving the nonalignment of methods and approaches used in microplastic research to consistently characterize risk. Environ Sci Technol 54:12307–12315

Kögel T, Bjorøy Ø, Toto B, Bienfait AM, Sanden M (2020) Micro- and nanoplastic toxicity on aquatic life: determining factors. Sci Total Environ 709:136050

Kor K, Mehdinia A (2020) Neustonic microplastic pollution in the Persian Gulf. Mar Pollut Bull 150:110665

Kroon FJ, Motti CE, Jensen LH, Berry KLE (2018) Classification of marine microdebris: a review and case study on fish from the Great Barrier Reef, Australia. Sci Rep 8:16422

Kuhn S, Bravo-Rebolledo E, van Franeker JA (2015) Deleterious effects of litter on marine life. In: Bergmann M, Gutow L, Klages M (eds) Marine anthropogenic litter. Springer, pp 75–116

Kukulka T, Proskurowski G, Morét-Ferguson S, Meyer DW, Law KL (2012) The effect of wind mixing on the vertical distribution of buoyant plastic debris. Geophys Res Lett 39:7601

Laganà P, Caruso G, Corsi I, Bergami E, Venuti V, Majolino D, La Ferla R, Azzaro M, Cappello S (2019) Do plastics serve as a possible vector for the spread of antibiotic resistance? First insights from bacteria associated to a polystyrene piece from King George Island (Antarctica). Int J Hyg Environ Health 222:89–100

Lattin GL, Moore CJ, Zellers AF, Moore SL, Weisberg SB (2004) A comparison of neustonic plastic and zooplankton at different depths near the southern California shore. Mar Pollut Bull 49:291–294

Law KL, Morét-Ferguson SE, Goodwin DS, Zettler ER, DeForce E, Kukulka T, Proskurowski G (2014) Distribution of surface plastic debris in the Eastern Pacific Ocean from an 11-year data set. Environ Sci Technol 48:4732–4738

Lebreton L, Slat B, Ferrari F, Sainte-Rose B, Aitken J, Marthouse R, Hajbane S, Cunsolo S, Schwarz A, Levivier A, Noble K, Debeljak P, Maral H, Schoeneich-Argent R, Brambini R, Reisser J (2018) Evidence that the Great Pacific Garbage Patch is rapidly accumulating plastic. Sci Rep 8:4666–4666

Lee OO, Wang Y, Tian R, Zhang W, Shek CS, Bougouffa S, Al-Suwailem A, Batang ZB, Xu W, Wang GC, Zhang X, Lafi FF, Bajic VB, Qian P-Y (2014) In situ environment rather than substrate type dictates microbial community structure of biofilms in a cold seep system. Sci Rep 4:3587

Lee YK, Romera-Castillo C, Hong S, Hur J (2020) Characteristics of microplastic polymer-derived dissolved organic matter and its potential as a disinfection byproduct precursor. Water Res 175:115678

Lehner R, Weder C, Petri-Fink A, Rothen-Rutishauser B (2019) Emergence of nanoplastic in the environment and possible impact on human health. Environ Sci Technol 53:1748–1765

Lenz R, Enders K, Nielsen TG (2016) Microplastic exposure studies should be environmentally realistic. PNAS 113:E4121–E4122

León VM, García I, González E, Samper R, Fernández-González V, Muniategui-Lorenzo S (2018) Potential transfer of organic pollutants from littoral plastics debris to the marine environment. Environ Pollut 236:442–453

Leslie HA, Brandsma SH, van Velzen MJM, Vethaak AD (2017) Microplastics en route: field measurements in the Dutch river delta and Amsterdam canals, wastewater treatment plants, North Sea sediments and biota. Environ Int 101:133–142

Li J, Liu H, Paul Chen J (2018) Microplastics in freshwater systems: a review on occurrence, environmental effects, and methods for microplastics detection. Water Res 137:362–374

Li J, Song Y, Cai Y (2020) Focus topics on microplastics in soil: analytical methods, occurrence, transport, and ecological risks. Environ Pollut 257:113570

Lima ARA, Costa MF, Barletta M (2014) Distribution patterns of microplastics within the plankton of a tropical estuary. Environ Res 132:146–155

Liu F, Olesen KB, Borregaard AR, Vollertsen J (2019a) Microplastics in urban and highway stormwater retention ponds. Sci Total Environ 671:992–1000

Liu J, Yang Y, Ding J, Zhu B, Gao W (2019b) Microfibers: a preliminary discussion on their definition and sources. Environ Sci Pollut Res 26:29497–29501

Liu Z, Cai M, Wu D, Yu P, Jiao Y, Jiang Q, Zhao Y (2020) Effects of nanoplastics at predicted environmental concentration on *Daphnia pulex* after exposure through multiple generations. Environ Pollut 256:113506

Liu Z, Li Y, Pérez E, Jiang Q, Chen Q, Jiao Y, Huang Y, Yang Y, Zhao Y (2021) Polystyrene nanoplastic induces oxidative stress, immune defense, and glycometabolism change in *Daphnia pulex*: application of transcriptome profiling in risk assessment of nanoplastics. J Hazard Mater 402:123778

Löder MGJ, Kuczera M, Mintenig S, Lorenz C, Gerdts G (2015) Focal plane array detector-based micro-Fourier-transform infrared imaging for the analysis of microplastics in environmental samples. Environ Chem 12:563–581

Lohmann R (2017) Microplastics are not important for the cycling and bioaccumulation of organic pollutants in the oceans—but should microplastics be considered POPs themselves? Integr Environ Assess Manag 13:460–465

Lorenzo-Navarro J, Castrillón-Santana M, Santesarti E, Marsico MD, Martínez I, Raymond E, Gómez M, Herrera A (2020) SMACC: a system for microplastics automatic counting and classification. IEEE Access 8:25249–25261

Lusher A (2015) Microplastics in the marine environment: distribution, interactions and effects. In: Bergmann M, Gutow L, Klages M (eds) Marine anthropogenic litter. Springer, Cham, pp 245–307

Lusher AL, Burke A, O'Connor I, Officer R (2014) Microplastic pollution in the Northeast Atlantic Ocean: validated and opportunistic sampling. Mar Pollut Bull 88:325–333

Lusher AL, Tirelli V, O'Connor I, Officer R (2015) Microplastics in Arctic polar waters: the first reported values of particles in surface and sub-surface samples. Sci Rep 5:14947

Lv L, Yan X, Feng L, Jiang S, Lu Z, Xie H, Sun S, Chen J, Li C (2020) Challenge for the detection of microplastics in the environment. Water Environ Res 93:5–15

Maes T, Van der Meulen MD, Devriese LI, Leslie HA, Huvet A, Frère L, Robbens J, Vethaak AD (2017) Microplastics baseline surveys at the water surface and in sediments of the North-East Atlantic. Front Mar Sci 4:135

Magni S, Binelli A, Pittura L, Avio CG, Della Torre C, Parenti CC, Gorbi S, Regoli F (2019) The fate of microplastics in an Italian wastewater treatment plant. Sci Total Environ 652:602–610

Martínez-Campos S, González-Pleiter M, Fernández-Piñas F, Rosal R, Leganés F (2021) Early and differential bacterial colonization on microplastics deployed into the effluents of wastewater treatment plants. Sci Total Environ 757:143832

Mateos-Cárdenas A, O'Halloran J, van Pelt FNAM, Jansen MAK (2020) Rapid fragmentation of microplastics by the freshwater amphipod *Gammarus duebeni* (Lillj.). Sci Rep 10:1–12

Mato Y, Isobe T, Takada H, Kanehiro H, Ohtake C, Kaminuma T (2001) Plastic resin pellets as a transport medium for toxic chemicals in the marine environment. Environ Sci Technol 35:318–324

Mattsson K, Hansson LA, Cedervall T (2015) Nano-plastics in the aquatic environment. Environ Sci Process Impacts 17:1712–1721

McCormick A, Hoellein TJ, Mason SA, Schluep J, Kelly JJ (2014) Microplastic is an abundant and distinct microbial habitat in an urban river. Environ Sci Technol 48:11863–11871

McCormick AR, Hoellein TJ, London MG, Hittie J, Scott JW, Kelly JJ (2016) Microplastic in surface waters of urban rivers: concentration, sources, and associated bacterial assemblages. Ecosphere 7:e01556

Meyns M, Primpke S, Gerdts G (2019) Library based identification and characterisation of polymers with nano-FTIR and IR-sSNOM imaging. Anal Methods 11:5195–5202

Miller ME, Kroon FJ, Motti CA (2017) Recovering microplastics from marine samples: a review of current practices. Mar Pollut Bull 123:6–18

Mitrano DM, Beltzung A, Frehland S, Schmiedgruber M, Cingolani A, Schmidt F (2019) Synthesis of metal-doped nanoplastics and their utility to investigate fate and behaviour in complex environmental systems. Nat Nanotechnol 14:362–368

Moore CJ, Moore SL, Leecaster MK, Weisberg SB (2001) A comparison of plastic and plankton in the North Pacific Central Gyre. Mar Pollut Bull 42:1297–1300

Moore CJ, Moore SL, Weisberg SB, Lattin GL, Zellers AF (2002) A comparison of neustonic plastic and zooplankton abundance in southern California's coastal waters. Mar Pollut Bull 44:1035–1038

Munno K, Helm PA, Jackson DA, Rochman C, Sims A (2018) Impacts of temperature and selected chemical digestion methods on microplastic particles. Environ Toxicol Chem 37:91–98

Murphy F, Ewins C, Carbonnier F, Quinn B (2016) Wastewater treatment works (WwTW) as a source of microplastics in the aquatic environment. Environ Sci Technol 50:5800–5808

Nakamura H, Watano S (2018) Direct permeation of nanoparticles across cell membrane: a review. Kona Powder Part J 35:49–65

Napper IE, Thompson RC (2016) Release of synthetic microplastic plastic fibres from domestic washing machines: effects of fabric type and washing conditions. Mar Pollut Bull 112:39–45

Obbard RW, Sadri S, Wong YQ, Khitun AA, Baker I, Thompson RC (2014) Global warming releases microplastic legacy frozen in Arctic Sea ice. Earth's Future 2:315–320

Oberbeckmann S, Loeder MGJ, Gerdts G, Osborn AM (2014) Spatial and seasonal variation in diversity and structure of microbial biofilms on marine plastics in Northern European waters. FEMS Microbiol Ecol 90:478–492

Oberbeckmann S, Osborn AM, Duhaime MB (2016) Microbes on a bottle: substrate, season and geography influence community composition of microbes colonizing marine plastic debris. PLoS One 11:e0159289

Ogonowski M, Motiei A, Ininbergs K, Hell E, Gerdes Z, Udekwu KI, Bacsik Z, Gorokhova E (2018) Evidence for selective bacterial community structuring on microplastics. Environ Microbiol 20:2796–2808

Ojha N, Pradhan N, Singh S, Barla A, Shrivastava A, Khatua P, Rai V, Bose S (2017) Evaluation of HDPE and LDPE degradation by fungus, implemented by statistical optimization. Sci Rep 7:39515

Panagiotidou E, Konidaris C, Baklavaridis A, Zuburtikudis I, Achilias D, Mitlianga P (2014) A simple route for purifying extracellular poly(3-hydroxybutyrate)-depolymerase from *Penicillium pinophilum*. Enzyme Res 2014:159809

Pärnänen KMM, Narciso-da-Rocha C, Kneis D, Berendonk TU, Cacace D, Do TT, Elpers C, Fatta-Kassinos D, Henriques I, Jaeger T, Karkman A, Martinez JL, Michael SG, Michael-Kordatou I, O'Sullivan K, Rodriguez-Mozaz S, Schwartz T, Sheng H, Sørum H, Stedtfeld RD, Tiedje JM, Giustina SVD, Walsh F, Vaz-Moreira I, Virta M, Manaia CM (2019) Antibiotic resistance in European wastewater treatment plants mirrors the pattern of clinical antibiotic resistance prevalence. Sci Adv 5:eaau9124

Pauly JL, Stegmeier SJ, Allaart HA, Cheney RT, Zhang PJ, Mayer AG, Streck RJ (1998) Inhaled cellulosic and plastic fibers found in human lung tissue. Cancer Epidemiol Biomark Prev 7:419–428

Pedrotti ML, Petit S, Elineau A, Bruzaud S, Crebassa J-C, Dumontet B, Martí E, Gorsky G, Cózar A (2016) Changes in the floating plastic pollution of the Mediterranean Sea in relation to the distance to land. PLoS One 11:e0161581

Peeken I, Primpke S, Beyer B, Gütermann J, Katlein C, Krumpen T, Bergmann M, Hehemann L, Gerdts G (2018) Arctic Sea ice is an important temporal sink and means of transport for microplastic. Nat Commun 9:1505

Peñalver R, Arroyo-Manzanares N, López-García I, Hernández-Córdoba M (2020) An overview of microplastics characterization by thermal analysis. Chemosphere 242:125170

Peng L, Fu D, Qi H, Lan CQ, Yu H, Ge C (2020) Micro- and nano-plastics in marine environment: source, distribution and threats – a review. Sci Total Environ 698:134254

Petersen EJ, Mortimer M, Burgess RM, Handy R, Hanna S, Ho KT, Johnson M, Loureiro S, Selck H, Scott-Fordsmand JJ, Spurgeon D, Unrine J, van den Brink NW, Wang Y, White J, Holden P (2019) Strategies for robust and accurate experimental approaches to quantify nanomaterial bioaccumulation across a broad range of organisms. Environ Sci Nano 6:1619–1656

PlasticsEurope (2020) Plastics – the Facts 2020: an analysis of European plastics production, demand and waste data. Plastics – the Facts 2020: an analysis of European plastics production, demand and waste data, Brussels

Prata JC, da Costa JP, Duarte AC, Rocha-Santos T (2019a) Methods for sampling and detection of microplastics in water and sediment: a critical review. TrAC Trends Anal Chem 110:150–159

Prata JC, Reis V, Matos JTV, da Costa JP, Duarte AC, Rocha-Santos T (2019b) A new approach for routine quantification of microplastics using Nile Red and automated software (MP-VAT). Sci Total Environ 690:1277–1283

Primpke S, Christiansen SH, Cowger WC, De Frond H, Deshpande A, Fischer M, Holland E, Meyns M, O'Donnell BA, Ossmann B, Pittroff M, Sarau G, Scholz-Böttcher BM, Wiggin K (2020) EXPRESS: critical assessment of analytical methods for the harmonized and cost efficient analysis of microplastics. Appl Spectrosc 74(9):1012–1047

Puglisi E, Romaniello F, Galletti S, Boccaleri E, Frache A, Cocconcelli PS (2019) Selective bacterial colonization processes on polyethylene waste samples in an abandoned landfill site. Sci Rep 9:14138

Rao JP, Geckeler KE (2011) Polymer nanoparticles: preparation techniques and size-control parameters. Prog Polym Sci 36:887–913

Rist S, Vianello A, Winding MHS, Nielsen TG, Almeda R, Torres RR, Vollertsen J (2020) Quantification of plankton-sized microplastics in a productive coastal Arctic marine ecosystem. Environ Pollut 266:115248

Roch S, Friedrich C, Brinker A (2020) Uptake routes of microplastics in fishes: practical and theoretical approaches to test existing theories. Sci Rep 10:3896

Rodrigues JP, Duarte AC, Santos-Echeandía J, Rocha-Santos T (2019) Significance of interactions between microplastics and POPs in the marine environment: a critical overview. TrAC Trends Anal Chem 111:252–260

Romera-Castillo C, Pinto M, Langer TM, Álvarez-Salgado XA, Herndl GJ (2018) Dissolved organic carbon leaching from plastics stimulates microbial activity in the ocean. Nat Commun 9:1430

Ruiz-Orejón LF, Sardá R, Ramis-Pujol J (2016) Floating plastic debris in the Central and Western Mediterranean Sea. Mar Environ Res 120:136–144

Russell JR, Huang J, Anand P, Kucera K, Sandoval AG, Dantzler KW, Hickman D, Jee J, Kimovec FM, Koppstein D, Marks DH, Mittermiller PA, Núñez SJ, Santiago M, Townes MA, Vishnevetsky M, Williams NE, Vargas MPN, Boulanger L-A, Bascom-Slack C, Strobel SA (2011) Biodegradation of polyester polyurethane by endophytic fungi. Appl Environ Microbiol 77:6076–6084

Saido K, Koizumi K, Sato H, Ogawa N, Kwon BG, Chung S-Y, Kusui T, Nishimura M, Kodera Y (2014) New analytical method for the determination of styrene oligomers formed from polystyrene decomposition and its application at the coastlines of the North-West Pacific Ocean. Sci Total Environ 473-474:490–495

Salvador-Cesa F, Turra A, Baruque-Ramos J (2017) Synthetic fibers as microplastics in the marine environment: a review from textile perspective with a focus on domestic washings. Sci Total Environ 598:1116–1129

SAM (2019) Environmental and health risks of microplastic pollution. Environmental and health risks of microplastic pollution. Publications Office of the European Union, Luxembourg

SAPEA (2019) A scientific perspective on microplastics in nature and society. A Scientific Perspective on Microplastics in Nature and Society, Berlin

Schell T, Cherta L, Dafouz R, Rico A, Vighi M (2020a) Bioconcentration of organic contaminants in fish in the presence of microplastics: is the "Trojan horse" effect a matter of concern? Paper presented at the SETAC Europe, 30th Annual Meeting, Dublin

Schell T, Rico A, Vighi M (2020b) Occurrence, fate and fluxes of plastics and microplastics in terrestrial and freshwater ecosystems. Rev Environ Contam Toxicol. https://doi.org/10.1007/398_2019_40

Schönlau C, Karlsson TM, Rotander A, Nilsson H, Engwall M, van Bavel B, Kärrman A (2020) Microplastics in sea-surface waters surrounding Sweden sampled by manta trawl and in-situ pump. Mar Pollut Bull 153:111019

Schultz SG (1976) Transport across epithelia: some basic principles. Kidney Int 9:65–75

Schür C, Rist S, Baun A, Mayer P, Hartmann NB, Wagner M (2019) When fluorescence is not a particle: the tissue translocation of microplastics in *Daphnia magna* seems an artifact. Environ Toxicol Chem 38:1495–1503

Schymanski D, Goldbeck C, Humpf H-U, Fürst P (2018) Analysis of microplastics in water by micro-Raman spectroscopy: release of plastic particles from different packaging into mineral water. Water Res 129:154–162

Seghers J, Stefaniak EA, La Spina R, Cella C, Mehn D, Gilliland D, Held A, Jacobsson U, Emteborg H (2021) Preparation of a reference material for microplastics in water—evaluation of homogeneity. Anal Bioanal Chem. https://doi.org/10.1007/s00216-021-03198-7

Setälä O, Magnusson K, Lehtiniemi M, Norén F (2016) Distribution and abundance of surface water microlitter in the Baltic Sea: a comparison of two sampling methods. Mar Pollut Bull 110:177–183

Shah AA, Hasan F, Hameed A, Ahmed S (2007) Isolation and characterization of poly (3-hydroxybutyrate-co-3-hydroxyvalerate) degrading bacteria and purification of PHBV depolymerase from newly isolated Bacillus sp. AF3. Int Biodeter Biodegr 60:109–115

Shruti VC, Pérez-Guevara F, Kutralam-Muniasamy G (2020) Metro station free drinking water fountain – a potential "microplastics hotspot" for human consumption. Environ Pollut 261:114227

Silva AB, Bastos AS, Justino CIL, da Costa JP, Duarte AC, Rocha-Santos TAP (2018) Microplastics in the environment: challenges in analytical chemistry – a review. Anal Chim Acta 1017:1–19

Skariyachan S, Setlur AS, Naik SY, Naik AA, Usharani M, Vasist KS (2017) Enhanced biodegradation of low and high-density polyethylene by novel bacterial consortia formulated from plastic-contaminated cow dung under thermophilic conditions. Environ Sci Pollut Res 24:8443–8457

Skjolding LM, Ašmonaitė G, Jølck RI, Andresen TL, Selck H, Baun A, Sturve J (2017) An assessment of the importance of exposure routes to the uptake and internal localisation of fluorescent nanoparticles in zebrafish (*Danio rerio*), using light sheet microscopy. Nanotoxicology 11:351–359

Solomon K, Matthies M, Vighi M (2013) Assessment of PBTs in the European Union: a critical assessment of the proposed evaluation scheme with reference to plant protection products. Environ Sci Eur 25:10

Song YK, Hong SH, Jang M, Kang J-H, Kwon OY, Han GM, Shim WJ (2014) Large accumulation of micro-sized synthetic polymer particles in the sea surface microlayer. Environ Sci Technol 48:9014–9021

Stephens B, Azimi P, El Orch Z, Ramos T (2013) Ultrafine particle emissions from desktop 3D printers. Atmos Environ 79:334–339

Suaria G, Avio CG, Mineo A, Lattin GL, Magaldi MG, Belmonte G, Moore CJ, Regoli F, Aliani S (2016) The mediterranean plastic soup: synthetic polymers in Mediterranean surface waters. Sci Rep 6:37551

Suhrhoff TJ, Scholz-Böttcher BM (2016) Qualitative impact of salinity, UV radiation and turbulence on leaching of organic plastic additives from four common plastics – a lab experiment. Mar Pollut Bull 102:84–94

Sullivan GL, Gallardo JD, Jones EW, Hollliman PJ, Watson TM, Sarp S (2020) Detection of trace sub-micron (nano) plastics in water samples using pyrolysis-gas chromatography time of flight mass spectrometry (PY-GCToF). Chemosphere 249:126179

Sultan I, Rahman S, Jan AT, Siddiqui MT, Mondal AH, Haq QMR (2018) Antibiotics, resistome and resistance mechanisms: a bacterial perspective. Front Microbiol 9:2066–2066

Sun X, Li Q, Shi Y, Zhao Y, Zheng S, Liang J, Liu T, Tian Z (2019) Characteristics and retention of microplastics in the digestive tracts of fish from the Yellow Sea. Environ Pollut 249:878–885

Talvitie J, Mikola A, Koistinen A, Setälä O (2017a) Solutions to microplastic pollution – removal of microplastics from wastewater effluent with advanced wastewater treatment technologies. Water Res 123:401–407

Talvitie J, Mikola A, Setälä O, Heinonen M, Koistinen A (2017b) How well is microlitter purified from wastewater? – a detailed study on the stepwise removal of microlitter in a tertiary level wastewater treatment plant. Water Res 109:164–172

Tetu SG, Sarker I, Schrameyer V, Pickford R, Elbourne LDH, Moore LR, Paulsen IT (2019) Plastic leachates impair growth and oxygen production in Prochlorococcus, the ocean's most abundant photosynthetic bacteria. Commun Biol 2:184

Teuten EL, Rowland SJ, Galloway TS, Thompson RC (2007) Potential for plastics to transport hydrophobic contaminants. Environ Sci Technol 41:7759–7764

Thiel M, Luna-Jorquera G, Álvarez-Varas R, Gallardo C, Hinojosa IA, Luna N, Miranda-Urbina D, Morales N, Ory N, Pacheco AS, Portflitt-Toro M, Zavalaga C (2018) Impacts of marine plastic pollution from continental coasts to subtropical gyres—fish, seabirds, and other vertebrates in the SE Pacific. Front Mar Sci 5:238

Tourinho PS, Kočí V, Loureiro S, van Gestel CAM (2019) Partitioning of chemical contaminants to microplastics: sorption mechanisms, environmental distribution and effects on toxicity and bioaccumulation. Environ Pollut 252:1246–1256

Triebskorn R, Braunbeck T, Grummt T, Hanslik L, Huppertsberg S, Jekel M, Knepper TP, Krais S, Müller YK, Pittroff M, Ruhl AS, Schmieg H, Schür C, Strobel C, Wagner M, Zumbülte N, Köhler H-R (2019) Relevance of nano- and microplastics for freshwater ecosystems: a critical review. TrAC Trends Anal Chem 110:375–392

Troost TA, Desclaux T, Leslie HA, van Der Meulen MD, Vethaak AD (2018) Do microplastics affect marine ecosystem productivity? Mar Pollut Bull 135:17–29

UNEP (2016) Marine plastic debris & microplastics – global lessons and research to inspire action and guide policy change. In: Programme UNE (ed) Marine plastic debris & microplastics – global lessons and research to inspire action and guide policy change, Nairobi

van der Hal N, Ariel A, Angel DL (2017) Exceptionally high abundances of microplastics in the oligotrophic Israeli Mediterranean coastal waters. Mar Pollut Bull 116:151–155

van Doremalen N, Bushmaker T, Morris DH, Holbrook MG, Gamble A, Williamson BN, Tamin A, Harcourt JL, Thornburg NJ, Gerber SI, Lloyd-Smith JO, de Wit E, Munster VJ (2020) Aerosol and surface stability of SARS-CoV-2 as compared with SARS-CoV-1. N Engl J Med 382:1564–1567 van Mourik LM, Crum S, Martinez-Frances E, van Bavel B, Leslie HA, de Boer J, Cofino WP (2021) Results of WEPAL-QUASIMEME/NORMANs first global interlaboratory study on microplastics reveal urgent need for harmonization. Sci Total Environ 772:145071

Verhaar HJM, van Leeuwen CJ, Hermens JLM (1992) Classifying environmental pollutants. Chemosphere 25:471–491

Verma R, Vinoda KS, Papireddy M, Gowda ANS (2016) Toxic pollutants from plastic waste – a review. Procedia Environ Sci 35:701–708

Vianello A, Jensen RL, Liu L, Vollertsen J (2019) Simulating human exposure to indoor airborne microplastics using a Breathing Thermal Manikin. Sci Rep 9:8670

Wang F, Wong CS, Chen D, Lu X, Wang F, Zeng EY (2018) Interaction of toxic chemicals with microplastics: a critical review. Water Res 139:208–219

Wang S, Xue N, Li W, Zhang D, Pan X, Luo Y (2020) Selectively enrichment of antibiotics and ARGs by microplastics in river, estuary and marine waters. Sci Total Environ 708:134594

Webb HK, Crawford RJ, Sawabe T, Ivanova EP (2009) Poly(ethylene terephthalate) polymer surfaces as a substrate for bacterial attachment and biofilm formation. Microbes Environ 24:39–42

WEF (2016) The new plastics economy: rethinking the future of plastics. World Economic Forum, Geneve

Wesch C, Bredimus K, Paulus M, Klein R (2016) Towards the suitable monitoring of ingestion of microplastics by marine biota: a review. Environ Pollut 218:1200–1208

Wright SL, Thompson RC, Galloway TS (2013) The physical impacts of microplastics on marine organisms: a review. Environ Pollut 178:483–492

Wright SL, Ulke J, Font A, Chan KLA, Kelly FJ (2020) Atmospheric microplastic deposition in an urban environment and an evaluation of transport. Environ Int 136:105411

Xu J-L, Thomas KV, Luo Z, Gowen AA (2019) FTIR and Raman imaging for microplastics analysis: state of the art, challenges and prospects. TrAC Trends Anal Chem 119:115629

Yakovenko N, Carvalho A, ter Halle A (2020) Emerging use thermo-analytical method coupled with mass spectrometry for the quantification of micro(nano)plastics in environmental samples. TrAC Trends Anal Chem 131:115979

Yang Y, Liu G, Song W, Ye C, Lin H, Li Z, Liu W (2019) Plastics in the marine environment are reservoirs for antibiotic and metal resistance genes. Environ Int 123:79–86

Zettler ER, Mincer TJ, Amaral-Zettler LA (2013) Life in the "Plastisphere": microbial communities on plastic marine debris. Environ Sci Technol 47:7137–7146

Zhang H, Kuo Y-Y, Gerecke AC, Wang J (2012) Co-release of hexabromocyclododecane (HBCD) and nano- and microparticles from thermal cutting of polystyrene foams. Environ Sci Technol 46:10990–10996

Zhang M, Zhao Y, Qin X, Jia W, Chai L, Huang M, Huang Y (2019a) Microplastics from mulching film is a distinct habitat for bacteria in farmland soil. Sci Total Environ 688:470–478

Zhang Y, Gao T, Kang S, Sillanpää M (2019b) Importance of atmospheric transport for microplastics deposited in remote areas. Environ Pollut 254:112953

Zhang W, Dong Z, Zhu L, Hou Y, Qiu Y (2020) Direct observation of the release of nanoplastics from commercially recycled plastics with correlative raman imaging and scanning electron microscopy. ACS Nano 14:7920–7926

Zhu L, Zhao S, Bittar TB, Stubbins A, Li D (2020) Photochemical dissolution of buoyant microplastics to dissolved organic carbon: rates and microbial impacts. J Hazard Mater 383:121065

Ziajahromi S, Drapper D, Hornbuckle A, Rintoul L, Leusch FDL (2020) Microplastic pollution in a stormwater floating treatment wetland: detection of tyre particles in sediment. Sci Total Environ 713:136356

Zubris KAV, Richards BK (2005) Synthetic fibers as an indicator of land application of sludge. Environ Pollut 138:201–211

Printed in the United States
by Baker & Taylor Publisher Services